IN WAR AND PEACE

IN WAR AND PEACE

MY LIFE IN SCIENCE AND TECHNOLOGY

Guy Stever

JOSEPH HENRY PRESS
Washington, D.C.

Joseph Henry Press • 500 Fifth Street, N.W. • Washington D.C. 20001

The Joseph Henry Press, an imprint of the National Academies Press, was created with the goal of making books on science, technology, and health more widely available to professionals and the public. Joseph Henry was one of the founders of the National Academy of Sciences and a leader in early American science.

Any opinions, findings, conclusions, or recommendations expressed in this volume are those of the author and do not necessarily reflect the views of the National Academy of Sciences or its affiliated institutions.

Library of Congress Cataloging-in-Publication Data

Stever, H. Guyford.
  In war and peace : my life in science and technology / Guy Stever.
      p. cm.
Includes bibliographical references and index.
  ISBN 0-309-08411-3 (hardcover)
  1. Stever, H. Guyford. 2. Scientists—United States—Biography. I. Title.
  Q143.S743 A3 2002
  509.2—dc21

                        2002007968

TO
Bunny
and our children,
Guy Jr., Sarah, Margo, and Roy

Contents

Preface

Science came into my life first in its purest form when as a young boy I saw the leaders of the California Institute of Technology come to my hometown to watch the casting of the Mount Palomar mirror. In the 1940s, as I walked through the wreckage of wartime London, I saw science in quite a different way, as central to the desperate struggle to survive. That life and death testing of science in turn after World War II ended propelled science in its institutions, style, and quality into the national enterprise that with technology has transformed our lives, our country, and our world.

I was blessed to be a part of that—to contribute to the science and technology that helped make us more secure in the unnerving decades of the Cold War, to see and be a part of the arrival of the Space Age, to help and empower institutions—from Carnegie Mellon University to the National Science Foundation to science advisor to the President—and to help in strengthening and enlarging the American science and technology enterprise.

Memoirs are by definition suspect. They are a melange of memories, of pieces of history selected and shaped by experience, and of imagination invested in events that happened long ago. Memoirs are neither history nor biography, not reflecting the judgments of the professional historian or the dispassionate biographer recounting a person's life, but rather of someone who was in the thick of the action, involved in many things but not everything, making judgments shaped by the prisms of personal values; by care for the opinions and reputations of colleagues, friends, and family; and by a watchful eye on the judgments of history.

In that sense this book is suspect. My only defense is that I knew it when I set out to write this book, and determined early on to be as honest as possible, to document where I could. I relied heavily in telling my story on daily pocket calendars I kept beginning in 1943 that recorded the topic, people, and place for meetings. The first draft of the story moved from the daily calendar by dictation and transcription. It retained much of its conversational nature right through lengthy editing. I also relied on taped interviews done over the past three decades, several conducted by professional historians interested in particular parts of my professional life. For example, the Air Force Office of Scientific Research has oral histories of its chief scientists. I also relied heavily on extensive records available, especially at government agencies such as the Air Force Office of Scientific Research, the National Science Foundation (NSF), and the National Aeronautics and Space Administration. Not least, I had the good fortune of comments and criticisms of many people who knew in rich detail part of the story I was telling. The clear leader of these is Philip M. Smith, who was with me at NSF and at the White House Office of Science and Technology Policy (OSTP), retaining a close relationship with me in his extended service with Frank Press at the OSTP and the National Academies. He was a lively participant as the editing of the book progressed and spearheaded the fund that made an editor possible. I am enormously grateful to him and to many others for their help and honesty in helping me tell my story. At the same time, any distortions, judgments that some consider unfair, or just plain mistakes are of course entirely my doing.

The brightest day of all the years of preparing this book came when Norman Metzger agreed to be the editor. With about 2,000 pages of early draft in 16 chapters on my hands, I had been floundering in that search and even had one abortive effort that discouraged me. Trained in chemistry, Norm has rich experiences as a science writer and as executive director of the Commission on Physical Sciences, Mathematics, and Applications at the National Research Council. In that position he became acquainted with a broad spectrum of issues in these sciences and with science and technology leaders. He has worked with much of the institutional infrastructure of academe, government, and industry to which many of the issues I have been involved in were vital. His work ethic is driving but flexible and understanding. He is dogged in corroborating information as well as enriching it with additional material. Most of all he is easy to work with and has a delightful sense of humor.

My wife, Louise Risley Stever, lived through and participated with me in everything in the last 60 percent of my life. She has kept priceless, often-used journals, diaries, and trip reports and has been close to the agony and ecstasy of writing the memoirs.

Guy Stever
Gaithersburg, Maryland

Acknowledgments

Writing a book inevitably leaves debts, and I certainly have mine. They include those who made the book financially possible, who helped in documenting my story and most especially in getting it as right as possible, and who reviewed in detail parts of the manuscript and corrected errors, deepened and broadened interpretations, and corrected the distortions of my memory. I was and remain deeply moved by their help.

First, I want to thank those friends and colleagues who backed with financial and institutional help their belief that my story should be told and that I would tell it well. I started making notes and dictating sections of this memoir in 1994, but it became clear that I would not succeed with the daunting task of organizing the eight exciting decades I describe here without retaining professional help for research and editing and additional funding to ensure the production of a well-illustrated book. Richard C. Atkinson, now president of the University of California, and Philip M. Smith, former executive officer of the National Research Council, who were with me at the National Science Foundation, proposed raising a fund that would help me go forward. William A. Wulf, president of the National Academy of Engineering, readily agreed to establish a Guy Stever Memoirs Fund at the academy. The following individuals and institutions contributed to the fund: Holt Ashley; Atkinson Family Foundation; Elizabeth and Stephen Bechtel, Jr. Foundation; Fletcher

Byrom; Robert W. Galvin; William T. Golden; Trevor O. Jones; Ralph Landau; Robert P. Luciano; Ruben F. Mettler; National Academy of Sciences, Bruce Alberts, president; Simon Ramo; Schering-Plough Corporation; Science Service, Inc.; Louise R. Stever; John F. Welch, Jr.; and William A. Wulf. I am deeply appreciative of their support.

That support enabled us to get the Joseph Henry Press to assign Stephen Mautner and his skilled associates the task of publishing the book. His and their prompt, knowledgeable, accurate, response was just what the author and editor needed.

I am also very grateful to the many people and institutions that helped in the research underpinning this book. The resources and people of the Library of Congress were of enormous help. Jeffrey K. Stine of the Smithsonian Institution was invaluable in guiding us to sources, offering counsel, and reviewing major parts of the manuscript. James David, also of the Smithsonian Institution, patiently advised on the availability, or frequently the lack, of military records and how to locate them. The staff of the National Science Foundation who aided in many ways included William Blanpied, Michael Sieverts, George T. Mazuzan (ret.), and Linda Boutchyard. The National Archives and Records Administration helped greatly, especially staff at its College Park branch and at the Gerald R. Ford Presidential Library. At College Park I had the help and guidance of the staff of the military records branch and of Marjorie Siarlante in locating records of my tenure as director of the National Science Foundation. Invaluable resources and hospitality were made available with patience and thoroughness at the Ford Presidential Library, led by David Horrocks and including Donna Lehman, Karen Holzhausen, Ken Hafeli, Geir Gundersen, and Richard Holzhausen. Janice Goldblum, of the National Academies, was cheerfully and very professionally responsive to many requests for information on my activities associated with the National Academy of Sciences and the National Academy of Engineering. In addition, I am pleased to acknowledge the help of Warren Kornberg, formerly of the National Science Foundation; Michael H. Gorn of the NASA Dryden Flight Research Center; Walter S. Poole of the Joint History Office of the Joint Chiefs of Staff; Dwayne Allen Day; Audrey Pendergast, writer and editor; Michael McGeary of McGeary and Smith;

Ben Patrusky of the Council for the Advancement of Science Writing; and Lauren Belliveau of the Embassy of Australia.

The people who read and critiqued parts of the manuscript, and who certainly should share the credit for what is worthwhile but not the blame for what is not, include Jeffrey K. Stine, Bart Hacker, Gregg Herken, and Michael Neufeld, all of the Smithsonian Institution; Dwayne Allen Day, an independent consultant; Roger L. Geiger of Pennsylvania State University; Robert Buderi of *Technology Review*; Jack Gibbons; Charles F. Larson of the Industrial Research Institute; Edwin Fenton of Carnegie Mellon University (ret.); Michael H. Gorn; Ezra Heitowit of the Universities Research Association, Inc.; Ruben Mettler of TRW, Inc. (ret.); Jacob Neufeld of the Air Force History Support Office; William Pickering of the California Institute of Technology (ret.); Alex Roland of Duke University; Marcia Smith of the Congressional Research Service; and Myron F. Uman and Donald C. Shapero of the National Research Council.

IN WAR AND PEACE

1

Starting Out

1916–1941. From Corning to Colgate to Cal Tech. King Tut and Mount Palomar turned me to science, balky mules, near disaster in the High Sierras, and a fateful decision to go east.

On a summer day in 1935 I stood on the "New Bridge" peering over the rail at the flooding Chemung River dividing the north and south sides of my hometown, Corning, New York. It was the biggest flood the locals remembered. It certainly impressed me as I walked along the north dike of the river watching trees, pieces of cottages, and occasional cows, dead and alive, float by.

On that summer day I was starting work on the 2 to 10 shift in the packing and shipping department of Corning Glass Works in its large plant right by the flooding river. When I got to the plant I found intense flood control under way, using as many pumps as could be found, including fire trucks, to keep ahead of the rising waters. The desperate work had two aims: to keep the water from the huge molten glass furnaces, where it would have instantly and explosively produced live steam, and to save the immense glass casting for the 200-inch telescope for the Mount Palomar Observatory. The telescope mirror was saved through extraordinary efforts to shift the large heating elements from below to above the mirror. These heaters controlled the slow annealing of the glass to prevent strains.

I was then 18 years old, an undergraduate at Colgate University in Hamilton, New York, with ambitions in science. And that flood and the

1

people fighting desperately to save the mirror added a powerful and durable memory to the many I had of Corning. My first memory was Armistice Day, 11 a.m., November 11, 1918, when I was just a bit over two years old. Church bells rang, factory whistles blew, cars honked, and my father, Ralph, came home with his friends to borrow grandfather's open car so they could join the parade, men hanging all over the fenders and hood. They shouted that they were going to burn the Kaiser in effigy. I'd burned myself by then, knew how much it hurt, and wondered why they would burn a person. Other memories before starting school were of my grandmother, Mattie, my grandfather, Horton, and my sister, Margarette, a year and a half older. Our mother, Alma, was hospitalized before I was a year old and later died of tuberculosis, then a dread disease. Our father continued to live with us at our grandparents' home, but he died before I started school, leaving our grandparents to bring us up. Grandmother particularly was the primary teacher of ethics and behavior to Margarette and me, in a very loving and understanding way. When my grandparents were not around, Margarette—bigger, stronger, handsomer, and smarter than me—enforced the principles, salted with sibling rivalry, which, she always said, was for my own good. We had a close relationship, full of mutual respect and admiration, that continues to this day.

As World War I ended, Corning already was a major glass-manufacturing city with three distinct downtown areas and school systems, Corning Southside, Corning Northside, and Painted Post, a separate but attached town. The largest employer of course was (and is) Corning Glass Works,[1] founded and mostly owned by the Houghton family. There were also specialty cut-glass manufacturers and a cottage industry of crystal glasscutters. With a population of about 15,000, Corning was big enough to have a strong character and small enough to feel part of. A good place to grow up.

A few of the great engineering revolutions of the twentieth century were beginning to be used. Nationwide electrification was beginning to light streets and power streetcars, but there was still much use of coal gas and coal. The mass production of automobiles was beginning to change daily lives, but there were still many horse-drawn vehicles—to deliver groceries and packages and fresh milk and ice for the kitchen ice box. I remember at a young age "helping" a family friend deliver groceries on a sleigh in the winter. An airplane was a rarity. Train travel was king. Motion pictures were changing from silent to sound to color. The telephone and radio revolutions were just getting started. The materials revolution was just starting in chemistry, but there were practically no plastics or artificial fibers like nylon. For us kids black cotton stockings were the choice, with many holes and darned patches. The modern medical and pharmaceutical revolutions had hardly started, and serious illness and early death were always close.

Many of the games we played after World War I were war games, and the most exciting were those based on then-new aerial fighting, an excitement that reached a crescendo when Lindbergh crossed the Atlantic in 1927. I knew all the aircraft types flown by both sides, and the Aces were my favorite. Then Houghton Stevens and I rushed to join Boy Scout Troop 23, sponsored by our church, the First Congregational, to which Grandmother belonged to ensure that Margarette and I attended Sunday school. Scouting became a passion for several years until I became a Life Scout. I won the Camp Spirit Medal at Camp Gorton, the Steuben County Boy Scout Council's summer camp.

Grandfather Horton designed and built a new house in Corning, with a rental unit upstairs. It became a gathering place for our relatives, all with nearby farming roots: Uncle Perry and Aunt Hattie Gaylord, twin sister of Grandmother Mattie, farmers from Moreland, near Corning; cousins Guy, produce broker, and Nell Ford from Philadelphia; cousins Phil, food machinery salesman, and Ethylene Considine, and their sons, Bob and Paul, from Cleveland; cousins James and Allie Moore, farmers, and their children, Melvin and Rhea. Grandmother was clearly the most influential in the family, but the men folk did most of the talking, mostly about the economy and politics.

Politics just before the Depression were very rough and didn't escape the notice of the young. In the bitter 1928 presidential campaign between Herbert Hoover and Alfred Smith, then governor of New York state, two events imprinted on my 12-year-old memory. My cousin, Nell Ford, took me to an outdoor speech by Al Smith, in Bath, the county seat. Her husband, Guy P. Ford, a well-to-do commodities broker in Philadelphia was the reason for my unique middle name, Guyford, one that my parents thought would pay off in the long run.[2] (Unfortunately,

the Depression was part of the long run, so it didn't.) On that day in
Bath, after a political lunch with the power brokers of Steuben County,
Governor Smith gave a campaign speech to the crowd. It was a boiling
hot day, and as he stepped to the podium, he wiped sweat off his red-
dened face. A friendly shout came from the front row, "Give the governor
a drink." My aunt instantly shouted from the middle of the crowd: "He
looks like he already had one too many." I was embarrassed, but the ayes
and nays were about equally divided.[3]   And there was something much
more sinister. The Ku Klux Klan, then a powerful political group, was
very much against having a Catholic, Al Smith, in the White House.[4]
The Klan marched in full regalia in nearby Elmira and burned crosses on
the hills around Corning. But we didn't have anything like those tensions
in our neighborhood. There were snowball fights, sometimes a little
rough, between the students from the Catholic and public schools. That
problem went away when a new public high school was built and the
Catholic school, on economic grounds, sent their students there, with
easy mixing.

Another feature of our home life in the 1920s was my exposure to
the renters in the upstairs apartment. The first couple was Hugh and
Marge Gregg. Hugh had just come to Northside High School to be the
chemistry teacher and assistant principal. When they left to start a family,
they recommended the place to Truman ("Jake") and Ruth Jacoby. Jake
was the new coach of football, baseball, basketball, and track at Northside
High School. Both the Greggs and the Jacobys took a great interest in
Margarette and me, and that interest continued through our grammar
and high school days on into college. Another couple was a government
employee and his attractive wife. He had rather odd hours, or odd until
we learned that he was a "revenuer" tracking down bootleggers. We
learned this when he landed in the hospital after being shot in a sting
operation. Our renters often dropped in for an evening of chat with our
grandparents, especially on hot summer nights. It was my first exposure
to college-educated people other than our teachers.

Grandfather made our living room attractive with one of the best
radios for picking up stations across the country and indeed the world.
Its only disadvantage to Grandmother's disgust was that it took its power
from large automobile wet batteries that spilled the acid on the living

room rug. And for special friends there was homemade elderberry wine, the berries picked along dusty country roadsides by Margarette and me. Occasionally, fermentation led to a small explosion of a glass container in the basement. Prohibition was still in effect except for home consumption.

In high school I began to come into my own, getting involved in many different extracurricular activities. These meant more to me than my studies, which I did easily. And I tried to emulate Margarette. She was a campus leader, with the lead in the senior play, a basketball star, head cheerleader, class officer, very popular socially, and at graduation received the best citizenship cup. I didn't like to take home any homework, so I concentrated on getting it done in the study periods. Quite often I did an assignment on one subject while listening to a teacher on a different one. I thought I had gotten away with this, but later, at graduation time, my civics teacher mentioned that he didn't understand how I got such a good mark in the New York State Regents exam on his subject when I always did my math assignments during class. In my junior year I was elected founding president of our Footlight Society and president of the Latin Club and also served as senior class president and had the lead in the senior play. Though skinny, I won three football letters, and in my senior year I was quarterback of our undefeated, untied, and almost unscored-on team. In those days the quarterback selected and ran the plays and played on both offense and defense. To Margarette's particular delight, I too got the best citizenship cup.

Although I remember those high school years as among the best of my life, the Depression shattered our close family life, then supported by Grandfather's job as a manager in a small Corning department store, Wing and Bostwicks. By the end of my sophomore year, his deteriorating health and the failure of the store forced a move to a small two-room upstairs apartment, leaving two units for rental. Margarette left to earn her way through college at Cortland State through scholarships, jobs at school, and summer work as a waitress at a resort in the Adirondacks. Like so many others during the Depression, we were now very poor. At the end of my junior year, Grandmother, her spirit broken from trying to make ends meet, died from an infection following surgery, a common occurrence in those days. In my senior year I did the cooking and housework, not brilliantly. Grandfather, by then virtually an invalid, kept spirits high by encouraging me in all my school activities. Family friends and relatives helped with food, laundry, and mending clothes. One of the best gifts was from Mr. Stanton, who owned a small hotel and "beanery." He gave me a hot lunch of my choosing every school day.

Adding to the Depression and to the poverty marking our daily lives in the mid-1930s was the "gathering storm" in Europe. We talked about it a lot in my high school days, as we "shot the breeze" at the "Four Corners" center of Corning Northside during balmy days and in the winter as we played pool and bowled in the recreation center at the First Congregational Church. Those sessions were enlightened by children of families newly arrived in America, bringing skilled glassblowers to Corning from the glass centers of Europe.

Amidst all this I got my first tastes of science. These came in many ways, the first being in the early 1920s when Howard Carter, supported by the Fifth Lord Carnarvon, discovered King Tut's tomb in the Valley of the Kings in Egypt. Just a beginning reader, I followed the stories of the discovery and the pictures in the Sunday rotogravure sections, intrigued by tales of lost treasures, the mysteries of ancient Egypt, and, when Lord Carnarvon died soon after the discovery, the supposed "mummy's curse." I read more deeply in an encyclopedia about Egypt and archeology and soon wandered on to pictures of galaxies beyond the Milky Way, descriptions of our solar system, and the "Martian canals."

But it was the 200-inch Mount Palomar telescope that led me to a life in science and more particularly physics. Corning suffered serious unemployment during the Depression. So the announcement in 1934 that the company had been selected to cast the mirror blank for the telescope was a real upper.[5] Our newspapers were full of articles about it for years, and we were very impressed when such leading science figures as Robert A. Millikan,[6] chairman of the Executive Council of the California Institute of Technology, visited the casting project. Many of us young high school students, watching our elders, were infected by local pride in Corning and its glass works division, pride I still have. Frank Hyde, later a famous chemist at Corning Glass Works, instructed our Boy Scout troop in grinding an 8-inch parabolic mirror, with which we constructed a fine telescope and got kudos at a scout jamboree.

COLGATE—"YOU CAN DO IT!"

Hugh Gregg, now school superintendent, who with his wife had rented our upstairs apartment, persuaded me in my senior year to go to his alma mater, Colgate University, in Hamilton, New York, about a three-hour drive from Corning. On Friday, June 13, 1934, Gregg drove me to Colgate to meet the dean of students and admissions. There was both good news and bad news. The good news was that my high school records and references were plenty good enough. The bad news was that scholarship money was in short supply. If I could find $175 for tuition and fees for the first semester, they would give me a tuition scholarship for the duration, beginning with the second semester, if every semester I did exceptionally well in my courses. They would also find me a job to pay for room and board, costing $15 a week in a cooperative dorm. I accepted the offer, knowing I had saved $200 from my *Corning Evening Leader* paper route. I had just missed out on getting one of ten Steuben County $100 scholarships. Our high school valedictorian, Alice Rose, and our salutatorian, Mildred Lindbloom, each won one.

Starting at Colgate in the fall of 1934 took me on an emotional roller coaster. It was sad to leave my grandfather, the only remaining member of my immediate family, even though I knew he was in good hands with Aunt Hattie and Uncle Perry. I cheered up during the drive to Colgate, but I almost cried when I saw the shabby state of the dormitory for impecunious students, which I soon learned was to be remodeled the following year. My spirits climbed again when I met other residents of the dorm, all in the same financial boat.

Colgate was a well-rated liberal arts college for males. Its remote location in Hamilton and small size, though, seriously cut down on social activities, such as the dancing and dates that I had enjoyed so much in high school. So I substituted work, getting my social life at home during vacations. In the first two years there were required core courses in the different liberal arts to ensure that we did not get swallowed up immediately in our major studies—courses in philosophy and religion, biological sciences, social sciences, physical sciences, and languages. Of all these core courses I enjoyed philosophy and religion the most, having my eyes opened to a broader world. Confident that I could handle the load, I started off the first month with a full load, trying out for drama, and taking on a job as a janitor's helper in one of the dorms. The job was hard, because the janitor I worked for had very high standards. I worked my tail off. I didn't get a part in the first play that semester. The heaviest blow of all was the "D" I got on my first mathematics exam, which had never ever happened to me and risked my tuition scholarship. In mathematics, the language of science! I was scared! Fortunately, my mathematics teacher, Professor Aude, a big ham-fisted Dane who came to Colgate after a successful career at Carnegie Tech, gave me a couple of special sessions in the evenings. He had me at the blackboard trying to solve problems, gave me a friendly poke on the shoulder with his big fist saying, "Come on, Stever. You can do it." It worked, and I wound up getting an "A" in that math course and in the next five.

I was determined never to get a low mark again—and I didn't. For the next seven years, four as an undergraduate and three as a graduate student, my studies came first. It wasn't only the fear of losing my scholarship that drove me, but also my rapidly emerging goal of becoming a college professor. That meant a doctorate as an entry ticket and that meant doing exceptionally well in my major and minor courses, physics for the major and mathematics and chemistry for the minor.

By the end of my freshman year, my confidence in my academic capabilities was restored. I made the Phi Society, a way station to Phi Beta Kappa. I lined up a better work program for earning room and board in my sophomore year. I had a summer job at Corning Glass Works in the packing and shipping department. And I had the summer to join my old gang of Northsiders[7] in the social life of 18 year olds. On the weekends I visited Grandfather, who was being very well cared for, although his health was obviously worsening. In the middle of the summer of 1935 he was brought to Corning Hospital, where he died of heart failure the day after he gave me his most prized possession, his gold Waltham watch and chain. Margarette came from her summer job waiting tables at a hotel at Saranac Lake. And we shared a sad but grateful remembrance of the life our grandparents had given us. Margarette and I had been making our own decisions and earning our way, so our lives didn't change after his death. Vernon Schoonover, with whom Grandfather had worked at Wing and Bostwicks, became our legal guardian until we were 21, and he managed the closing of the estate, which was not much once all the bills were paid.

By the end of my junior year I had taken all the physics, math, and chemistry courses needed for a major in any of the three. I declared a major in physics, and along with some other physics majors got to work with professors on individual research projects. The professors were very good, with doctorates from first-class research universities, thanks to the efforts of the President of Colgate, George Barton Cutton. He set high standards and capitalized on the Depression, when jobs were hard to get and good people were available at reasonably low salaries. The head of the physics department, Paul Gleason, asked me to assist him on his research on photoelectricity, a continuation of the work he had done at Harvard. Preparing samples for Dr. Gleason meant carefully depositing in a vacuum a thin film of selenium onto a copper-oxide-coated iron wafer about the size of a quarter. That was done through sputtering, in which an electric field was established between a selenium cathode and a coated iron wafer anode. If the vacuum, anode and cathode positioning, and electric field strength were all just right, selenium flowed from cathode to anode, depositing as a thin film on the wafer. I became reasonably expert with the technique and could provide Dr. Gleason with samples for studying photoelectricity.

The physics faculty informally selected some of us as likely candidates for graduate school and a research career and exposed us to meetings of the American Physical Society and its journal, *The Physical Review*. They also assigned us additional and very well-chosen special projects in which one did the sort of background analyses essential to starting research. One was reading papers on the theory of auroras, caused by the ionization of air molecules in the upper atmosphere by solar-emitted particles. This study was to link directly to my graduate research on cosmic rays, also due to charged particles entering the earth's atmosphere from space. A second study was on the theory of metals then flowering in the United States and seeded by theoretical and experimental work in Europe. This helped me understand the silicon semiconductors used as detectors of microwave radar, a field I got into during World War II. A third special project was the ultracentrifuge work of Jesse Beams, to become very important in the development of nuclear materials.

That high-powered science stuff mixed with some very "real-world" experiences in my summer jobs, first at Corning Glass Works and then for two summers in Detroit at the Ford Motor Company, which during the summer hired about a dozen or two college students from around the country. I got some lucky assignments at Ford, first with an expert tool-maker who taught me the use of metal-working tools at least for simple jobs. Then I worked with a machine repairman. He and I fixed machines on the spot all over the plant. The largest job we did was replacing a bearing on a three-story-high metal stamping machine.

And I picked up the labor politics of the days when I worked with a dyed-in-the-wool communist in the electric repair shop. Our routine was interrupted one day to set up a trial of a new technology for heat treating cylinder blanks, by putting them in a coil and heating them with eddy currents from an alternating magnetic field. Observers as-sembled, wearing the white shirts and ties of higher authority. The cylin-der got red hot but not evenly, and the man with the whitest shirt pronounced the technique useless. My radical work mate then told me what every one of the white shirts did and whispered scurrilous remarks about them.

I saw in close proximity some of the labor battles of the mid-1930s. Henry Ford, abetted by Harry Bennett, fought the unionization of his plants. It got very rough. There was a fight on the overpass to the factory gate through which we entered between a union organizer and company goons. The organizer was shot. All of us workers had to walk past a line of goons at each entrance to the plant. My social science studies were sharply augmented.

CAL TECH

The great drive of the spring semester of my senior year at Colgate was to be accepted to graduate school and get a fellowship or scholarship to pay for room, board, and tuition. I applied to Harvard, the California Insti-tute of Technology, the University of California at Berkeley, the Univer-sity of Illinois, and the University of Rochester. All accepted me and all offered financial aid. I accepted Cal Tech, partly because the 200-inch mirror was being ground, polished, and coated there and partly because Robert A. Millikan had made a great impression on me when he visited Corning. And that Cal Tech offered both a teaching scholarship for room and board at the Athenaeum (the faculty club) and a teaching assistant-ship in the physics department was more than I could imagine.

In 1997, when I was given the Vannevar Bush Award of the National Science Board, Victor Neher, who became my graduate thesis advisor, told an interesting story about my acceptance at Cal Tech, a story which I had never heard until I was 80 years old. He told this story at the end of the ceremony after I had given my prepared "extemporaneous" acceptance remarks and the cheers of my many friends were dying out in expectation of the end of the occasion. He asked to be heard and said that he had been on the graduate admission and scholarship committee of the Cal Tech physics, astronomy, and mathematics division in the spring of 1938, and there arrived an application from me with many letters of recommendation. No one on the committee knew any of the letter writers, and they had never had a Colgate student apply to Cal Tech. After some discussion one of the members said, "Oh, let's take a shot in the dark." Richard N. Zare, presiding at the dinner, answered, "Blessed are the risk takers."

I was Colgate valedictorian and graduated summa cum laude. I was even more excited that my classmates voted me "Most Brilliant," without even an honorable mention as "Most Studious." I like to think they were very perceptive! Although I had planned to return to Corning to get a job for the summer of 1938, I received a personal letter from Dr. Millikan inviting me to come to Cal Tech early to join the cosmic-ray research group, starting on thesis research right away. Paul Gleason urged me to do it and gave me a $300 loan out of his pocket to make up for missed summer earnings. So with hurried good-byes at Colgate and Corning, I was off by train to Chicago, the farthest west I had ever been, and on to Pasadena and Cal Tech.

Cal Tech was ideal for me. To be immersed completely in a community where scientific research was the prime interest was my first reward. And with room and board at the Athenaeum, my living standards rose sharply. I remember the lively luncheon discussions of a variable group of graduate students—for example, Charley Townes[8] and Pief Panofsky[9] and occasionally a faculty member or two, like Willie Fowler—right under a large painting of the founding fathers, the "Trinity," the professors from the effete East: Millikan, the physicist; Noyes, the chemist[10]; and Hale, the astronomer. And I enjoyed the climate and other amenities of Southern California that had attracted those founding fathers.

HIGH-ALTITUDE MULES

In the spring of 1939, near the end of my first pressure-cooker graduate year at Cal Tech, Victor Neher, my thesis advisor, came by my research room in the subbasement of the physics building with a startling offer: that I do a cosmic-ray experiment in the summer in the High Sierras of California. I accepted immediately. Dr. Neher had planned to do the experiment himself, but he and Millikan, like a lot of people, knew the world was on the verge of war and that time was running out on experiments they wanted to do, based mostly in India, to send balloons with cosmic-ray detectors into the upper atmosphere.

I took a week's vacation with some friends, stopping at Yellowstone, Bryce, Zion, and the Grand Canyon, and put on hold my original thesis problem: probing the discharge mechanism, or quenching, of Geiger counters, a now familiar instrument for detecting and measuring ionizing radiation. The experiment I was to undertake, the hunt for cosmic rays in the High Sierras, was an important one. Cosmic rays were not well understood in the 1930s when Millikan organized three lines of attack on them, using three different detectors: Carl Anderson using cloud chambers, William Pickering using Geiger counters, and Victor Neher using electroscopes.[11] Millikan had taken detectors all over the world so that he "could use the earth as a giant magnet to analyze the energies of primary cosmic rays."[12] The stuff of cosmic rays is mostly protons with an admixture of other particles, traveling at very high speeds and probably coming from outer space. The intense interest in cosmic rays since their discovery in the early twentieth century by Victor Hess was both in understanding them and in using them, since, until the advent of accelerators later in the 1930s, they were the only source of high-energy particles for probing matter. Indeed, in 1937, Carl Anderson and his graduate student Seth Neddermeyer had concluded from cloud chamber experiments that most of the secondary cosmic-ray particles in the earth's atmosphere were particles intermediate in mass between electrons and protons.[12] Hideki Yukawa, a Japanese theoretical physicist, had postulated their existence earlier. We called them mesotrons, but today those particles detected by Anderson and Neddermeyer are called muons.[13] Yukawa also postulated that the mesotron once outside the nucleus had a finite lifetime, yielding an electron

upon decaying.[14] It was measuring that lifetime that our experiment was all about.

The experiment was conceptually simple. Mesotrons created when cosmic-ray protons hit the earth's atmosphere make up most of the cosmic radiation passing through the atmosphere onto and into the ground. Cosmic rays lose their intensity as they course through the atmosphere, pass through matter, are absorbed by other particles, and as mesotrons decay into electrons. Cosmic-ray mesotrons travel almost at the speed of light but have a lifetime of about a millionth of a second, which means a steep drop in their number for every mile of travel down through the atmosphere.

So the experiment to measure the lifetime of mesotrons becomes "quite obvious"—the sort of statement that has forever irritated students. We would measure the intensities of the radiation at two different heights above sea level and compare them. These would differ because of both absorption and decay of mesotrons between the two heights. All we had to do was measure the intensity at the two different altitudes, separating out the difference due to absorption by the intervening atmosphere. That would give us the difference in intensity owing solely to decay of mesotrons into electrons and therefore the lifetime.

That explains why we went for our measurements to two different elevations in the High Sierras but not why we put our detectors, recording electroscopes, into a lake at each site. That was so the extra water depth in the higher lake equaled exactly the mass that the mesotrons penetrated to reach the lower lake, canceling absorption effects and making the difference in intensities due simply to mesotron decay. We actually sank the detectors in both lakes to even greater depths to eliminate any transition effects as the mesotrons went from air to water. There were other complications that had to be dealt with in our calculations: the relativistics of particles traveling near the speed of light and the different angles of the cosmic rays entering the atmosphere, with intensities dropping as their angle from the zenith dropped.

Getting the equipment ready during the summer was a great experience in working with Vic Neher, who was a master experimental physicist. He had designed and built the electroscope to detect charged particles. It had a slight Rube Goldberg flavor: Bombarding mesotrons ionized gas particles in a chamber that created charged particles that in turn discharged a quartz fiber. A battery-driven clock arrangement recharged the fiber and also advanced the film recording the position of the fiber at regular intervals. The rate of discharge of the fiber gave the intensity of the mesotrons, which is what we were after. And there was a nice little optical port for looking at the discharging fiber to make sure the instrument was OK.

This rather delicate instrument was going to be suspended from buoys in mountain lakes. So we needed a container that could protect the electroscope against water and against the mules bearing it up the mountains. We tested the container, which looked like one of those large milk cans farmers use, for watertightness in a pool in the courtyard of the physics lab. We could not think how to simulate riding on a pack mule, so we trusted to luck. More on that later.

Robert Millikan had secured money for the experiment from the Carnegie Corporation, all $300 of it. We had to scrounge, beg, and borrow more. We needed a boat from which to submerge and retrieve our instruments. We were lucky. Dr. Goetz, a research associate in physics, was the proud owner of a "faltboot,"[15] a sort of large kayak with a rubberized cover stretched over a wooden frame, which could be disassembled somewhat like a tinker toy. It weighed about 95 pounds, which seemed just right to balance the load for the mule of the electroscope in its watertight container. We learned later that meant a pretty strong mule. We promised Dr. Goetz that we'd be especially careful with his boat. To my great pleasure, two of my fellow graduate students, Hugh Bradner and Bob Hoy, joined the expedition. Bob was in geology and Hugh in physics, and, even better, Hugh offered the use of his handsome Ford convertible. We packed our food, 60 pounds of it, including such essentials as a pound of raisins, 3 pounds of chocolate, a half-pound of peanut butter, and one jelly-jar key.

Our first lake was the Kerchkoff Reservoir, at 6,000 feet, in the foothills of the High Sierras, west of Mount Whitney. The faltboot assembled with no problems. The electroscope when wound for the day ticked away happily. All we had to do each morning was row out to the buoys, haul up the gear, get it to shore, open the watertight can, take the electroscope out, wind the clock, check that it was recording, reverse the whole process, and, after the rig was back in the water, relax for the rest of the day.

It was pleasant for us but a disastrous time for the world. As we did our measurements from September 1 to 3, 1939, Nazi Germany invaded Poland, and Britain, France, Australia, and New Zealand declared war on Germany. Even though the United States declared its neutrality a couple of days later, we knew we were on the verge of a world war. It was depressing but not a surprise. The three of us had lived with Hitler's rise to power as we made our way through high school, college, and now graduate school.

Now we had to get to the upper lake, Lake Tulainyo, at 12,865 feet. To get to it we had to climb over Whitney Pass just south of Mount Whitney, on the ridge line of the dramatic cliff of the High Sierras, then drop down on the west side of the pass to Crabtree Meadows, where there was a ranger station at the junction of the Whitney Pass and John Muir trails. From Crabtree Meadows north on the John Muir Trail, we would use a fisherman's path along Wallace Creek to Wallace and Wade Lakes, to our base camp at about 11,000 feet. Lake Tulainyo was 2,000 feet higher and three miles east. It took four days before we could put the instrument in the lake.

Bob Hoy knew something about pack animals, and he negotiated with the hostler for our pack animals, two mules and a horse. Listening to Hoy and the outfitter, I learned a few things—for example, that the horse should be a mare because it had a quieting effect on the mules. Not quieting enough, it turned out. One of the two mules was a male, a big fellow and strong enough to handle the electroscope in its watertight can and the faltboot. A pleasant-tempered female mule rounded out the train. As we learned about diamond hitches, hobbles, and the like, I also learned, in a sort of bees and flowers conversation, that even though mules were sterile they were not uninterested in sex.

We made it to Crabtree Meadows, where the ranger and his wife were closing the station and pulling out after a five-day reconnaissance of their territory. We would be the last party in that summer. He told us his route in case we needed help. We found an almost perfect spot at the base camp area at 11,000 feet. At a lunch stop en route, the ornery male mule suddenly lay down and began pounding the pack with the delicate instrument against a boulder. We had to cut off his pack, and with my heart in my mouth I immediately looked through the porthole to see if the electroscope was working. It was!

Wales and Wallace Lakes were just above the tree line. And just below where their outlet streams came together, there was a pleasant grassy field for the animals to graze, a rocky outcropping to protect us from wind and rain and, not least, plenty of firewood. I reconnoitered our route to Lake Tulainyo. Cliffs that in most places seemed to call for an experienced climber blocked access. And even where an ordinary climber like me could do it, a mule could not. Luckily, one of the physics professors at Cal Tech, Ira Bowen, foresaw our problems and had given us an old clipping from the *Los Angeles Times*, with pictures of a fish-stocking expedition to Lake Tulainyo, including an aerial photo on which a dotted line showed the route taken. It wasn't easy, but we made it, not without pushing, cursing, beating metal on rocks, and whatever else we could do to harass our animals into scrambling over the worst places.

Lake Tulainyo nestled on the west side of the steep eastern slope of the High Sierras less than a hundred yards from the divide. A steep slope of boulders ringed it. It was deep and cold. No living plant was in sight except lichens. And we found no signs that the fish stocking a few years back had been successful.

We got our apparatus into the faltboot and then into the lake, with a lot harder work than at Kerchkoff Reservoir, over 6,000 feet below. Our instruments checked out fine the next day and the following morning dawned beautifully. But there was something missing, our two mules to be exact. There was consternation, especially since a search of our immediate surroundings didn't turn up any mules. We finally found tracks and agreed that Bob would set out after them while Hugh and I did our daily turn with the gear. Bob returned later in the day without mules but with a big smile. He had tracked the mules down to the John Muir Trail and found a penciled note on a large stone cairn in the middle of the trail. It was from the ranger telling us that our mules had joined his pack train and that he would keep them until Friday when we could meet him and the mules at the cairn.

After two more pleasant days we had enough data. Bob picked up our mules. And we planned a day of retrieving the instruments and retreating from the mountains. But a great storm blew up during the night, and the weather turned cold, cloudy, and very windy. Still, Hugh and I set out to retrieve the instruments and the faltboot. The higher we climbed, the colder and windier it got. The lake was covered by nasty whitecaps, and occasionally little swirling "dust devils" whipped across the water blowing up spray. It looked scary but not nearly as much when we found our faltboot missing. We had wedged it between large boulders to keep it from blowing away, but it wasn't enough. There were rubber scrape marks on the boulders, pieces of gear, and a broken strut or two. Hugh and I were shocked, I even more so because I was responsible for the whole thing. We found the faltboot across the lake floating upside down and being battered against the boulders by the high waves. It was very hard, but we got to the boat, righted it, emptied the water, and carried it back to the launch site rather than risking the lake.

But we still had to get our gear out of the lake. Perhaps the supposed loss of mental acuity with altitude accounts for what we did next, which was to get into the faltboot and paddle out to the buoys. We never made it. About 15 yards out we were flipped by one of those little twisters. The cold water was a terrible shock, and we both bellowed when we surfaced. Our heavy wool clothing and climbing boots were hard to manage, but holding on to the boat we struggled to shore. We were too frightened to try again, and after storing our boat, we ran toward camp. We threw off our wet and freezing clothes except for our boots and socks and passed naked by a startled Bob Hoy, who had harnessed the mule and was en route to meet us to pick up the equipment and the faltboot. We kept going toward camp for some warm clothes and our sleeping bags.

The instruments and boat still had to be retrieved. The storm continued for two more worrisome days and nights. We were running out of food and dreaded a retreat without the instruments and data. Finally, the winds calmed and with record speed we managed to retrieve our equipment, break camp, and get down to Crabtree Meadows that night to the now-closed ranger's cabin. Winter was returning to the High Sierras. We had a light snowfall for our return over Whitney Pass, but otherwise the trip went well except that the horse was becoming increasingly lame from a rope burn. The hostler lost his temper about that and asked for

$45 in damages. We had already used the $300 from the grant. I was later chastised by an administrator at Cal Tech and had to verify the claim.

Steak! Mashed potatoes and gravy! Chocolate milkshakes! Fresh bread and butter! A *warm* night's sleep in the desert near Lone Pine! We were new men. We returned to Kerchoff Reservoir for three pleasant days of more measurements to confirm our earlier ones and then to Pasadena and Cal Tech. I spent a tense evening in the darkroom developing 2 feet of 35-mm movie film. The whole record was on it, and it came out very well. The next night I dined at the Athenaeum with Bill Pickering[16] and his wife, Muriel, who were sailing to the Orient to join Robert Millikan and Victor Neher. They were so interested in my account that they asked me to go with them to the station to continue the story. As I got into the little car that was taking them to the train station in Glendale for the trip to Vancouver where they would embark, I stepped right in the middle of their farewell basket of fruit. They said the stories made up for the loss.[17]

Then the real work started. I had to plot the data, correlate the data with the barometric pressure during the entire time, and work up the accompanying theory and calculations. We did not have a computer programmed to receive data and then spit out the results nicely graphed and diagrammed. The theory was quite complex, and I had to work hard at it. And there was competition for my time, for I was in the second year of graduate study when the toughest of the required courses had to be passed.

William Fowler,[18] a future Nobelist but then a young instructor, just out of his own graduate school days, told me to get going before I was scooped. I was learning something about the pressures of "publish or perish." Robert Oppenheimer,[19] who spent a term each year on the Cal Tech faculty and who was especially interested in our results, saw me at lunch one day and asked how it was coming. When I told him some of our difficulties, he came to my research room and spent an entire afternoon with me discussing the problem, rapidly filling in and erasing my small blackboard as I took notes. He was a magnificent lecturer, but he had one disconcerting habit. He was a chain smoker, and at the blackboard he always had a cigarette in one hand and chalk in the other. We were always waiting for him to write with the cigarette and smoke the chalk, but he never did.

The research work, including the trip into the Sierras to get the data, took up the summer of 1939 and much of the time through the spring of the 1939–1940 academic year. With a heavy course load it was a very intense, high-pressure time. By the end of the spring term in 1940, I felt a tremendous load lifting, for the toughest courses were well passed, the theory for the cosmic-ray work was done, the paper by Neher and Stever for *The Physical Review*[20] had been submitted, and a letter to the editor of the *Review* summarizing our results had been published. I had punched my ticket for acceptance into the research world. Although my doctoral work wasn't done, I was clearly on a higher plane in my career scramble.

My spirits soaring, I set out with three graduate school friends, George Wheeler, Bob Wells, and Byron Havens, in Bob's car on one of the best auto tours of my life. We camped out every night for three weeks from Southern California to Seattle, where we dropped Byron, to Edmonton, Alberta, and back to Yellowstone, where George and Bob continued on east and I bussed back to Pasadena. It was totally effective in renormalizing me after months of battering by the metronomic beat of seven days a week of studying, solving homework problems, going to classes, taking exams, and somehow finding time for research. I was particularly grateful to George, not only for his friendship but also for a loan of $600 for incidentals. This, added to Professor Gleason's $300, came to $900 debt to be repaid at the end of my education, which I did in three months.

INQUISITION

The trip rested and refreshed me for the 1940–1941 school year. But what was there to do? I had already taken the "unrequired required" courses, the core courses for a Ph.D. in physics and a minor in mathematics: electricity and magnetism with Smythe, optics with Bowen, mechanics with Zwicky, atomic physics with Millikan, mathematical physics with Fowler, complex variables with Ward, thermodynamics with Epstein, quantum mechanics with Houston, and nuclear physics with Lauritsen and Oppenheimer—a powerful package with a galaxy of star teachers.

That qualified me to take my oral exam, for which I immediately began to study. And I could take some electives, including theory of numbers with Bell, relativity with Tolman, and an extended course with Epstein that brought together at a higher plane some of the earlier courses.

I returned full force to the problem of what makes a fast Geiger counter quench. Victor Neher and I agreed that I should take a new tack on the problem. I hadn't gotten far when Victor came to my office one day to explain that he was taking a temporary assignment as a staff member of the newly formed Radiation Laboratory at the Massachusetts Institute of Technology (MIT). It was at this time that I learned of the National Defense Research Committee, an organization put together at the request of President Roosevelt by Vannevar Bush to mobilize academic scientists and engineers to work on military needs. Victor assured me that I would be well looked after by Bill Pickering in my pursuit of the Geiger counter work.

All this neither surprised nor upset me. Like many Americans, shocked by the quick capture of France, Belgium, and the Netherlands and frightened by the seemingly impossible job the British had in the Battle of Britain, I was convinced we had to start helping the British. Fortunately, President Roosevelt was working as fast as allowed by the practicalities of politics to get us engaged. And even many who did not want us to get into the war were still supportive of a military buildup, including having a draft. The draft bill passed by the narrowest of margins.

Soon after Victor Neher left to join the MIT Rad Lab, Professor Ira Bowen dropped by my research lab to inform me that the experiment on the lifetime of the mesotron was a good thesis and that I should concentrate on taking and passing the oral exam to finish by June 1941. I agreed but said that I would also like to work on the Geiger counter problem. He agreed, so the shape of my academic year was determined. The highest priority was studying for the oral exam. A standard start in that process was to get the "bone book" from the last candidate who took the exam. It was a record going back several years, with each candidate describing the questions that the various faculty members asked and giving the candidates' answers to them at the time and, if necessary, the correct answer worked out later. Since the faculty knew of the bone book, this was more of a challenge to them to think up different questions or to cleverly conceal the real nature of their questions. More important was to go over all of one's notes from all the important courses and to study alternative texts on the same subject. As the time approached, I found myself getting more than uptight. I solved that by going ice skating at the Pan Pacific Rink in Los Angeles or to a "happy" movie. I found Olivia de Havilland just right.

The three-hour exam with a quarter-hour coffee break was as tough as I expected. I stood up quite well, even though there were moments when I did not have the foggiest notion of an answer. The worst question was from Fritz Zwicky: You have a string with a ball of wax at the end and a beaker full of water into which you dip the ball of wax. What is the electric charge on the surface of the ball? Zwicky had taught me mechanics, and I could not think of anything he taught me that would solve that question, so I searched for some things I had learned from Epstein in thermodynamics and wrote a couple of basic equations on the board and turned and found that Epstein had a great big smile on his face. I continued and somehow out of those equations found an answer. Epstein then asked me some more questions, and I answered them correctly. I had a feeling Zwicky was disappointed that I answered him with someone else's stuff. I waited outside with some friends after I finished. When the faculty came out with big smiles on their faces, my friends cheered and then carried me on a stretcher over to the Athenaeum for lunch, to cheers from the diners.

With the exam behind me, I tore into the Geiger counter problem, and solved it.[21] I was able to show the deadtime of the counter, that is, the time when it cannot register another incoming ionizing particle, and the recovery time, the time it took for the count register to regain full strength.[22] I also discovered that a directional counter could be made by putting a small number of glass beads on the central wire to prevent the initial ionization breakdown, confined to a thin sheath right around the wire, from spreading along the wire unless the ionizing particle traveled lengthwise through the counter, giving a directional measurement. By measuring the strength of the discharge, one could tell whether the particle came along the wire or from the side. Cal Tech insisted that I apply for a patent, which I did and then forgot about until after the war, when

I began to receive small checks periodically, my 7 percent of the take by the institute. So I had completed two parts of a thesis, either of which would qualify me according to my graduate committee. And it kept me hopping right up to the end of my third year.

SPENCER TRACY'S EYES

Although I worked briefly for two corporations, Corning Glass Works and Ford Motor Company, my first contact with scientific consulting, later a substantial part of my professional life, came when Professor Bowen asked if I could spend a few days consulting for a movie company, Metro Goldwyn Mayer. MGM was making a movie of Robert Louis Stevenson's *The Strange Case of Dr. Jekyll and Mr. Hyde,* with Spencer Tracy in the title role and Ingrid Bergman and Lana Turner as the leading ladies. They would show him changing from Jekyll to Hyde by painting his face with a grease paint that fluoresced in ultraviolet light, so-called dark light. The concern was that ultraviolet light might damage Tracy's eyes, recalling the problems early movie stars had with "klieg eyes" from the burn—in effect, a retinal sunburn—by the ultraviolet light from the powerful arc lights then (and still) used.

I dressed in the only suit I had, a white shirt, and tie and drove to the MGM studios in a borrowed car. The head cameraman took me to the set and handed me over to the assistant cameraman, who was very helpful, as were the technicians who understood my questions and were very forthcoming. It didn't take many questions and measurements before I was sure of the answer, based on my optics studies with Bowen and on a talk I had with him. They offered to show me around the studio, an offer I declined lest it appear that I was not a busy man.

Back at Cal Tech, I spent a day writing up my report and sent it to MGM along with a bill for $100. I stated that not only would Spencer Tracy be safe from "klieg eyes" in that scene, he would be safer than if the scene were shot with normal arc lights. The industry had already solved the "klieg eyes" problem by fronting the carbon arc light with a Pyrex lens, which wouldn't crack with the heat but would filter out the extremely high-frequency portion of the dark or ultraviolet light, the burning part of the radiation. Adding an ultraviolet filter to block visible radiation striking Tracy's face left the nonburning, lower-frequency ultraviolet for the fluorescence to turn Jekyll into Hyde. QED for $100.

I had more consulting ventures, none as glamorous as that one. But it was spring 1941 and my time at Cal Tech was ending. Physics Ph.D.'s were in great demand. I had three offers: one from the Research Corporation, which developed patents assigned to it by universities; one from Stanford University, to be a physics instructor; and one from MIT, to join the Radiation Laboratory. Had times been normal I would have leaped at the Stanford offer to start a career in physics in an outstanding department. However, my strong desire to defeat Nazism turned me to the Rad Lab. Professor I. I. Rabi of Columbia University came through Cal Tech to lecture and to recruit for the Rad Lab, dropping by my research room to convince me to go east. But convincing wasn't necessary.

Saying goodbye to my many friends and colleagues at Cal Tech brought tears. I boarded a Santa Fe train for the east, stopping in Corning. I didn't stay long. I had to start work.

2

War

1941–1945. From the Rad Lab via a stormy Tagus to London, the Blitz, radar, and almost losing Britain's most important secrets. On to the beachhead in Normandy, capture by the Germans, witness to the liberation of Paris, ghastly discoveries in Germany.

I boarded a Pan American Airways clipper for London at 6 a.m. on New Year's Day 1943. I was to serve as scientific liaison officer for the London Mission of the Office of Scientific Research and Development. It was my first trip on a flying boat, my first Atlantic crossing, and the longest flight in my life. What was scheduled for 3 to 5 days, with stopovers in Bermuda, the Azores, Lisbon, and Poole in England, ending with a short train ride to London, actually took 19. We spent 13 days in Bermuda, first delayed by foul weather in the Azores and then by "maintenance needs." The "needs" were in fact the use of our aircraft to help support the summit in Casablanca from January 14 to 24 between Franklin Delano Roosevelt and Winston Churchill, where they agreed on unconditional surrender as the terms for ending World War II.

It was an epochal time. The day after I left for London the Germans began their retreat from the Caucasus and later that month the Soviets began their drive to clear Stalingrad, Montgomery's Eighth Army took Tripoli, and American bombers hit Wilhemshaven, the first U.S. raid on Germany. A month earlier Enrico Fermi and his colleagues on a former squash court at the University of Chicago achieved "the first sustaining chain reaction and thereby initiated the controlled release of nuclear energy."[1]

A BRITISH GIFT

Things were different in the summer of 1940. In June, Norway surrendered and Hitler toured Paris, and then on September 7, 1940, at about 4:00 p.m., 348 German bombers escorted by 617 fighters blasted London until 6:00 p.m. [2] The Blitz had started. In September 1940 a British group in great secrecy bore "gifts" to the United States, including a palm-sized device inside a black japanned metal box.[3] The device, a "cavity magnetron," solved head-on the major obstacle to radar images sharp enough for precision location of aircraft, especially at night, for directing artillery fire, and for finding submarines. Until the magnetron, "seeing" objects at great distances was severely limited: the beam width of the signals sent out from a radar device and echoed back from the target was too wide to provide enough detail and the power or strength was too small for sufficient range. The cavity magnetron changed all that. Invented by physicists at the University of Birmingham only eight months before it was carried to the United States, it emitted highly focused microwaves[4] at great power. The British team told a U.S. group[5] gathered in a Washington hotel that the device "could generate ten kilowatts of power at ten centimeters, roughly a thousand times the output of the best U.S. tube on the same wavelength."[6]

That was stunning. Radar then was long wave with meter-long signals. The British used the technology throughout the war for its Chain Home system to detect and track planes coming across the coast and to guide interceptors close enough for visual contact. While useable for daytime attacks, it lacked the resolution to guide interceptors to planes attacking at night, a tactic the Germans quickly turned to. The British desperately needed better resolution to guide artillery fire and fighter aircraft. Narrower beam widths—radar with wavelengths in centimeters not meters—meant greater accuracy, better resolution, less clutter from extraneous echoes, and less atmospheric interference. It would also yield better range information, better height readings, and more discrimination of closely spaced targets. As early as 1935, Robert Watson-Watt in Britain had pushed hard for the development of microwave radar, an urgency then shared neither by the United States or Germany[7] nor the many other countries working on radar in the 1930s, including

Japan, Italy, France, and the Soviet Union. The Germans, for example, developed several radar systems in the 1930s, but they were of limited designs and intended mainly for gun laying[8] (for example, the 53-cm Würzburg radar I encountered later).

The problem was getting centimeter radar with sufficient power. The cavity magnetron did that, a claim the Americans confirmed in tests a few weeks after the Washington meeting. More magnetrons were built and tested in short order, with components built by AT&T, RCA, Sperry, Westinghouse, and General Electric. But as remarkable as the cavity magnetron was, microwave radar also needed further development of its components—receiver, detector, and so forth—to be a workable device. On October 18, about a month after the magnetron arrived in the United States, the notion of a central laboratory to work on radar was approved by the Microwave Committee of the National Defense Research Committee (NDRC). The laboratory was charged with making the magnetron useable for insertion in aircraft, for gun laying or directing and timing artillery fire, and for guiding bombers. It was to be administered by civilian scientists and not report directly to the military, roughly the model used by the British. The Radiation Laboratory—the Rad Lab—was set up under the NDRC.[9] Vannevar Bush, the NDRC director, started the lab with a budget of $455,000, approved the site (the Massachusetts Institute of Technology) and named as director Lee Dubridge, a physicist from the University of Rochester.[10] By December 1, some 20 physicists had signed on, a full year before Pearl Harbor.[11] A few weeks after the cavity magnetron came to the United States, a major laboratory to exploit it was up and running. And in March 1941 a device using the cavity magnetron enabled over Cape Cod the first air-to-air detection by Americans, coincident with a similar achievement by the British.[12] At its peak, some 4,000 people worked at the Rad Lab from 69 academic institutions, and the products that flowed out of it by 1944 totaled $640 million.

I was given badge 149 when I arrived at MIT in June 1941, with a fresh doctorate in physics from the California Institute of Technology, having been recruited for the Rad Lab by I. I. Rabi and Victor Neher. Victor was among the 20 "charter members" of the lab. Rabi was to win the Nobel Prize in 1944 for his work on magnetic properties of atomic nuclei. I was first put to work on modulators that formed the pulses and powered the magnetrons but was soon asked to help establish MIT's Radar School. In the summer of 1942 I returned to the Rad Lab to work on K-Band, or 1-centimeter, radar components.

TO LONDON

On a Saturday evening in early fall 1942, Lee Dubridge called me at the Neher's rented home where I lived to ask if I was interested in joining the London Mission of the Office of Scientific Research and Development for radar liaison with the British and our growing American forces in Europe. I listened hard and asked a lot of questions. But I had decided in microseconds that I was going hard after this opportunity. I was young, single, and feeling pressured by my counterparts who were going into the armed services. I wanted to get closer to the real war, so I accepted. My spirits leaped.

For the next three years I was part of what was then the world's biggest effort in technology transfer, a term not used then but in vogue now. The exchange between Britain and America of science, technology, and engineering in radar and other technology for modern warfare had already begun, but it was about to become enormous. So too was the flow of ideas and equipment from laboratories and industry to our fighting forces. The part I was to play also foreshadowed much of my postwar life.

I spent two months getting ready for London. Thanks to Lee Dubridge, Rad Lab division heads briefed me. And thanks to Vannevar Bush, who arranged the introductions, leaders of industrial and military laboratories briefed me on radar work. Occasionally, it went beyond that. At Westinghouse, given the cachet of the Bush introduction, I was also told about recent work on the separation of uranium isotopes, even though I didn't have the necessary clearance. I kept quiet. I visited Bell Laboratories to be briefed on magnetrons by James Fisk, later the head of the laboratories, and General Electric, to be briefed on vacuum tube developments by its head of research. And I talked with scientists who had visited British laboratories and with Britons assigned to the Rad Lab, notably Denis Robinson,[13] the British Air Commission and Royal Air Force representative for aircraft-to-surface work.

Rough Passage

I also learned some lessons on the 19-day trip to Britain, one being the limits at the time of flying boats and navigation. On the leg from the Azores to Lisbon, I went forward to talk to the navigator of the Boeing 314 flying boat, who was unable to use celestial navigation because of a severe Atlantic storm and had to rely instead on compass and dead reckoning. Not too accurate. We missed Lisbon by several tens of miles but finally turned and landed on its Tagus River in a raging storm. A launch came out with enough space to take off some but not all of us. Huge waves pinned the launch against the plane. The boat and plane heaved up and down at different frequencies, with very heavy waves breaking over the boat. We finally maneuvered away from the plane, started toward shore, and were treated to foul language from the captain and crew when a big wave hit just as the captain opened a window to see his position. Cold and wet, we were forced to wait in a steel-corrugated customs shed and were told that no one would clear customs until everyone was off the plane. But the captain wouldn't go out again until the storm let up. So the remaining passengers bobbed up and down in the plane all night. When they were finally taken off, the three women were limp with seasickness and had to be carried off. To top it, the plane steward had collected our passports and then dropped the bundle in the sea. Fortunately, it was retrieved, but for the rest of the war I traveled with a badly stained passport. During two more days of delay, I had a chance to telephone greetings from Sam Goudsmit of the Rad Lab to his sister, who had escaped the Nazi invasion of the Netherlands and was now in neutral Lisbon. We finally got from Lisbon to Ireland's seaplane base at Foynes, circling far out in the Atlantic to avoid German interceptors based in France. Some weeks later another Clipper was intercepted and shot down with the actor Leslie Howard of *Gone with the Wind* aboard. The land plane that was to take us from Foynes to England had flown into ducks and damaged its wood-surfaced wings. More delays, although our stay in Limerick was considerably less tense than in Lisbon.

"Nonsense!"

I finally made it to London, checked into the Park Lane Hotel, and set out for the American Embassy in Grosvenor Square, to see Bennett Archambault, the science attaché, whom I had met in the States. I went past the Connaught Hotel into Grosvenor Square to the embassy and once past the Marine guard to Archambault himself, where I excitedly told him that "there's an air raid on!" "Nonsense! There haven't been any sirens." "But," I argued, "I heard both bombs and antiaircraft fire." "Nonsense!" The air raid sirens sounded some 15 minutes later. The newspapers reported the next day that the German bombers went undetected by radar by flying in very low above the Thames estuary. Another hole in Britain's radar defenses, adding to that of the Chain Home long-range radar against nighttime attacks.

The first blitz was from September 1940 to May 1941. London was hit the hardest:

> Almost half of the 60,595 civilian dead were killed in the London region. The maximum tonnage of high explosives dropped on any provincial city was 2,000: London received more than 12,000 tons. It was bombed continuously for a far greater period than any other city, being raided for seventy-six continuous nights, with only one exception, in the autumn of 1940. . . . In central London only one house in ten escaped damage.[14]

The British security clearance for working with Americans was simple. Our top officers would introduce us to their top people, who would in turn introduce us to the people working for them down the line. Introductions done, we could work directly with our British counterparts or indeed with anyone at any level. Our London office for radar liaison, all of two people, David Langmuir, the senior member, and I, worked with all the British armed services—the Royal Navy, Army, and Air Force. Each had legitimate and complex interests in radar. Each had established laboratories to bridge civilian and military communities, many led by university people and populated by younger faculties and graduate students.

The morale and spirit of the military were up, with recent Allied victories at Midway, Stalingrad, El Alamein, and Guadalcanal. Ahead were increasing air attacks on Germany and the return to Europe. Although air attacks on Britain had dropped substantially since the 1940 Battle of Britain, night attacks remained a problem. And attacks by German submarines on Atlantic convoys continued to be devastating, the allies having lost 70 ships totaling over 400,000 tons by February 1943.[15]

Better radar was desperately needed against nighttime attacks and the U-boats. That meant microwave radar, and I got to work on continuing the exchange of information on a critical component, the vacuum tubes essential to building successful microwave radar systems. Since the tubes, or valves as the British called them, were developed primarily by industry, I had many opportunities to visit British industrial laboratories and compare them with the American ones I had visited earlier. I took over from Dave Langmuir, attending meetings of the Committee on Valve Development, established early to coordinate British work on these critical components. I took notes all the while, although it was difficult for me in the beginning to understand the very rapid British speech. I always had the feeling that their brains and speech were running in a higher gear than mine.

Out of London

The principal feature of my second week in London was a four-day trip to Great Malvern, my first chance to get out of London to where the real work of research and development in radar was being done. Great Malvern—on the border between England and Wales and well inland—was chosen a couple of years earlier because of Britain's deep concern with German air raids and the possibility of a German coastal raid to capture radar secrets. A raid on Bruneval, a French coastal town near Le Havre, had captured a German 53-centimeter, or Würzburg, radar. If the British could do it, why not the Germans?

Great Malvern housed two major radar centers—the Telecommunications Research Establishment, TRE, and the Army Defense Research and Development Establishment, ADRDE. The Telecommunications Research Establishment, established in 1940 under the Air Ministry, was where the first cavity magnetron was sent—to improve its power output, to build microwave receivers, and the like. And it was at TRE that air-to-air microwave radar was developed and tested on the same day the Rad Lab tested its device over Cape Cod. The ADRDE, the principal research and development agency for the army, was more operationally attuned, focusing, for example, on building radar-controlled antiaircraft guns and gun-laying radar, for bearing, range information, and fire control for artillery.

I spent a good bit of time from early 1943 to mid-1945 on the London-Great Malvern route. The Paddington train went through Reading along the Thames to Oxford and then across the beautiful Cotswolds to Worcester, where we boarded a local for the last leg to Great Malvern. The TRE had taken over a "public" boys' school; Americans would call it private.

The Royal Air Force was beginning to install sets using the cavity magnetron technology for ground radars and for airborne interception, bombing, and navigation systems. Priority was on air-to-surface radars critical to antisubmarine warfare. One tanker convoy lost seven of nine ships just south of the Azores. About 100 German submarines were hunting in the Atlantic at any given time.[16]

Early on the TRE and the Rad Lab had divided the work, with the United States working hard on the components for 3-centimeter radar, the X-band, while the British continued their strong development of 10-centimeter radar, the S-band. Indeed, that latter work was already paying off. By February 1943 there were initial successes against German submarines with 10-centimeter air-to-surface radar.[17]

Bombs and Coded Love Notes

Our work in London was occasionally interrupted by air raids, which we mostly ignored except when the explosions got too close. We weren't alone in our insouciance. Even in the ferocious assaults of the 1940 blitz, about 60 percent of Londoners didn't go to the shelters. As E. K. Chamberlin noted: "The majority of the population under attack seems to have gone to bed as though nothing untoward had happened, or was likely to happen—a statistical demonstration of the power of sheer habit and of an ingrained optimism that seems indistinguishable from fatalism."[18] The air raid shelters were usually in hotel basements or subway stations. Traveling late at night on the "tube," I saw families, old folks, and children, unrolling their blankets on the hard pavement just a few feet from the trains, unable to sleep. I decided that wasn't for me. Still, we were kept awake by air raids, and I quickly found that British coffee—muddy and strong—magnified the effect. I had to choose between air raids and coffee. I quit coffee.

I greatly admired the air raid warning crews on the roofs. They would stay there with their hard hats and gas masks, ready to douse fires. And they would bellow at anyone pulling aside a blackout curtain and letting light out.

American organizations and their personnel were coming to London in increasing numbers, and they all wanted to work and live close to the American Embassy in Grosvenor Square. Our mission was also growing quickly, and we had to move from the embassy to a very good location one block from the Marble Arch off Hyde Park. Our new quarters were excellent. They were on the fifth floor, and we quickly discovered that the sixth and seventh floors were occupied by the government in exile of the Netherlands. We often met Dutch officials on the elevator, including the very tall Prince Bernhard in a fine uniform. A pleasant way to start the morning was to get on the elevator with him looking down at us from his great height. We would say "Good morning," and he would respond with "Guten Morgen." We were then living at a posh place, 40 Berkeley Square, in an apartment large enough to house five of us. We shared costs, pooled ration coupons, hired a cook, and on many Saturday nights had parties. The idyll was short lived. The Office of Strategic Services, which in time became the Cental Intelligence Agency, was also looking for space, and it had a lot more clout. So after a final and very noisy party, it was back to the Mount Royal Hotel, which fortunately was right across Oxford Street from our new offices.

In summing up my life in Britain those first few months in 1943, I realized I had grown in my capability to help in large scientific affairs. I had worked with impressive people, particularly great leaders in science, engineering, and government in both countries. I enjoyed my work but also found time to enjoy my stay, despite the frequent air raids. I dined and danced with very attractive women—British, American, Swedish,

Belgian, and Canadian. Our search for good food often ended up in a Chinese restaurant in Soho. I remember especially a cold Saturday night in early 1943. Dave Langmuir and I were far from home and not going to any parties. So we walked after dinner down to Piccadilly Circus to the cable office so Dave could cable his new wife, Nancy, in Washington. They had worked out codes from lines in poems that best expressed their feelings at the moment; for example, K1 for some lines by Keats or S34 for Shelley. The message sent, we had a drink to usher in the weekend, and hearing some singing outside went out to join in. The two of us wandered down the Strand in the blackout singing at the top of our voices when suddenly a tall figure emerged out of the dark. We looked up at a policeman, who said, "'Ere, 'ere, that will not do for the Strand in the middle of the night." We apologized and went our way.

The next day was Sunday, and the security officer at the American Embassy called Bennett Archambault, the head of our office, at home asking if "you have a person by the name of Langmuir working in your group." He went on that the British censors had discovered a message Langmuir sent that was clearly in code. "That is a security violation, and they want to know about it." They called Dave in, and he described it as a love note to his wife. They did not smile. And they were quite emphatic that one did not use coded messages. Still, I bet those very serious censors relished telling the story to their wives and girlfriends, even "decoding" the message by reading the poems.

I continued to meet and work with a widening group of the three military services, with the major laboratories, and with companies working on microwave tubes—in London and the provinces. Before I left for London, I'd been told by Jerrold Zacharias at the Rad Lab that the British manufacturer of vital semiconductor crystal rectifiers had a much better yield than ours did, even though there had been information exchange on the process. I visited the manufacturer, the British General Electric Company,[19] and was cordially received and given a tour of the tube manufacturing facilities, ending at the assembly line for the rectifiers. I compared every step of the process with Zach's description of the U.S. one, and they were identical. Then I noticed that the women in packing each unit gently tapped it on the wooden table. I asked why and was told: "Oh, we find we get a better yield if we tap it just so." That very firmly set the tungsten "cat's whisker" part of the mixer on the silicon crystal semiconductor. Problem solved and reported to Zach.

THE COMPTON MISSION

The intensifying war effort made even more urgent the effective coordination of British and American radar work. The American radar program had at the time nothing like the tight coupling by the British of research and development to military needs and operations. British laboratories were in close contact with their front-line airfields, naval ports, and army units. Their need to defend Britain from air attack was still great. And Bomber Command taking the war to the Germans was at airfields all around them. They had daily working relationships, some scientists were even involved in operations, and there was a "Sunday Soviet," in which the research and development leaders would meet with military leaders for informal talk on immediate- and long-range problems.

All this activity led to the Joint Chiefs of Staff setting up an ad hoc Committee on Radar Research and Development headed by Karl T. Compton, president of the Massachusetts Institute of Technology (MIT). The British counterpart was a high-level radar committee, led by Sir Stafford Cripps, the minister of aircraft production. These committees were in some ways creatures of success, the realization that research and development programs in each country were exceptionally successful in developing radar components for a widening set of military uses. Needed now was integration of British and American priorities for different radar sets, production goals, and the calibration of new applications.

"Badly Shaken"

In April 1943, three months after I arrived in London, Karl Compton came to England as head of the Committee on Radar Research and Development[20] to discuss, if not create, bilateral radar policy. Dave Langmuir, in preparation for this visit, came up with an ingenious idea— to create a pocket reference book for the mission members summarizing every radar system the British had in research, development, or production and including for each set performance data, development schedule, and intended use. The British had not done this, fearing quite reasonably that their proximity to the enemy made the security risk too great. I demonstrated one day how great the risk was when I took a taxi from our office to the Air Ministry to confer about a couple of British radar sets for inclusion in the notebook. The taxi took a pleasant route along Park Lane, over to Hyde Park Corner, past Buckingham Palace, up the mall through the Admiralty Arch, then down Whitehall to the turn into the Air Ministry. I studied the notebook for a bit and then just enjoyed the scenery. Arriving at the ministry, I paid, bounded up the steps, and entered. I was just getting cleared when I realized I didn't have the notebook, the guide to some of Britain's most valuable radar secrets. I rushed out. No taxi. No traffic at all.

I ran the 100-yard dash in record time to Whitehall where it joins Parliament Street. There was only one taxi. It was my only hope and so I intercepted it. It was my taxi! It had gone down to Parliament and getting no fare had turned around. The notebook was on the back seat. I returned to the Air Ministry badly shaken. The finished notebooks made a great hit with the Compton mission.

The mission had a major impact. One of its first recommendations was to intensify the scientific and technical cooperation on radar with specifics on who should do what. Thus, the United States should focus on longer-range development work, given the likelihood that the Pacific War would last longer; on the further development of tunability, lower voltages, and higher power in the magnetrons; and on the technology for the 3-centimeter, H_2X, radar, as well as 1-centimeter, K-band, radar.

Lee Dubridge and Louis Turner, also from the Rad Lab, stayed on in Britain after the Compton mission left to begin implementing its recommendations. That kept us plenty busy, helping them and the flood of other Rad Lab visitors who came after them and who needed help pairing with their British counterparts. Lee sought especially to understand the decision by the British to continue their emphasis on 10-centimeter radar, the H_2S system, when the Americans were working hard on 3-centimeter, H_2X, developed under the leadership of George Valley. The British of course knew that the H_2X would give better target definition. But they had already invested much of their limited capacity on an H_2S crash program under heavy pressure by the Royal Air Force Coastal Command for its battle against German submarines and by the RAF Bomber Command for its attacks against German arms industries, synthetic fuel plants, and other targets beyond the range of the bomber guidance systems then in use, GEE and OBOE.[21]

EUROPEAN WEATHER

The U.S. Eighth Air Force as it intensified its operations in 1942 discovered the limitations of daylight bombing. It didn't have the long-range fighters to protect its bombers inland against German fighters and was working hard to correct that. And there wasn't all that much daylight. The European continent is famous for its cloudy weather, and there were many days when daylight visual bombing was impossible. The Eighth Air Force in late 1942 tested British equipment for blind bombing, with one or two aircraft in each formation equipped with a British grid navigation system, either the GEE or the more precise OBOE, both using flight patterns set down by in-flight radio communications from ground stations. But both systems were limited to line-of-sight transmissions and could not go beyond the Ruhr.

The Eighth Air Force search for blind bombing technology then went to the 10-centimeter, or H_2S, system. It was carried on the plane and showed cities, coastlines, rivers, and other large features of the terrain. But there were signs that the Germans had countermeasures to the 10-centimeter radar, having retrieved a set from a bomber that crashed near Rotterdam. And in any case the air force was anxious for even more precise guidance for its bombers and successfully tested, in a November 1943 raid on Wilhemshaven, crash production models of the 3-centimeter system, which offered finer details. That target was missed by eight previous bombing missions. By year's end, H_2X radar was guiding 90 percent of American bombing runs. And in time it became the Allies' major bombing radar, and a stellar Rad Lab success story.[22]

Window

Luis Alvarez, a brilliant physicist, inventor, and experimentalist, was responsible for two Rad Lab projects of especial interest to the RAF, both using 3-centimeter radar: the Eagle blind bombing system and a blind landing system. Pilots then would not admit that a completely blind system was safe, so to accommodate the doubters the name of Alvarez's system was changed from ground control landing to ground control approach. It is now standard in commercial airports.

I arranged for Luis to meet at an airfield with Air Commodore, later Vice Marshall, Donald Bennett, commander of the RAF Pathfinder Force, responsible for marking targets for the main bombing force. Bennett was Australian and the youngest air commodore in the Bomber Command.[23] Luis was very late. Suddenly, while the air commodore and I were talking, Luis burst in along with his pilot, both disheveled. They had crash landed their plane. After they recovered, the discussion went on, ranging from the night bombing techniques of the Pathfinder force, especially its technique for marking targets, to Luis's blind landing system. It turned out Bennett himself had just tested in Scotland the first set sent to Britain.

When it was time to leave, the air commodore said, "If your plans would permit you to stay overnight, we have a very large raid on, in fact our largest ever, and you can hear the crew briefings before takeoff, and their debriefings after their return tomorrow morning." Luis and I jumped at the chance. And then the air commodore added, "By the way, we'll be using Window for the first time tonight." Our eyes flew wide open; this was the first airborne use of a highly classified countermeasure to confuse enemy radar. It meant dropping bundles of thin strips of aluminum foil, cut to just the proper length to resonate with the wavelength of enemy radar. The idea was that a bundle of strips would spread out and look like an aircraft on radarscopes.

This raid was the first of three days of bombing the Port of Hamburg, with 739 planes used the first night and 3,000 over three days, including Eighth Air Force daylight bombers. It was a frightfully destructive raid with many civilian casualties, largely from the enormous firestorm touched off by incendiary bombs.

The predawn debriefing of the crews after the first raid differed dramatically from the preraid briefing the night before when the crews were told to use Window, by dropping the foil bundles serially and regularly. They argued strongly and loudly that each crew member already had more than enough to do. They changed their tune when they learned at the debriefing that they had lost no planes from their airfield and very few from other fields. The Germans had a hard time with Window. Radio operators listening to the German broadcasts heard their ground-to-air controllers complaining: "The planes are having babies. What's happening?"

The Germans came up with countermeasures to Window. And we did our part to make Window even more effective. The British Countermeasures Board, with which I liaisoned in Dave Langmuir's absence, worked on keeping the initial successes up. For example, illustrating the coordination between British and U.S. efforts, there was a proposal that the British use American machines for stiffening the aluminum foil to keep them at full length since those that folded did not reflect as well, attenuating the false spoofing.

"LAND OF MILK AND HONEY"

The post-Compton mission stream of visitors and bilateral work intensified. There was intense collaboration on LORAN, a long-range navigational system similar to the British GEE but working in a different frequency and with longer-range capability. Both systems laid down overlapping hyperbolic grids, enabling, through two closely coordinated radio beacons, an airplane or ship to position itself very accurately. And we had joint coordination on K-band, or 1-centimeter radar, which I had worked on at MIT just before I left for England and which in principle could show bombers details as minute as railroad tracks.

In September 1943 I returned to the United States. It was a culture shock even though I'd been away only eight months. Despite the rationing of gasoline and shortages of sugar, coffee, and the like, home was a land of milk and honey compared to Britain. Yes, the United States had been at war for almost two years when I returned and there had been many killed and wounded, but the civilians hadn't experienced the war firsthand.

Using my copy of the notebook on British radar sets I had shipped in a classified pouch, I gave a talk to the Rad Lab staff almost immediately upon my return. My theme was that it was time for radar to go on the offensive rather than being used in the main to defend against attacks. The example I used was height-finding radar at Appledore in Devon, England, which the British were using for offensive air sweeps over the European continent. It was critical for RAF fighters engaging German fighters to know their altitude and to get above and behind them for maximum advantage. The tracking radar being worked on at the Rad Lab, using microwaves versus the longer-range Appledore radar, gave excellent positioning information but not height. Height finding was added after my talk.[24]

Concurrent with my visit, Lauriston ("Larry") Marshall was getting ready to leave for London to be the first director of the newly established British Branch of the Radiation Laboratory, the BBRL, another product of the Compton mission and Lee Dubridge's follow-up. It was solid recognition that a group was needed onsite to install, test, maintain, and modify as needed Rad Lab radar sets, especially the H_2X systems. The British branch grew from about 30 people at the start to over 100 after D-Day.

After the Rad Lab meeting and the departure of Larry Marshall for London, I went to Washington for meetings at the OSRD Foreign Liaison Office, my home base. Bennett Archambault, my London boss, and Lee Dubridge had come in for meetings. I also met with people in the countermeasures business and visited major laboratories at Bell Laboratories and Columbia and other universities working on magnetrons and 10-centimeter and 1-centimeter radar. When I got back to the Rad Lab toward the end of October, new ideas were percolating for me to take back to London, including the Microwave Early Warning (MEW) system, a very high-powered system conceived by Luis Alvarez.

Turf

I returned to London on November 27, 1943, the day before Churchill, Roosevelt, and Stalin met for the first time, in Teheran, where they agreed to invade Western Europe in 1944. I came back to major changes. One was that the establishment of the BBRL meant realigning some of the work done until then by the London OSRD radar office, namely, Dave Langmuir and I. The second change was the reorganization of the

war effort to establish Britain as the base for landings in Europe. A new
tactical air force, the U.S. Ninth, had been created to attack targets close
to the invasion site and to transport paratroops. It complemented the
U.S. Eighth Air Force, which would continue joint attacks with RAF
Bomber Command on German air bases and cities and the plants making planes, tanks, and munitions for the German war effort. About this
time I got a letter from Victor Neher that included a message from I. I.
Rabi that I should return to the United States lest I get damaged in the
fighting. No way, with London on the frontlines and about to be part of
the greatest event of the war. Still, it was clear my portfolio would change.

Larry Marshall returned to the United States in January 1944, and
Sam Goudsmit became acting head of the British branch of the Rad Lab.
Goudsmit had a second and highly secret task as scientific head of the
ALSOS mission to learn whether the Germans were successfully developing an atomic bomb.[25] There was a second reason for Larry's return to
the United States: a fierce turf battle had broken out between OSRD
and BBRL, more especially its two strong and talented leaders,
Archambault and Marshall. I kept as aloof from that battle as I could.
Soon John Trump replaced Marshall. John had a foot in both camps,
being secretary of the radar division of the NDRC, now a part of OSRD,
and having his office in the Rad Lab.

Before he left, Larry helped the Ninth Air Force acquire OBOE, the
British blind landing system to cope with Europe's foul weather. Sam and
I finished the job by gathering the pertinent data, presenting it to the
Ninth Air Force, and recommending that it use OBOE, while the Eighth
Air Force, bombing deep in Germany, used the H_2X system. First used in
1942, OBOE, while accurate, had a limited range and the Germans had
begun jamming it. However, by late 1943 it had become much less vulnerable to jamming. OBOE depended on distance measurements from
two ground stations, enabling the plane to locate its position very accurately and to time its release of bombs or markers for the follow-on bombers. Joe Platt[26] came to England with a strong Rad Lab and BBRL team
to handle the equipping of the Ninth Air Force with the latest microwave
OBOE, which was credited by General Sam Anderson with about half of
the Ninth Air Force raids over France before and after the invasion.

I also helped a Rad Lab team[27] arrange the emplacement of the very powerful MEW system near Dartmouth, for tracking large bomber and fighter sweeps over the Channel, Normandy, and Brittany. The MEW system was installed in February 1944, and while managed by the Ninth Air Force, it served all of the Allied air forces. It gave a superior overview of the land and air armadas during the Normandy invasion. A MEW set was also on Omaha Beach to direct fighter operations six days after the Normandy invasion began.

MISSILES

In late December 1943 Dwight D. Eisenhower became commander of the Supreme Headquarters of the Allied Expeditionary Force, SHAEF, based in London. The U.S. Air Force's presence in Britain was already heavy, but now the number of U.S. Army people coming into the country was overwhelming. "The American invasion struck the British Isles like a huge Technicolor bomb, scattering nylons and cigarettes and goodwill over the whole country. . . . The sheer prodigality of American equipment and supplies dazzled a population which had been cozened into accepting carrot flan as a kind of luxury."[28]

My customer base widened enormously with the complex hierarchical structure and many new levels to negotiate that necessarily came with SHAEF and its invasion planning. A third unit was added to the London mission of the OSRD and BBRL: a civilian radar unit,[29] sent by Eddie Bowles, special assistant to the secretary of war, to advise, with British colleagues, including Robert Watson-Watt, who headed it, on optimal uses of radar and countermeasures during the first critical days of the invasion.

The Germans took note of the buildup in their own fashion with "The Little Blitz," a series of raids in 1944, the most intense during the week of February 19. One of these was the worst air raid I experienced in London. I was in my fifth floor room at the Mount Royal Hotel, and on the top floor was Colonel Arthur Warner, an academic, who had been at the Rad Lab serving as reserve officer in the Coast Artillery and Antiaircraft Command. He came over to look at radar preparations. As many times before, I opted to stay in my room rather than go to the shelter during the raid. Art Warner joined me at its peak, badly shaken by what was happening, with many nearby explosions and fire bomblets[30] hitting our roof followed by commands to extinguish them from fire marshals. We kept up each other's spirits. We walked outside the building after the raid to see the building across the street in flames and many fires lighting the sky. Art Warner's report on antiaircraft activities would now have the ring of authenticity. And later that year in June the Germans first used Vergeltungswaffe-I, their Vengeance Weapon-1, or V-1, a flying bomb. Britishers called the V-1 "buzz bombs" because of the frightening noise they made before they struck.

The buildup by the U.S. Army included antiaircraft units, with a major part being the integration of SCR-584 radar into the preinvasion defense of Britain. The SCR-584, developed at the Rad Lab under Ivan Getting, was a gun-laying radar that came to be heavily used for antiaircraft batteries in Britain, for the European invasion, and for the war in the Pacific. It was then the largest investment in radar sets ever made but also very "profitable" since it was used to detect enemy aircraft and to control antiaircraft guns and, offensively, to guide aircraft to air and ground targets.

By spring 1944 I had two new assignments. First was planning technical intelligence missions to the continent to follow our troops as closely as we could to capture intelligence on the Germans' technical capacities. This meant briefings on intelligence already in hand, planning specific missions, and working with other Allied units and our own three services. There were also safety briefings on the booby traps used by the Germans. As Technical Observer #300, I was given the simulated rank of major, to help get the equipment I needed and, not least, to be treated as an officer if captured.

The second assignment was guided missiles, the precursor to my professional life after the war. The stimulus of course was the intelligence coming in on the V-1 and V-2 German missiles. And the British had started the GAAP program, on a guided antiaircraft projectile. Vannevar Bush assigned me as liaison to the project. I soon became pretty knowledgeable about those new technologies, with frequent visits to the National Physical Laboratories, where they had a supersonic wind tunnel and a brilliant theoretical team; to the Royal Aircraft Establishment, where they were working on missile guidance and control; and to the other governmental and industrial laboratories working on rocketry. I was now heavily involved in the guided-missile business and less in radar, which in any case was being well handled by the new BBRL leadership and by Bill Breazeale at the London Mission.

"A FLOATING DOCK"

It was May 1944, and the time for invasion was close. The mood in Britain was extraordinary. "Living on this little island just now uncomfortably resembles living on a vast combination of an aircraft carrier, a floating dock jammed with men, and a warehouse stacked to the ceiling with material labeled 'Europe'. . . . The fight everybody is waiting for hasn't started yet, but all over England, from the big cities to the tiniest hamlet, the people, at least in spirit, seem already to have begun it."[31]

My first mission to the continent was beginning to take shape. It was to be joint with Howard P. ("Bob") Robertson,[32] a theoretical physicist from Princeton, who, although I didn't know it at the time, had been sent to England by Vannevar Bush in part because of the intelligence leaking through on the Germans' missile programs, especially the V-1. Odd construction sites, called ski sites, had appeared along the French coast in an arc stretching from Cherbourg to Belgium and Holland. They were "ski sites" because of a jog at the end of a long and narrow building. The odd design, easy to see in reconnaissance photos, was copied from a building at Peenemünde on the Baltic coast that had a little jog for truck loading. The Germans soon learned why we spotted their sites so easily and went to a straight design. But the secret was out.

On the morning of June 6, D-Day, shortly after we heard that the invasion of Normandy had begun, Bob Robertson, Bennett Archambault, and I met with R. V. Jones,[33] the superb intelligence leader for the British Air Ministry, to begin planning in detail the Robertson-Stever mission to Normandy. Bob and I were to land on the invasion beaches on D-Day plus 16 and then follow the army up the Cotentin Peninsula toward Cherbourg. Jones and his colleagues shared with us their technical work and the vast amount of intelligence on the German missile program they had. Our office routinely received copies of the weekly intelligence s ummary of the British war cabinet. For some time the summary comment was that "the Germans are not ready to use the V-1 operationally this week." Then, six days after the invasion, the summary reported that "the Germans are ready to use the V-1s operationally this week." And, sure enough, they did. Four V-1s were fired at London a week after D-Day.

The V-1 was a forerunner of cruise missiles, a pulsejet that got its thrust from burps of hot, high-pressure gas at the rear of a cylindrical tube.[34] A simple, inexpensive, and tremendously effective weapon, it would not have been a great weapon if it had to be very accurate. But it had a target, London, 15 to 20 miles wide, and it had to fly only 100 to 200 miles.

Each V-1 snarled impending disaster from the buzzing of its pulsejet, followed by an eerie silence as the engine cut off and the weapon fell. Many were shot down over open country or sea, but those that got through killed or wounded over 24,000 people. There was no use shooting one down once over London because it was just as lethal. Antiaircraft guns were set up over the south coast, using the SCR-584 radar, a Bell Laboratories gun director aiming where the missile would be when a proximity fuze reached it. It surely startled the Germans looking out over the channel to see some of their V-1s shot down by a single shell rather than the usual barrage. The reason of course was that the shells didn't explode unless they had run in close enough to touch off the fuze. These fuzes used small, continuous-wave radar to sense that they were close enough to the target to explode. Those that missed their target exploded on the sea surface.

Curiously, on the same day that the first V-1 was fired, a V-2 shot from Peenemünde went astray and landed in a farmer's field in southern Sweden. British intelligence successfully arranged with Swedish officials to pick up the pieces. One of my colleagues in the Air Ministry, Squadron Leader Wilkinson, flew to Sweden with another squadron leader in the bomb bay of a Mosquito bomber. The pieces were brought back to the Royal Aircraft Establishment in Farnborough, where, just as curators do with dinosaur bones, they were suspended from a ceiling until they could be partially pieced together. It didn't give all the answers, but it was an important step toward understanding the next weapon to hit Britain.

And it soon did—on September 8, 1944. The V-2 was the first rocket-propelled, medium-range ballistic missile.[34] It came with no warning and hence was more terrifying. Just a flash, the bang of a supersonic shock wave, and an explosion. If you heard nothing, all was well; if you heard two bangs, a shock wave and an explosion, you were ok; if you heard only one big bang, you were gone! When one final explosion on March 29, 1945, ended five years of aerial bombardment of Britain by airplanes, buzz bombs, and rockets, "the enemy had dropped a grand total of 70,995 tons of high explosives and uncountable thousands of incendiaries, killing 60,595 civilians and seriously injuring another 86,182, destroying 222,000 homes and damaging nearly five million more."[36]

NORMANDY

Bob Robertson and I flew to France about 20 days after D-Day, delayed by a strong storm at the Normandy beaches. Our C-47 had one other passenger, an army nurse, a two-man crew, and one very heavy replacement engine for a P-47 fighter. Since the invasion, the nurse had shuttled every day between France and Britain with a load of severely wounded men and then had returned to France with badly needed cargo. She was in a very bad state, her morale understandably low after having dealt with badly wounded men in tough flying conditions. We talked to her about other things and it may have done some good.

After an aborted first try, we landed on a small dirt field on the cliffs above Omaha Beach. We got an instructive tour of the beach, reminders of the terrible fighting a few days before, and then drove through Carentan toward the main road north up the Cotentin Peninsula toward Cherbourg. And we drove quickly, because the Germans were still shelling Carentan. The battle was on to break through to the west coast of the peninsula and to seize Cherbourg to the north. We passed through Ste.-Mère Église, where paratroops had landed on D-Day and saw the aftermath of the vicious fight, especially in the fields where gliders carrying troops landed—or tried to, some smashing into trees, walls, or earth berms.

We were hosted by an engineer's unit encamped in a field surrounded by the typical Norman bocage of small trees and bushes, solidified by centuries of accumulated rock and dirt. The engineers built a new dirt airfield every day, returning every night to tell us that another had been built. Remarkable. We slept in pup tents to the nightly sounds of German shells heading north and east and American shells south and west. Still, the field, circled by very high hedges, gave one at night an illusion of security as the bombardment went on overhead.

Tit for Tat

Our standard work routine was to leave right after breakfast, mostly in one jeep when Bob Robertson and I went together and our escorting officers had other missions and two jeeps when we went together. We had a bunch of different targets. For radar and guided-missile targets, we would take as many pictures as needed, make notes of both the effects of bombardment on the targets and of the demolition by the Germans as they retreated, and try to find something new in the equipment the Germans left behind. We seldom found any intact radar equipment, although the nature of the antennae, the electronic equipment, the communications, and the raid plots at the radar centers were revealing. I was quite impressed that our earlier intelligence briefings on what kinds of radars were located at given sites were so good. We had a good idea of what was there, from detecting the emanations from the radar sets and from photo reconnaissance observers picking out from the photos the antennae and layouts of the station. The very large long-wave antennae used to lay down beams for German bombers in their raids over Britain at night or during bad weather were located and identified by this sort of work by the British, who then used the information to jam the stations. The engineers who operated those beams constantly found countermeasures to the jamming, and then new kinds of jamming were introduced. An interesting tit-for-tat electronic battle. We arrived at one of those beam sites so soon after the fighting that we found German bodies still unburied. This made our site inspection unpleasant, even more so by our first discovery of booby traps.

Also, the V-1 launch sites had been bombed for some time before the invasion. Bob Robertson visited V-1 sites to study the effects of the bombing on them and to check if they were as advanced as intelligence said. They were. Also, aerial photography in France found several structures, dug deep into the ground with thick concrete walls and roofs, which were candidates for V-2 use, either assembly or storage. One of these V-2 sites was on the Cotentin Peninsula not far from Cherbourg in a town called Sottevast. It was on our target list and of most interest to Bob Robertson with his missile intelligence portfolio and also a second portfolio: operational research, with his particular interest being the effectiveness of bombing raids and bomb penetration against concrete structures such as submarine pens and the V-2 sites.

Bob and I drove our jeep to the Sottevast site, and while we found considerable damage, none of it was lethal to the operation, and it was mostly repairable. Even a direct hit on the concrete bunkers would have left a crater 5 feet deep and 12 to 15 feet wide but wouldn't have penetrated. Bigger and more penetrating bombs were needed. At the V-2 site we were looking at, we came upon an office with a safe circled by a lot of dried blood. Someone had forgotten about booby traps.

About the same time we were visiting the V-2 storage site at Sottevast, American troops finally captured Cherbourg, secured almost the entire Cotentin peninsula, and captured the rest of it, Cap de La Hague, within a few days. Bob Robertson's high credentials easily got us to the commanding general, who arranged for us to tour the vast fortifications and technical systems the Germans had built in the city, particularly in the port. The army hoped to capture the Cherbourg port intact, but the Germans had demolished the place, including sinking ships in the harbor to block navigation. However, a fuel pipeline was laid within several weeks from England to Cherbourg to supply airplanes, trucks, cars, and tanks.

We drove to the west coast of the Cotentin Peninsula to look at technical emplacements. Near Barneville on the coast we found huge antennae used by the Germans to communicate with their submarines and ships in the Atlantic. We didn't stay long because just south of us troops were still fighting hard to capture La Haye-du-Puits, finally taken on July 5. On the coast north of Barneville, we found another concrete bunker. We probed it with flashlights and soon found inside bundled explosives with fuses attached. We left.

The Generals

Driving leisurely in sparse traffic down the main Cotentin road to an airport for our return to England, suddenly there were sirens and flashing lights. We were pulled off the road by military police. Then we saw a long armored column led by an open command car with tanks rumbling behind. In the command car stood General George Patton. He was in full "travel costume," with a brass-plated helmet and his famous special pistols on his belt. He stood straight, looking like he was in a victory parade, although the only audience was us. He was leading his Third Army to St. Lo to exploit a breakthrough by the First Army. Later in Brittany I saw General Omar Bradley, commander of the U.S. land forces on D-Day, with a driver, a single guard, one jeep, no battle regalia, and no tank column.

I was happy to be back in London and spent much of my time there preparing for my next trip to France, this time to Brittany. Much of Brittany was open, although there was still intense fighting in Mont St. Michel to the east and in Brest to the west, and there were scattered pockets of Germans cut off from escape who might still put up a fight. It would stay that way for some time because Patton's Third Army headed to Paris after the breakthrough at St. Lo.

This time our team consisted of Colonel Eric Bradley, commander of the U.S. Army Air Force Technical Section in London; Bennett Archambault, science attaché at the U.S. Embassy; Major Carl Lindstrand on the staff of the U.S. Army Technical Section; and me. We tried first to go to Lorient on the Brittany coast to get to the German submarine pens. Nothing doing. The Germans still held the town. Our route then took us inland to the northern Brittany coast, from Rennes through St. Brieuc and Guingamp to Lannion. We must have been the first Americans to get into Lannion because the crowd was so huge the police had to lead us through. There were flowers on the jeep, hands to shake, and hugs and kisses from women of all ages. Pleasant, but I felt guilty because I hadn't done the fighting to get us there.

Capture and Surrender

We drove on without stopping to ask about the local situation at one of the Free French barriers. Not wise, it turned out. We drove a little farther north to Perros-Guirec, a small coastal town, to have a look at a major radar control center in a farmhouse on top of a hill, which was used to direct fighter aircraft. We drove up the hill along a small and narrow sunken road bordered by hedges. As we got closer, we suddenly realized there were many people on both sides wearing German helmets and carrying rifles. German soldiers in the road with rifles ready stopped us. Our senior officer, Colonel Bradley, a calm veteran of World War I, told us in a very low voice, "Don't say anything." That was wasted on me because I was speechless!

The Germans led us through a minefield to the farmhouse. There the German commander insisted that we use only German or French, which made us grateful for our school language studies. He wanted to know how close our tanks were, thinking that colonels commanded divi-

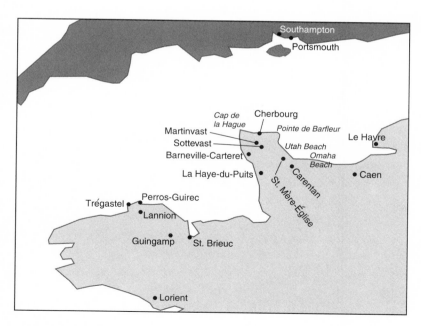

Normandy and Brittany sites.

sions and that we had come forward to parley, leaving our tanks hidden down the road. Colonel Bradley temporized for some time but then realized that this group wasn't anxious to fight but to surrender. We got over that shock—we were after all technical not combat personnel—and suggested to the Germans that they surrender to us. They agreed, with two conditions: that we honor the Geneva Convention for the treatment of military prisoners, although we didn't know much about it, and that they be made prisoners of either Americans or British. About this time three Free French officers were brought in under a flag of truce along with a Canadian officer who had been dropped by parachute well before the invasion to help the Free French work with our troops. We agreed to the Germans' surrender terms, sealed at the German commander's request with a glass of schnapps, and after some more palavering, some 500 German troops were carted off in rickety trucks, the engines fueled by natural gas kept in a bag on the top.

Whispered Conversations

Then we were back to Lannion for an impressive celebration dinner with the mayor and all the leading citizens of the area. Sad Brittany songs were sung with tears—"Ni zo bepred Bretoned, Bretoned, tud kaled," "We are the Bretons forever, a robust people."[37] We responded first with the "Star-Spangled Banner" and then, after much wine, with "I've Been Working on the Railroad." We stayed the night in a large and well-appointed chateau, finally going to sleep after a lively talk with the leader of the Free French group we had met after we "captured" the Germans. He was a thoroughly civilized fellow whose politics we disagreed with. It turned out he was a communist sent from Paris to direct local operations—a harbinger of French politics after the war.

Exploring the abandoned site the next day we found the entire raid-reporting center housed in several spacious rooms underneath the farmhouse. And we found an ingenious device for tracking aircraft on a visual map to control the air battle. Eric Bradley and Bennett Archambault returned to London with this haul.

Carl Lindstrand and I drove on to Trégastel on the Brittany coast, to look at a German Würzburg radar, but the Germans had blown it. We were invited to stay for dinner and the night by the local pharmacist, Monsieur W. Jouille, and his wife. The dinner, contributed by many neighbors, was memorable, and our rooms were pleasant, but we were kept awake by nearby machine gun fire and whispered conversations just outside our rooms. The next morning we were told that the local Free French unit had guarded both our house and jeep. The danger was real. Another unit of our technical intelligence group working about 10 miles to the west came upon some Germans, including SS. They opened fire with no warning, killing an American officer and wounding another. And we met the wife of a Free French leader who had delivered supplies to a Free French unit. She and her child were held hostage by the Germans and their house was burned, but the leader did not give himself up. He was caught and executed. Carl and I stopped at a beautiful stone church to pay our respects to him. It was a moving moment with many of the local citizens taking part.

On the fiftieth anniversary of D-Day, I wrote letters to M. Jouille and the leader of the Free French unit that had guarded us and got back wonderful letters from Madame Jouille, the pharmacist's widow, from the Free French fellow, and from the current mayor of Lannion, who had an account of the entire episode published. The pharmacist and his wife moved to Paris after the war and set up a very successful pharmaceutical laboratory there. The Free French fellow had gone with his colleagues, after we left, to help American troops in their attack on Brest. In 1946 he returned to his job working on a fishing boat off the Brittany coast.

Paris

Then it was back to London to prepare for our next technical intelligence mission, this time to Paris and this time as part of a joint British-American technical intelligence team, coordinated by a new SHAEF organization, the Combined Intelligence Operations Section. We landed on a grassy airfield just outside Chartres, where we were held up because of the heavy fighting in Paris. The spectacular view of the cathedral across the flat fields of the Beauce with its uneven steeples, one twelfth-century Romanesque and the other thirteenth-century early Gothic, rebuilt after a disastrous fire, became for me the symbol of France. It was the first time I saw a part of France not torn by war.

We entered Paris when it was liberated by the French Second Armored Division and by the American Fourth Division. On August 29, 1944, General Eisenhower, in part worried about further German resistance or counterattacks and in part to reassure Parisians that the taking of Paris was an Allied victory, ordered a march through the city by the Twenty-Eighth U.S. Infantry Division, newly landed in Normandy and transported to Paris on its way to the front. I was so proud to see those Americans dressed for battle marching down the Champs-Elysées, the tanks and artillery caissons rolling along in great numbers three abreast. Paris celebrated as never before.

My colleagues and I billeted at the Ambassador and had immediate invitations, one from Sosthenes Behns, the head of International Telephone and Telegraph, who had flown in with us to reestablish the company's Paris office, and the other at the Hotel George Cinq, where the host was Jacob P. den Hartog, then a professor of mechanical engineering at Harvard and after the war at MIT. Den Hartog, a Navy Reserve captain, was leading a delegation looking at the same things we were: captured technical information and equipment.

Our work was very different from our first two trips to France, for we were visiting operating industrial plants and laboratories and intact German facilities. But that doesn't mean that we got any more information. Industrial officials mostly claimed they had done no technical work for the Germans. I'm sure this was in part true. Paris quickly replaced London for technical operations supporting our troops, both for the BBRL and for the London liaison office of OSRD. Many more scientists arrived from the United States, including Charlie Lauritsen, who had taught me nuclear physics at Cal Tech and had been on my graduate thesis committee. And Sam Goudsmit came, to join me on my radar visits in Paris intended to hide his real purpose, the ALSOS mission described earlier to determine German progress, if any, on building an atomic bomb.

Of all my technical intelligence trips, I got the least useful information out of Paris. But it was a fabulous two weeks walking in a beautiful city, relatively undamaged and in a festive mood. A. J. Liebling told his

American readers that "for the first time in my life and probably the last, I have lived for a week in a great city where everybody is happy."[37]

The Allies made rapid progress. Patton's Third Army reached the German border and freed Luxembourg. To the north and along the coast, the British and Canadians had a difficult time of it as the Germans fiercely defended the port towns of Le Havre and Antwerp. And to the east the Russians and Poles reported evidence that the Germans had murdered 1.5 million people at the Majdanek concentration camp. It was the first of many such horrifying discoveries by liberation troops.

Disaster

Our next trip was more promising—to Eindhoven in Holland, home of Philips Electronics. But getting to Eindhoven depended on the success of a major Allied operation, "Market Garden," an airborne assault on September 17 on Holland. The goal was to seize canals and bridges to enable ground troops to enter Germany over the Rhine at Arnhem and into the Ruhr Valley. The rationale for the attack, flawed as it turned out, was that the German forces were so weakened that maintaining the pressure through a focused rather than broad attack would lead to quick collapse. The German resistance in fact turned out to be fierce, especially at Arnhem.

We returned to London to prepare for the Eindhoven trip and found a mood change in the city that was dramatic if short-lived. The Germans' loss of their V-1 sites and airfields in northern France and more effective defenses—new high-speed RAF fighters together with SCR-584 directed guns and proximity fuzes—virtually stopped air attacks on England. The happiness in removing the blackout curtains quickly faded when in September the first V-2 landed in London. The arrival of that silent frightening weapon, against which there was no defense, made even more urgent the thrust into Germany and the capture of the V-2 sites.

We stopped in Brussels and then continued north while the Market Garden battles were intensifying. We were often delayed by German attacks on our long line of tanks, artillery, and troops concentrated on a single highway. We continued to the outskirts of Eindhoven, one of three sites for paratroop drops. More German aerial and tank attacks, but we finally made it to Eindhoven and a welcome by Dutch scientists at the Philips plant. Our team was led by Robert Watson-Watt, A. P. Rowe, and W. B. Lewis from the British side and Bennett Archambault from the American. In a few days, we got a superb overview of electronic componentry—oscillators, cathode-ray tubes, and the like. I did notice, however, that there were areas of their own research about which they were reticent. Quite understandable, since with the end of the war they would face intense international competition for markets. And it was a time not only of technical exchanges but also long talks with our hosts at whose home Dave Langmuir and I were staying, a Philips engineer and his wife, and with their neighbors, of how hard their life had been under the Germans.

And then there was the daily evidence out the window that Market Garden was a disaster. Every day an almost unbelievable number of C-47s would fly north in great formations to drop supplies to American troops at Nijmegen and to the British at Arnhem. American paratroopers and British tanks got across the bridge at Nijmegen but meeting strong German resistance could only get a few miles farther, and they couldn't get close enough to help British and Polish paratroops at Arnhem. Many of these had to surrender, but some escaped to be sheltered by the Dutch. The Germans were pretty rough on the Dutch people, who had counted on being freed. Civilians again suffered in the awful ways of war.

DEEPER INTO MISSILES

Back to London and two weeks later in mid-October to the United States. I met with Vannevar Bush, who surprised me with his intense interest in guided missiles. I told him of the British work on their guided antiaircraft projectile, GAAP, and our intelligence on German guided missiles, including, of course, the V-2. He was interested in the potential of guided missiles as a major war weapon although he didn't think they would have a great effect on the war.

Bush asked me before I returned to London to work with various groups engaged with guided missiles. Arrangements were also made for me to visit the Air Force's research facility at Wright Field and guided-missile facilities in Southern California. I visited Cal Tech and the Dou-

glas Aircraft Company, the latter working on a rocket-powered glide
bomb.

At Cal Tech I talked with William Pickering, my substitute thesis
advisor when Victor Neher left for the Rad Lab before I finished my
thesis. Bill was advising on electronics, telemetering, communications,
and tracking for ground-to-ground bombardment rockets being developed for the Army. That project became the foundation for the work of
the Jet Propulsion Laboratory with the National Aeronautics and Space
Administration, NASA, when it emerged a decade later.

As pleasant as all this was, I returned to New York rather quickly to a
meeting of the Vacuum Tube Development Committee, our counterpart
to the British Committee on Valve Development, where I reported on
what we had seen at Eindhoven and on new British magnetron work. But
the most important reason for hurrying east was to join a committee put
together by Vannevar Bush to advise him on the future for the guided
missiles then being worked on. The chairman was MIT Professor Joseph
Boyce and included Richard Courant, the exceptionally brilliant mathematician from New York University; Jesse Beams, a very able physicist
from the University of Virginia; and Ivan Getting, who was developing
radar guidance for missiles. We discussed a report by Fritz Zwicky of Cal
Tech, who had taught the hardest mechanics course I ever took. He proposed inserting very powerful rockets into the upper atmosphere high
enough to orbit the Earth. It was the first time I had contact with someone serious about injecting satellites into the Earth's orbit. That was 13
years before the Russians did it.

Before my scheduled return to London I also met with the group at
the Johns Hopkins Applied Physics Laboratory developing an antiaircraft missile to defend naval ships in the Pacific against kamikaze attacks.
Merle Tuve, a brilliant physicist from the Carnegie Institution of Washington, led the group. Tuve, with Gregory Breit, had used radar pulses to
measure the height of the ionosphere, which starts at about 30 miles
above Earth and has a big role in propagating radio waves. Merle invited
me "to sit with us while we put the finishing touches on the proposal
that we're going to make to the Navy for a guided antiaircraft projectile
program." First, he asked everyone to propose the work they were going
to do—on air-breathing ramjets, rockets, supersonic aerodynamics and

control surfaces, electronic controls, guidance, warheads, and so on. Then he asked each "to give me the amount of money you will need for your share of development." Merle toted up the estimates, which came to "two and a half million dollars. But you guys are always underestimating the amount of money you're going to spend, so I'm going to multiply that by four and ask the Navy for ten million dollars." He did, and some years later with about a *billion* dollars spent, they produced the Bumblebee[39] program for the Navy.

Double Bang

I was supposed to return to London by mid-December, but my intestinal pains were sharpening. I was soon having my appendix removed at St. Luke's Hospital in New York. I was there on December 16, 1944, when the Germans attacked in the Ardennes Forest in Luxembourg, trying to retake Antwerp and split the British and American armies. The German attacks in this Battle of the Bulge were initially successful, partly because bad weather made air support tough, and it took the British and Americans some seven weeks to regain the initiative. There were great acts of heroism, particularly at the Battle of Bastogne, on the Belgium-Luxembourg Border. And there were great military maneuvers, as when Patton's Third Army, coming through from the south, was routed right through the communications and supply lines of other army units, usually a disastrous maneuver, to reach the front and break through to Bastogne. Bastogne was relieved by the Fourth Armored Division on December 26, and the battle in the Ardennes was effectively over on January 26.

I finally got back to London in January. Confidence in victory was very high, although V-2s were still being fired. On my first day back the Germans put on a great show for me by landing one close enough that I got the double bang of the shock wave and explosion. My main interest now was guided missiles, and I dealt with many visitors on that front, from the National Advisory Committee for Aeronautics, in time to become NASA, and from different army and navy groups. And I made contact again with the team for the British Guided AntiAircraft Projectile and their work at the Applied Physics Laboratory. And I was just about to leave for the GAAP test range at Aberporth by Cardigan Bay off the Welsh coast when we got news that our troops were close to the next intelligence objective, Cologne. I led a technical intelligence team to Cologne, then to nearby Bonn, before returning to London. We didn't get all that much information, especially compared to our next trip. The destruction in and around Cologne was immense. Cities such as Jülich and Düren were flattened. We climbed inside the steeple of the Cologne cathedral, its roof gone, and could see streets made virtually impassable by the bombing; the terrible state of the railroad yards, tracks broken and twisted; and, east across the Rhine, the main bridge in the water, destroyed by the retreating Germans. The danger was still real: German snipers a few hundred yards across the Rhine fired at steeple visitors who stuck their heads out too far. And the Germans would occasionally lob mortar shells on the plaza in front of the cathedral hoping to hit GIs visiting the site.

FINAL SWEEPS

In mid-April 1945 I finally got to Aberporth to see a major test firing of an antiaircraft missile. I returned afterwards to my hotel where several British colleagues and I were to meet for beer and dinner. When I entered the pub, my British friends had long, drawn faces. They told me that President Franklin Delano Roosevelt was dead of a massive stroke in Warm Springs, Georgia. Harry Truman was now president. I was shocked and saddened, the British even more so. Roosevelt had been a great friend to them and critical to their war efforts.

The final sweeps into Germany were under way. And I was directly reminded that the end was near when an OSRD historian interviewed me for the first time on the work of our London Mission.[40] Even more interesting to me was a telegram from Theodore von Kármán in Washington, asking me to join a mission he was leading to Germany, as background for *Toward New Horizons*, intended as a postwar technical blueprint for the Air Force. Von Kármán did the report at the request of General Henry H. "Hap" Arnold, commanding general of the Army Air Corps during the war. Arnold was alarmed by the Germans' rapid development of advanced jet aircraft toward the end of the war. The Germans were already flying operational aircraft in the transonic range, powered by turbojets, and they soon expected to have turbojets powerful enough for supersonic flight. Von Kármán was then arguably a major figure in the United States in aeronautics. He came in 1930 to direct the Guggenheim Aeronautical Laboratory at the California Institute of Technology, which spawned NASA's Jet Propulsion Laboratory. He trained many of the world's leaders in aeronautics, including the director of research for the German Air Force. In fact, the head of the German Air Force, Hermann Göring, had offered the job first to von Kármán after he was already at Cal Tech. As von Kármán loved to tell it, "All I did was to send him a photograph of my face in profile." Von Kármán's nose was very prominent.[41]

I had little time to get ready but was there when von Kármán's group went to the Hermann Göring Air Force Research Establishment, near the small village of Volkenrode, near Braunschweig in Lower Saxony. The place, planned by a former colleague of von Kármán, was immense:

> Some fifty-six buildings, housing facilities for research in ballistics, aerodynamics, and engines, were built below the tree level, so they could not be identified or seen from the air. Some buildings were designed as farmhouses; others were underground, and still trees embedded in their concrete roofs camouflaged others. There was an airfield that was concealed by means of a cover of ash so it would not present a smooth surface from the air. The whole thing was incredible. Over a thousand people worked there, yet not a whisper of this institute had reached the ears of the Allies.[42]

The engineers and scientists at this remarkable place worked on transonic and supersonic aerodynamics, both theoretical and experimental, using wind tunnels for aircraft and firing ranges for artillery projectiles and rocket-propelled missiles. The wind tunnel data got the immediate attention of George Schairer, Boeing's chief aerodynamicist designing the B-47 bomber. He was so excited when he saw the Volkenrode data on a swept-back wing design near the speed of sound that he fired off a telegram to Boeing, which in essence said: "Sweep back the wings of the B-47. Data follow." That telegram gave Boeing the jump on designing high-speed subsonic aircraft. One famous German aerodynamicist refused to discuss his work until at a pleasant outdoor lunch the GI cook served up loaves of bread, real butter, and fresh milk. Turned out he hadn't eaten for a long time, and his hunger won out over his silence.

Ghastly Story

To expose them to less bombing, the Germans had moved their V-2 and some V-1 production sites to an abandoned salt mine in the Harz Mountains, near Nordhausen. We flew to that site to see the manufacturing processes and view missiles in various states of assembly. But much more startling and shocking were their labor practices. They used as slave labor great numbers of captured or displaced foreign workers, cruelly overworking and starving them.[43] The death rates were high. Nordhausen, its nearby Dora Concentration Camp, and its missile production site were liberated by American troops on April 11, 1945. The history of one of the liberating divisions, the 104th Infantry, tells the ghastly story:

> Two miles northwest of Nordhausen a huge underground V-bomb factory was discovered. It was two miles in length, with two large tunnels approximately fifty feet in width and height, connected laterally by forty-eight smaller tunnels. From 1943 to 1945, 60,000 prisoners had toiled here in the production of V-1 and V-2 bombs. Of these, 20,000 had died from various causes including starvation, fatigue, and execution. The SS was in charge of the factory and the camp, with German criminals as straw bosses. Workers were executed at the slightest suggestion of sabotage. No workers had ever been allowed to leave the camp and when they became too weak to work, they were abandoned to die and their bodies burned at the crematorium within the grounds. Reports indicated that approximately one hundred bodies were cremated per day, and there were about thirty corpses piled on the ground awaiting such treatment when the 104th arrived. These bodies showed many signs of beating, starvation, and torture.[44]

After that ghastly experience, we flew to the university city, Göttingen, where von Kármán had studied under Ludwig Prandtl, the foremost aeronautical engineer, though now that mantle had passed to von Kármán. A few of us visited Prandtl, ill and at home. The two giants could hardly speak and the rest of us could hardly guess what they felt, knowing our own torn emotions.[45]

We had climbed a tower shortly after we arrived at Volkenrode and saw rockets over a nearby airfield. It wasn't rockets but signal pistols fired in celebration: the war in Europe was over. And after more technical reconnaissance in Jena and other places, I returned to London for reports, for interrogation of German prisoners of war expert in guided missiles and rockets, and for another visit to Aberporth.

I thought about my four years in the war effort and of the immense debt we owed the soldiers, sailors, and airmen who faced death. My second thought was of gratitude to the British people for their courage through years of air raids and missiles. And my third thought was of sympathy for the wounded, the homeless, and the displaced who had to struggle on. I thought of what my colleagues from Britain and the United States had accomplished in our:

> secret war, whose battles were lost or won unknown to the public, and only with difficulty comprehended, even now, to those outside the small high scientific circles concerned. No such warfare had ever been waged by mortal men. The terms in which it could be recorded or talked about were unintelligible to ordinary folk. Yet if we had not mastered its profound meaning and used its mysteries even while we saw them only in the glimpse, all the efforts, all the prowess of the fighting airmen, all the bravery and sacrifices of the people would have been in vain.[46]

I returned to the United States in July 1945. I was 28.

3

MIT, Missiles, and Marriage

1945–1950. Chaos, a "new world," and a new life. "Chaos" in the many but short-lived missile programs started by the three services; "new world" in the pattern of postwar research and development in the United States formed by the experiences and people of World War II; and the "new life" being my own as I married, started a family, and began the professorial climb at MIT.

President Truman announced the end of the war in the Pacific on the evening of August 14, 1945. The celebrations were well earned, exuberant, and joyous—but not for all. Over 400,000 American soldiers were dead, almost 700,000 wounded. Fifty million military and civilians were killed worldwide in World War II, 6 million in the Holocaust alone.[1]

For those who made it through, the "what now?" question loomed. I did not have a job to return to. Had I gone to Stanford to teach when I finished my doctorate at Cal Tech some four years earlier, I might have worked into a teaching and research job. I had decided strongly against it, to get started on war-related science and technology. Now I could build on a fine physics education and experience and expertise in new technologies gained at a very rapid clip indeed. The London Mission of the Office of Scientific Research and Development (OSRD) acquainted me with many of the leaders in science and technology. Now was the time to think about where I wanted to live, how I wanted to live, what kinds of institutions I wanted to tie into, and what kind of work I wanted to do at those institutions. Not least, I was almost 29 years old and single, and I wanted to change that.

61

COLD WAR

Franklin D. Roosevelt died on April 17, 1945, one month before the war in Europe ended and three months before nuclear weapons were dropped on Hiroshima and Nagasaki to end the war in the Pacific. Truman met with Stalin in July and with Churchill from July 17 to August 2 in Potsdam, just outside Berlin. He came armed with the successful test of the new weapon on July 16 at Alamogordo, New Mexico, told Stalin about it, and got seeming indifference in return.[2] The Soviets of course knew about the program from their spies.

Any hope that the American nuclear monopoly would temper Soviet behavior was short lived. The Soviets tightened their control of Eastern Europe—in Bulgaria, Romania, and Poland. They turned North Korea into a satellite, with bloody consequences five years later. They pressured Turkey, Greece, Iran, and other countries for greater influence.[3] In early 1946 a young diplomat in our Moscow embassy, George Kennan, sent his famous "Long Telegram," which launched the policy of containing the Soviets that was to be the heart of U.S. defense policy for 50 years. And two weeks after Kennan's message, Winston Churchill made the Cold War official. In a speech on March 5, 1946, in Fulton, Missouri, with President Truman in attendance, he declared:

> From Stettin in the Baltic to Trieste in the Adriatic, an iron curtain has de-scended across the continent. From what I have seen of our Russian friends and allies during war, I am convinced that there is nothing they admire so much as strength, and there is nothing for which they have less respect than weakness, especially military weakness.

We possessed nuclear weapons, but the "military weakness" was never-theless telling. The inevitable decline in military budgets when a war ends was intensified by Truman's absolute determination to cut the total federal budget. Between 1943 and 1945, between my departure for Lon-don and my return to the United States, the federal research and develop-ment (R&D) budget grew almost sixfold, from $280 million to $1.59 billion.[4] That was brutally reversed, beginning in 1946, as Truman drove hard for an austerity budget. Defense spending hit a double black dia-mond slope, sliding from $81.5 billion in 1945 to $44.7 billion in 1946 to $13.1 billion in 1947. The Army Air Force in 1946 had 28 missile programs; half of these were cancelled in 1947 while others continued as component development programs.[5]

This paradox of defense budgets coming down with the onset of the Cold War was seasoned by the uncertain state of our military science and technologies at the end of the war. Yes, we possessed the Bomb, and we shared global leadership in radar with the British, but we couldn't be as sure about our position in jet engines and in missiles and rockets. The first turbojet engines had been tested in Britain and Germany in the late 1930s. A German turbojet fighter flew successfully in August 1939, a few days before the invasion of Poland. In 1941 the British flew a jet.[6] By the summer of 1944, both Britain and Germany had turbojet fighters in action.[7] The first American development, a Navy project at Westinghouse, passed its first 100-hour test in July 1943.

The 1920s were a time for rocketry, when technological enthusiasts—Hermann Oberth in Germany, Robert Goddard in the United States, and Konstantin Tsiolkovsky in the Soviet Union—all published on rockets and their potential for space travel. There was little governmental interest, except for the Soviet Union, where a disciple of Tsiolkovsky got a short audience with a receptive Lenin.[8] Robert Goddard launched liquid-fueled rockets in 1926, but for various reasons, including Goddard's secretive nature, the work at the time gained little attention.

In December 1944 the nascent Jet Propulsion Laboratory (JPL) fired a missile at a California test range. It was 92 inches long and had a range of 11 miles.[9] The work done by JPL and by General Electric, Western Electric, and Bell Laboratories toward the end of the war proved a good base for postwar missile development in the United States.

"Democratic Principle"

And not least, the work done through the OSRD set the pattern for postwar science in the United States: federally financed research done privately—that the nation maximizes its gain from research investments by funding but not directing the best work in universities by professors and their students. Vannevar Bush's great achievement was having the military see that it needed academic science to win the war, a recognition that was seconded by acceptance that the command and control style of the military would not work with academic science. Rather, Bush somehow persuaded the military that what Merle Tuve called the "democratic principle" was of the essence in successfully harnessing the very best academic science to war needs. The principle according to Tuve meant:

> Tell the worker or the people of the community what the need is, invite them to contribute in the best way they can, and let them help you and help each other meet that need. Any society or any group always selects men to handle certain tasks, by elections or by hiring them or by some other system. But notice that a boss using the democratic principle does not depend on others, he asks his men, his workers to participate. This means that they help him with the whole job, they don't just do what they are told to do. This system of asking people to help with the whole job was what I used in running the proximity fuze development. It worked so well, the whole team took hold so vigorously, that during most of the work it was a struggle to keep up with them. I often felt like a short-legged donkey trying to keep from being run down by a stampede of race horses.[10]

Some donkey! Some stampede! The Germans in contrast "made almost every conceivable blunder."[11] Convinced that they would win the war quickly, they didn't organize for wartime research, excluded their academic scientists, and gave the military direct control over research and development.

Bush saw to it that the scientists themselves figured out how to do the work to get to a goal and he created the instruments to do that in the form of flexible cost-basis contracts, highly decentralized research, and enormous variation in the types of agreements and research operations put in place, from a university simply providing space and management to that of the Rad Lab, where an institution provided space and management and recruited the scientific personnel. Bush also split procurement from research, telling a congressional committee that "new developments are upsetting to procurement standards and procurement schedules. . . . Research, however, is the exploration of the unknown. It is speculative, uncertain. It cannot be standardized." Or as a historian of the OSRD put it: "A procurement unit that is also responsible for research tends to think all its geese are swans."[12]

INTO GUIDED MISSILES

Yes, the V weapons the Germans fired at London, other English cities, and Antwerp were primitive. But they were also pretty destructive. And now there was portentous technological convergence: the atomic bomb that could make a V-2-type weapon thousands of times more destructive wedded to aeronautical advances in high-speed bombers and fighters and greatly improved rockets that might even gain intercontinental range. Soon the notion and then the reality would emerge that intercontinental guided missiles and nuclear weapons were to be the dominant technology of the Cold War.

That was the setting when on my return to the United States Vannevar Bush asked me to stay on the payroll on assignment to the Committee on Guided Missiles of the Joint Research and Development Board of the Joint Chiefs of Staff (JCS). That's a mouthful for an organization, but it was the focal point for determining the development of guided missiles and their military applications after the war. It was created in May 1942[13] to set up a direct civilian link to military planning of the war at the highest levels, a crisp expression of Bush's blunt assertion that "scientists should sit with the heads of the Army and Navy in planning the overall strategy of the war."[14] Happily, it was right up my alley for I already had so much experience with the German, British, and American guided-missile programs of World War II.

It was very stimulating to be in Washington working with people in these emergent and growing fields. Dave Langmuir, my senior colleague on our two-man radar liaison group in London from 1943 to 1945, was also on the staff of the guided-missile panel. And there were many others, both new acquaintances and those I'd worked with at the Rad Lab at the Massachusetts Institute of Technology (MIT) or in Europe.

But in thinking about where to work, it wasn't going to be for the OSRD.[15] It was going out of business. Bush insisted on it, to the furious resistance of the military, of Roosevelt's and then Truman's budget director, of Lee Dubridge, the director of the Rad Lab through the war, and to Bush's cost in political power.[16] Why was Bush so insistent on closing down the OSRD? Partly, he may have thought that, without the exigent conditions of a war, the organization, which Bush ran with an iron hand, was not sustainable, especially as its best scientists and technologists

returned to their regular jobs in universities and industry. Another speculation is that Bush was pushing the shutdown to force the military, now
passionate in its faith in civilian science, to figure out how military research should be done after the war. Whether or not that's true, that's
what happened: for example, the Navy created the Office of Naval Research and the Air Force created its Scientific Advisory Board.[17]

In August 1945 the Committee on Guided Missiles of the JCS Joint
Research and Development Board was trying to decipher its own role.
There was guided-missile work in all the services. The service rivalries
were intense, between the Navy and the Army but also within the Army—
between the Army Air Force (AAF) and the rest of the Army, the Army
Services Forces (ASF). The Army Air Force regarded missiles as an extension of aircraft technology and therefore within its portfolio, while the
Army Services Forces saw missiles as an extension of its artillery. That
issue got a solomonic solution, by giving the AAF air-to-surface missiles
and the ASF missiles launched from the ground without any aerodynamic lift.[18] All this came to a head in the National Security Act of 1947,
which put the services under a National Military Establishment, what
became the Department of Defense, established the post of secretary of
defense, and in September 1947 established the AAF as a separate service, the U.S. Air Force.

All that was to come. Van Bush argued in 1945 that it was important
even before the future fit of nuclear weapons became clear to get started
on missile development work, both in the laboratories and on the test
ranges. That led to a staff paper for the guided-missile panel that made
two points: (1) mesh very closely developments in nuclear weaponry and
guided missiles and (2) establish a strong organization with the proper
stature and talent to both coordinate and control guided-missile budgets.
We also urged that the panel's report should get to the president before
the whole issue of development of a coherent guided-missile program got
into the hands of Congress and its committees.

At the end of August 1945, Bradley Dewey, chair of the Guided
Missile Committee, asked Dave Langmuir and me to examine issues
for the panel to deal with, this in the context of an uncertain future for
the panel and indeed the Joint New Weapons Committee. Bush, for
one, was skeptical about whether the panel, like the OSRD, had much

of a future in peacetime, observing that "in time of war almost any organization will work because in time of war men will agree; that is what happened. In times of peace, it is a very different thing."[19] Yet there was certainly no lack of policy issues for the panel, or whatever came after it, for the postwar missile program. For example, the panel was supposed to establish the immediate weapons needs of the three services and where they should emphasize long-term research and development; set out how much of the military research and development budget should be for guided missiles; figure out the technological and institutional means for marrying nuclear weapons and guided missiles; and, not least, examine countermeasures to guided missiles. Tall order.

Dave and I were asked to get more definitive information by the end of August and to report to Dewey before the end of September when he was to meet with Bush. We got busy, arranging meetings and trips. We talked with Theodore von Kármán, then finishing up a report for the Air Force, *Toward New Horizons,* setting out the technological future for the Air Force. A central recommendation was that over the next decade the Air Force vigorously develop new technologies, including long-range guided missiles, or what the Air Force, which then didn't much like the missiles title, called "pilotless bombers."[20]

We did as thorough a canvass as we could of missile programs at the time. That was no small job. The Army Air Force alone had 11 surface-to-air missile programs going. At the Applied Physics Laboratory (APL) of Johns Hopkins University, we met up again with Merle Tuve, who was creating for the Navy a very substantial surface-to-air guided-missile program, Project Bumblebee. We also visited General Electric, working on a surface-to-surface guided missile for the Naval Bureau of Ordnance. There I met Simon Ramo, who had gotten his Cal Tech doctorate just ahead of me and who later became the "R" in TRW. Many of the military officers Dave and I talked with were very well informed about the science and technology underlying the missile programs, and while each understandably defended his service's interests, each could also be quite statesman-like in his opinions. We visited the JPL, which had been working for sometime on surface-to-surface missiles and which was established by von Kármán as a unit of the Guggenheim Aeronautical Laboratory at Cal Tech in Pasadena.

In 1945 JPL was a fairly small place, with about 300 people, in the Arroyo Seco, a dry canyon wash north of the Rose Bowl in Pasadena. It was there that von Kármán's graduate students fired their first rocket in October 1936.[21] JPL was moving forward strongly in rocketry. H. S. Tsien, who had been with the Von Kármán mission to Germany, which I accompanied, and who like me was astonished by the enormous German Air Force research establishment at Volkenrode, talked to us on the metallurgical problems of handling extremely high temperatures in rocket combustion chambers and nozzles, the supersonic[22] aerodynamics of ramjet [23] motors, and of missile fuselages and control surfaces. A supersonic wind tunnel was being planned. We also stopped at the Naval Ordnance Test Station in Inyokern, California, in the middle of the Mojave Desert, not too far from Mount Whitney in the High Sierras, where I had done my Ph.D. thesis on cosmic rays. There I had a reunion with Bill McLean, a fellow graduate student at Cal Tech. Bill was emerging as one of the finest guided-missile designers, his great triumph being the Sidewinder, an air-to-air missile with a pioneering heat-seeking homing device, first fired successfully in September 1953.[24] We also visited several companies, such as Boeing Aircraft in Seattle. It became evident that the aeronautical industry, heavily concentrated on the West Coast, was easily and quickly moving into the guided-missile business. Dave's later career would be centered there, and mine would take me there often. We left the coast for Cleveland and the Lewis Laboratories of the National Advisory Committee for Aeronautics.[25] This laboratory concentrated heavily on research and development on propulsion and was very experienced in rocketry, turbojets, and ramjets and in developing test facilities such as wind tunnels. It had several supersonic tunnels under design or construction, for testing air-breathing engines and liquid-fuel rockets.

Vannevar Bush's Offer

In our travels Dave Langmuir and I had many talks about our own futures, recognizing that our employment at the JCS guided-missile subcommittee was transitional. I thought that my most saleable item for other employment was my knowledge of the total guided-missile program from its technology to its organization, gathered from the British, the Germans, and now the JCS guided-missile subcommittee. Not long after I returned from the West Coast trip, I was told Dr. Bush wanted to see me. He treated me to a long conversation in which he outlined some of his own plans, which included a strong role in the management of military research and development, including guided missiles. He asked me to join him in this effort as one of his assistants and made a pretty strong case for it. I was overwhelmed by the offer but asked if I could think about it for a couple of days. At the second meeting I said: "Dr. Bush, your offer is the greatest honor I have ever received and I would have liked to take it, but I really want to get on the faculty of a research university and spend my life as a professor." His face broke into a broad grin and he said: "Well, that's a straight statement and I certainly will honor it. Let's see if I can help." The next morning Julius "Jay" Stratton called me. He was a professor of physics at MIT and the newly appointed head of the Research Laboratory for Electronics (RLE), a phoenix-like rebirth in 1946 of some of the elements of the Rad Lab but with characteristics that would better fit into a research laboratory. Today it is MIT's oldest interdisciplinary research laboratory. Stratton was RLE head in its start-up years from 1946 until 1949 when he became provost of MIT and later its president.

Jay Stratton asked me to become executive officer of a new guided-missile program that MIT was negotiating with the Navy. The Navy Bureau of Ordnance wanted at least two major guided-missile contractors on the East Coast. One was already in place, Project Bumblebee at the Applied Physics Laboratory; APL had arrangements with some 20 subcontractors in industry, academia, and federal laboratories.[26] MIT was to fit into the Ordnance Bureau programs to the tune of about one-quarter of the financing. The Navy was very firm that MIT and APL shouldn't compete for the same contractors and that we needed to coordinate with Army missile programs at Cal Tech; Project Hermes at General Electric, developing a series of missiles starting with V-2 technology; and Bell Laboratories, developing the Nike, a surface-to-air missile.

I leaped at the chance, mostly so I could reenter the academic world and get closer to becoming a real professor. I was to report December 1, 1945, and spent the intervening three months with Dave Langmuir and other colleagues writing a comprehensive review of U.S. guided-missile programs based on our visits and interviews. Bradley Dewey had his meeting with Vannevar Bush, who used the information in constructing a picture of the rapidly changing military research and development effort. Those three months were the first time since January 1943 that I could actually relax, have free weekends, attend the Army-Navy game in Philadelphia as the guest of our military colleagues on the guided-missiles staff, and even take a vacation in New England's White Mountains.

A NEW LIFE IN MORE WAYS THAN ONE

In December I began two decades of the happiest, most productive, most rewarding part of my life. The most important reason for that was that I finally put down roots and established a strong and lasting family life. From the summer of 1934 when I left for Colgate University, the Depression, the deaths of my parents and grandparents, my doctoral studies, and the war had torn up my life. I had lived at Colgate, summered twice in Detroit and once back in Corning, then on to Cal Tech for three years, to MIT's Rad Lab for a little over a year, and then three years in London and Washington in many apartments and hotels. The latter 12 years of my 29 were without a home—exciting and enjoyable but nomadic. All my possessions were in a steamer trunk and a couple of large suitcases. I had few possessions, except for a few thousand dollars I had saved. No debt, no property, no house, no land, no car, no wife.

I resisted temptations to marry, although I had dated girls from Corning to California, from Boston to Washington to London, and very much enjoyed my social life. That was about to change.

One of my more fortunate acquaintances was Bradley Dewey, chair of the guided-missile subcommittee David Langmuir and I had worked on. The Deweys had four children, none then married, and they had a tradition of parties for the friends of their children. In December, shortly after I arrived in Cambridge, I was invited to a white-tie dinner at the Deweys, to precede one of the popular waltz evenings at the Copley Plaza Hotel. Not long after I arrived I wandered over to talk to Mr. Dewey who was sitting with a very attractive young woman, dark haired, wearing a blue gown to match her blue eyes. Mr. Dewey introduced us, and I immediately fell in love. We didn't get a chance to talk or dance that evening, but I made inquiries afterwards. She was Adelaide Louise Risley Floyd, born in Cambridge, nicknamed Bunny. She had married a Bostonian, Cleaveland Floyd, a Harvard graduate, shortly after he joined the Army Air Force at the beginning of the war. Captain Floyd was killed when the airplane he was piloting crashed over the Himalayan "Hump" from India to Burma to China. After his death she entered Simmons College for a graduate degree in medical social work, her undergraduate work having been at Smith College.

But I soon learned there were at least three other suitors. I worried a lot, especially on a couple of extended trips when I thought of her all the time. My courting of Bunny continued apace in early 1946. We skied in New Hampshire, went to Boston Symphony concerts with Serge Koussevitsky conducting, and spent the Easter weekend in 1946 at the Dewey vacation home on Lake Sunapee in New Hampshire. We talked seriously on the drive home, and I asked her to marry me. She hesitated, but the answer was *yes*! I didn't crash the car but stopped and kissed her. I went home in a fog, and it was only when I got an early call from her the next morning in which she said "darling" that matters became clear and bright. We were married on June 29, 1946, at the Episcopal church in Newton Center. We honeymooned in Canada and soon moved into Bunny's apartment on Beacon Hill. When we learned that Bunny was expecting our first child in August 1947, we looked for larger quarters and established a home in Belmont, not far from Cambridge. It was there where all four of our children were born and started school and Sunday school. And it was there that we established lifelong friendships.

With all these wonderful changes in my life, I was also embarking on my MIT career. One of the first things I did was to establish a strong link between Project Bumblebee at the Applied Physics Laboratory and our Project Meteor. Both were navy projects, although Bumblebee was a surface-to-air program and ours was air-to-air. Next we recruited MIT faculty. We had a real enticement: equipment, supplies, and the salaries of research assistants—preferably graduate students for any proposing faculty member whose project fitted the needs of Project Meteor—would be paid for. We got immediate and excellent returns because graduate education required both course work and a research project, and graduate enrollments were growing in all departments. And MIT was in the catbird seat in electronics because many of its Rad Lab members either came from the MIT faculty or joined it immediately after the war. Lan Jen Chu from the Research Laboratory for Electronics soon proposed an ingenious radar guidance component. Albert Hall from electrical engineering was developing a large platform for marrying guidance and control apparatus with the simulated dynamics of a missile in flight. Hoyt Hottel and Glenn Williams from chemical engineering worked on combustion for air-breathing ramjets and rockets. Eddie Taylor from aeronautical engineering and head of the Gas Turbine Laboratory was constructing a wind tunnel.

The new programs weren't universally loved. A combustion facility is much like a stripped-down jet engine. It burns a lot of fuel, has awfully hot gases roaring out the end, and is very loud. It turned out that the combustion facility set up by Hottel and Williams created a stir in the neighborhood, particularly with the physics department. The combustion facility was across Albany Street from the main MIT campus, where there were other MIT facilities, including an accelerator used by Professor Robley Evans to accelerate nuclear particles to extremely high energy. Robley Evans told the top brass at MIT that, if they didn't do something about the combustion facility noise, he would point the beam of his accelerator at it. We cut the noise.

BREAKING THE SOUND BARRIER

MIT's aeronautical engineering department was strong in the subsonic range, below the speed of sound. Not so in the transonic and supersonic ranges emerging for both aircraft and missiles. The MIT top brass tried to lure Theodore von Kármán. Failing that, we did recruit as a full professor his colleague, H. S. Tsien, who as a graduate student and participant in the missile program at the JPL, had marched hand in hand with von Kármán into the transonic and supersonic era. Some of his best papers were on the supersonic aerodynamics of cones, cylinders, and flat plates, all key elements of supersonic craft. In that hyperactive era, a university could not claim first rank without high-speed wind tunnels for aerodynamic research and design testing. Using Project Meteor's needs and pulling out all stops, MIT obtained Navy support for a supersonic wind tunnel. Designed by John Markham, head of MIT's Guggenheim Aeronautical Laboratory, it was started in 1947 and finished in 1949. It was operated very successfully in the Mach 2 to 4 ranges[27] and proved a valuable instrument not only for Project Meteor but also for many other programs around the country.

We needed test vehicles to try out our components, given that, as von Kármán put it, "supersonics is one branch of aviation where theory and speculation preceded practice."[28] With Navy urging, we contracted with the United Aircraft Corporation, more particularly its Pratt and Whitney Engine Company, then converting from piston to jet engines. For rocket development we worked with the Bell Aircraft Company in Niagara Falls. We were hassled incessantly by any number of military officers and civilian officials checking up on this or that. They were in part reflecting the intense jockeying by the services facing an emergent Department of Defense, declining budgets, and uncertain roles in missile warfare. And the new Research and Development Board, under Vannevar Bush, was taking a hard look at the missile programs of all the services. At each budget cycle the Navy asked for new programs, new write-ups, new cost estimates, and new schedules.

By the end of 1946 we were doing pretty well. I even progressed toward my real goal of becoming an MIT professor. I was made assistant professor of aeronautical engineering in spring 1946, went to departmental meetings and the like, was asked to teach an advanced course on gas kinetics for mechanical engineering students, and lectured on jet propulsion. Still, for most of my first two years at MIT, I had my nose to the Project Meteor grindstone. I did a lot of traveling as a consequence. Bunny got used to it. And I worked hard to minimize time away from home, especially weekends. That started a pattern of business travel that lasted my entire life, of being home on the weekend and multitasking, so I could take care of several tasks on one trip.

In 1947 I got a chance to repay Vannevar Bush for the help he gave me starting my career, although I'm sure he would not have looked at it that way. I was asked to join an ad hoc study of the current status of all the guided-missile programs and to predict if and when they might become operational.[29] Our group of "planning consultants" reported to the Research and Development Board, an agency that, like the National

Security Council and the Central Intelligence Agency, was created by the National Security Act of 1947.

To this point in my life I had viewed Vannevar Bush through the rosiest of rosy glasses. And why not! He had assigned me to the London Mission of OSRD for radar liaison; he had increased my portfolio to include guided-missiles technologies; he had temporarily assigned me to the staff of the JCS guided-missiles panel; and he had secured me a job on the MIT faculty. Working easily with President Roosevelt, he had brilliantly organized and led civilian science, particularly from academe, in World War II. His report, *Science, the Endless Frontier*, had secured the principle of government support of basic research. For all these contributions, Vannevar Bush is revered. With the end of the war new pressures on the organization of government-supported research emerged. Many leading scientists had thrived under his wartime arrangements, although many thought them autocratic, almost dictatorial, and they now wanted more influence. Bush wanted our committee to look at several interrelated questions. First were the technical feasibilities of all the different types of guided missiles : air-to-air and surface-to-air anti-aircraft missiles; rocket-propelled, surface-to-surface bombardment missiles, long-range artillery in effect; pilotless aircraft, an extension of manned bombers; and longer-range intercontinental missiles being talked about at the time. For each type and for each contractor, Bush wanted to know the feasibility and the problems. He even asked us to estimate when each missile type would be operational.

We finished our report in about nine months. It said this: "It seems highly unlikely that it will be possible before the following dates to make detailed and reliable predictions of the extent to which high-speed (supersonic) guided missiles will meet the military characteristics that have been established for them: Air-to-Air, 1952; Surface-to-Air, 1950; Air-to-Surface, 1951; Surface-to-Surface, range up to 1000 miles, 1952, range greater than 1000 miles, 1952–1954." The report went on to estimate that operational uses would not be possible in less than three to four years beyond those dates. That meant the operational date for intercontinental ballistic missiles (ICBMs) would be about 1960, a date we hit on the nose. We thought air-to-air missiles would be operational ahead of surface-to-air missiles, both in the early 1950s. But the order was reversed, air-to-air missiles having proven to be more difficult than we thought, given added problems of smaller size, lighter weight, and harder targeting. The problems were solved spectacularly by the Sidewinder. We set the mid-1950s for operational cruise missiles— that is, air-breathing subsonic weapons, with a lineage back to the V-1s. The big problem was their guidance, with three options: celestial navigation, radar map matching, or inertial guidance, an idea we picked up from the Germans. Inertial guidance would win out, but we didn't know that in 1948. And the Air Force was demanding in its targeting, requiring that a cruise missile land within 5,000 feet of its target.[30] The comparable accuracy for the V-1 was 4 miles.

While I think we did something quite useful for the Research and Development Board—the operational dates we gave for the guided-missile programs and their pros and cons were used in planning further work—the reality was chaotic. "Complex and confusing," we said. On the other hand, some of the briefings and discussions were eye opening. For example, one of my Cal Tech teachers, Fritz Zwicky, presented his proposal to eject some material into the earth's orbit or possibly to escape the earth. His proposal was to mount a shaped charge on the nose of a rocket fired vertically and to fire the charge at the peak of the trajectory. The very high speed of the ejected material would add enough energy for it to escape. Another example was the discussion of possible uses of orbital vehicles, many of which came to pass within a few years. Still another example was the briefing on rocket-propelled manned aircraft to break the sound barrier and go supersonic. This material would provide enrichment to a rocket course I was soon to assume responsibility for. Once again, government studies strengthened my subject bank for teaching.

However, Truman's severe postwar budget cuts were amplified for the missile programs, as they matured from relatively inexpensive research to much more costly development. That, together with the budget crunch, forced some very sharp decisions. We pointed out in our report that, of the 16 missile contracts of the Army Air Force, only one was more than a year old and that as these went into development their costs would rise substantially, to not less than $3 million to $5 million annually. With a budget then of $20 million, only five projects could be supported.[31]

And the Research and Development Board was failing. For Bush "the RDB post was like a slow-motion automobile wreck. Bush saw the wreck coming but was powerless to stop it."[32] The root problem was that he was determined to coordinate the research programs of military services that were fighting bitterly with each other for dwindling dollars. And to Bush "coordinate" also meant the right to approve budgets. That was denied. He got approval to attend all JCS meetings dealing with the interests of the Research and Development Board. He was never asked.[33] And Bush—ironically, for someone who chaired the Research and Development Board from 1946 to October 1948, when he resigned—ridiculed the notion of long-range ballistic missiles and military satellites, doing so in words and style that made plain his contempt for many of his military colleagues:

> We are . . . decidedly interested in the question of whether there are soon to be high-trajectory guided missiles, spanning thousands of miles and precisely hitting chosen targets. The question is particularly pertinent because some eminent military men, exhilarated perhaps by a short immersion in matters scientific, have publicly asserted that there are. We have been regaled by scary articles, complete with maps and diagrams, implying that soon we are thus all to be exterminated, or that we are to employ these devilish devices to exterminate someone else. We even have the exposition of missiles fired so fast that they leave the earth and proceed about it indefinitely as satellites, like the moon, for some vaguely specified military purposes. All sorts of prognostications of doom have been pulled from the Pandora's box of science, often by those whose scientific qualifications are a bit limited, and often in such vague and general terms that they are hard to fasten upon. These have had influence on the resolution and steadiness with which we face a hard future, and they have done much harm, vague as they are.[34]

I was never sure whether Bush's dismissal of the notion of intercontinental missiles was an emotional one, denying that warfare with intercontinental missiles tipped with nuclear weapons was possible. He was normally bullish on the future of technologies. I think he was growing tired of the Research and Development Board job. Perhaps, he was also sharing some of the guilt that affected the lives and judgments of many involved in making the bomb. And his contempt for the military's sophistication in science was quickly becoming dated. He did not recognize changes in the services, particularly in the younger officers who were rapidly taking over from some of Bush's enemies, such as Admiral Ernest J. King, chief of naval operations during World War II. Many of these younger officers had gotten or were getting advanced degrees in the sciences and engineering, often from Bush's own Department of Electrical Engineering at MIT.

To be fair, the Air Force at the time wasn't a big fan of ICBMs either, preferring to put its chips on bombers.[35] That was understandable. In their lifetime its current leaders saw the biplane turn into huge jet bombers, as strategic bombing became a major part of World War II. The linchpin of the argument for creating a separate Air Force was that no missile in the works could carry the very heavy and large nuclear weapons then available and that no missile could approach the accuracy of bombers. It wasn't until 1953 that the Air Force became a believer in ICBMs, for several reasons: the Soviets had an ICBM program going, the hydrogen bomb was a reality, the much greater destructive power of a missile was recognized, and inertial guidance, an old technology,[36] was looking more promising for long-range missile guidance.

The year 1947 was very busy for me. I was occupied with the intense work for the Research and Development Board, some beginning work for the new Air Force Scientific Advisory Board,[37] and Project Meteor, the latter spiced by pressures to get a missile designed, built, and flying. But I was frustrated. I had not gotten any closer to becoming a genuine university professor, doing research, personally working with undergraduate and graduate students, teaching, and participating in university life instead of in a special project. I went to see Jay Stratton and told him that I wanted to change, even though I recognized that Project Meteor was becoming more intense with respect to organizing the contractors. Jay asked me if I knew "the difficulties of being a real professor. You'll have to find support for graduate students, for their and your own research, never mind teaching and supervising theses." I replied: "That's exactly what I'm interested in doing. I don't think I can become a real professor until I do it." Jay then said and I realized later that this was great wisdom: "Well, you've got to recognize that you've not really done this kind of research and academic work since you left graduate school in 1941. And you've made a good name for yourself in a field where experts respect you. You'll now have to prove that you can do a professorial job."

I still wanted to do it. A successor for the Project Meteor executive directorship was appointed, Ed Schneider, a former Rad Lab staff member. I agreed to part-time consulting, especially to use my outside connections. Later Ed returned to industry, and Bob Seamans took over as the program was readying its first test vehicle firings, but the tightening missile budget closed the project down except for remaining component work.

Early in 1948 I was free to move on to my new life as a professor. I moved out of Temporary Building 20[38] to the Department of Aeronautical Engineering in the Daniel Guggenheim Aeronautical Laboratory. And I was appointed chair of the Athletic Board of the faculty, which meant modest overseeing of MIT's athletic activities. MIT was continuing to build a fine intramural athletic program, was competitive in most extramural sports except football, and excelled in some sports such as crew. There was a boathouse right in front of MIT and the beautiful Charles River to row in the fall and spring. There were no athletic scholarships. There was a joke that I loved to tell: that MIT was the only university in the country that could afford to have the director of athletics and the director of admissions next door to one another.

PROJECT LEXINGTON

In early February 1948, Walt Whitman, head of the chemical engineering department, asked to see me. I knew that when a department head asks to come see you instead of vice versa, it probably means he's going to ask you to do something. It did. MIT had been approached by the Atomic Energy Commission[39] (AEC) to look at the feasibility of nuclear-powered aircraft, particularly its nuclear reactor, being pursued under major contracts from the Air Force. Although I was quite interested in learning about another new technology, I was somewhat dismayed by the timing of the task. I asked if I could think about it. I talked with James Killian, soon to be MIT's president, and Jerome Hunsaker, my department head, and both encouraged me to join in.[40] The highly classified project was going to be done mainly in the summer of 1948, but some preparatory work had to be done. The study was to be housed in a windowless bunker, an abandoned antiaircraft gun site, on a hill over-

Plate 1

My grandfather, Horton Stever, who with my grandmother, Mattie, raised my sister and me after our mother was hospitalized when I was one and Margarette three.

My parents. Both died when my sister and I were very young.

My sister Margarette, my life-long supporter with me. That determined look on my sister's face was not without cause: She took a firm hand in seeing I behaved myself.

Quarterback for Northside High in Corning – "unbeaten, untied, almost unscored upon": 266 – 8.

Note: All photographs © H. Guyford Stever except as otherwise noted in parentheses following captions.

Plate 2

My Colgate graduation picture. I was voted "Most Brilliant," sharing honors with class-mates who were voted "Biggest Week-Ender" and "Biggest Tall Story Man."

The casting of the Mount Palomar mirror at Corning Glass. The Cal Tech scientists who came to Corning to check on the work, especially Robert Millikan, were a huge boost for me to enter science. *(Ayres A. Stevens)*

Plate 3

Two of us in our faltboot on Lake Tulainyo working with our cosmic-ray apparatus. In the foreground is the snow glacier we had to get across carrying our equipment and our faltboot.

Taking a break after hauling mules and equipment up the High Sierras. Left to right, Bob Hoy, I, and Hugh Bradner.

On the move toward Lake Tulainyo, at almost 13,000 feet in the High Sierras. Our rented animals were probably figuring out what to do to us next.

Plate 4

Victor Neher, who was my doctoral adviser until he left for the Rad Lab at MIT. I followed him immediately after I got my degree.

Cal Tech was booming when I got there, which had its downside, such as graduate students, me included, being "accommodated" in the loggia of the faculty club, the Athenaeum. *(Bernard Hoffman/TimePix)*

The triumvirate that transformed Cal Tech from the Throop College of Engineering into one of the premier technological and scientific institutions in the world. Left to right, Albert Noyes, chemist; George Ellery Hale, astronomer; and Robert Millikan, physicist. *(Courtesy of the Archives, California Institute of Technology)*

Plate 5

Temporary Building 20 at MIT, home of the Rad Lab. Paraphrasing Churchill, "Some Temporary!" It was home between 1943 and 1998 not only for the Rad Lab, but also after the war the Research Laboratory of Electronics and the Media Lab. *(MIT Museum)*

From my London office I had a pretty good view of the damage done by the Blitz and even without looking out I was shaken by the bombing blasts.

Plate 6

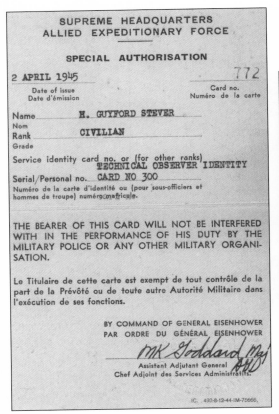

London, 1943, shortly after I arrived to become the second person in the two-man OSRD office to help coordinate the US-British work on radar.

SUPREME HEADQUARTERS
ALLIED EXPEDITIONARY FORCE

SPECIAL AUTHORISATION

2 APRIL 1945   772

Date of issue   Card no.
Date d'émission   Numéro de la carte

Name   H. GUYFORD STEVER
Nom
Rank   CIVILIAN
Grade

Service identity card no. or (for other ranks)
            TECHNICAL OBSERVER IDENTITY
Serial/Personal no. CARD NO. 300
Numéro de la carte d'identité ou (pour sous-officiers et
hommes de troupe) numéro matricule.

THE BEARER OF THIS CARD WILL NOT BE INTERFERED
WITH IN THE PERFORMANCE OF HIS DUTY BY THE
MILITARY POLICE OR ANY OTHER MILITARY ORGANI-
SATION.

Le Titulaire de cette carte est exempt de tout contrôle de la
part de la Prévôté ou de toute autre Autorité Militaire dans
l'exécution de ses fonctions.

BY COMMAND OF GENERAL EISENHOWER
PAR ORDRE DU GÉNÉRAL EISENHOWER

                    MK Goddard Maj
            Assistant Adjutant General
            Chef Adjoint des Services Administratifs

IC. 493-8-12-44-IM-75666,

Shortly after the Normandy landings till the end of the war I was "Technical Observer #300," authorized to go anywhere and even to take "custody of Prisoners of War." I elected not to exercise that privilege.

My first night in Normandy, about three weeks after D-Day.

Plate 7

Hennebont, a town in Brittany. Only one example of the destruction we saw in the aftermath of the fierce battles from town to town after the Normandy invasion.

In Lannion, en route to our "capture" of a Wehrmacht unit. Though we hadn't done the fighting and didn't deserve it, we didn't mind the enthusiastic welcomes all that much.

Plate 8

Sottevast, France, 1944. Bob Robertson, my colleague in my first technical intelligence mission after the Normandy landings at a captured – and destroyed – V-2 site.

In front of the pharmacist's house in Trégastel where we were invited to stay, guarded through the night by Free French forces to protect us, with reason, against SS attacks.

Volkenrode, 1945, the site of the immense and very well hidden research laboratory for the Luftwaffe. Hugh Dryden on the left, and Theodore von Kármán, third from the left.

looking the town of Lexington, hence the name for our task, Project Lexington. I helped Walt Whitman organize it, choose its members,[41] arrange clearances, and visit the AEC's major laboratory in Oak Ridge, Tennessee. I remember my astonishment at the size of the diffusion pumps for separating uranium isotopes.

As with missiles, we were operating in a chaotic time, technically and politically. The AEC and the Air Force saw the project from different, often-conflicting perspectives and priorities. As the Cold War intensified, the Air Force saw a critical need for long-range bombers, given the uncertainty and risks of basing nuclear weapons on foreign ground. The AEC was intent on developing peacetime uses for atomic power, especially the generation of electricity, but was also given control of the design and production of nuclear weapons. The Air Force launched its feasibility study on Nuclear Energy for the Propulsion of Aircraft (NEPA) in May 1946 with a budget of $1.3 million that grew to $8.3 million in 1951.[42] At first only Air Force funds were used, but by 1949 the AEC was financially involved.[43] The Air Force wanted to develop nuclear-powered aircraft quickly and chose "concurrency" as the way to get there. That meant tackling all parts of a weapons system at the same time, hoping for a quicker result. Many felt that was a more costly approach; the response was that development would be done more quickly and one could get on to other things. The AEC felt strongly that reactor development had to come first; no new reactors for producing fissionable uranium and plutonium had been built since the Manhattan Project.

The technical problems were formidable. The idea behind the plane was simple enough; air for jet propulsion was heated not by chemical burning—gasoline or kerosene with air—but by a nuclear reaction. The heat from a nuclear reactor can heat the air directly, direct cycle, or through a heat transfer agent, such as liquid sodium, an indirect cycle. Direct cycle was the route taken for the nuclear-powered airplane. The problem was that at the time of our study, 1948, the materials to do this in a reactor flying in a manned airplane weren't available—that is, materials that "would (1) stand up to the high-intensity nuclear radiation which necessarily existed throughout the interior of the reactor, (2) resist corrosion by the very hot air which passed through the reactor at great speed, and (3) not leak any of the highly radioactive fission products into the exhaust airstream."[44] Shielding was also a formidable issue. It had to be heavy enough to protect the crew but light enough to fly. And the whole thing had to be porous enough to be air cooled without radiation leakage.

And on the airplane side of things, we were moving from the well-known regime of subsonic flight to transonic flight, where the control issues were much less well known. We talked a great deal about the sound barrier. A few rocket planes had penetrated it so fast that control problems were ignored. And rockets, of course, were one-way trips. We wanted the nuclear-powered airplane to come back. Yes, a nuclear-powered airplane would fly for a very, very long time for a limitless range. How to keep it together that long was the problem. Turbojet technology was in its infancy; it was a long time before we could get to 1,000 hours, then 10,000, a rough equivalent of a year. Today's jet engines are very reliable and last as long as the airplane flies.

We had our first major meeting with the NEPA leadership in May 1948. Air Force representatives reminded us how important the nuclear-powered airplane was to them. At the time the Air Force was developing aerial refueling and setting up bases around the world for planes equipped with nuclear weapons. And, very tentatively, the Air Force was beginning to explore long-range ballistic missiles. A few days later we met with AEC representatives, including Robert Bacher, a physicist, head of one of the Manhattan Project divisions at Los Alamos and an AEC commissioner; Robert Oppenheimer, chairman of the AEC General Advisory Committee; and several others. Bob Bacher gave a general review of NEPA but pointed out that it was not well tied to the other problems the AEC had a primary interest in. And he noted differing opinions in the commission on the feasibility of a nuclear-powered airplane and that the short timescale the Air Force talked about was not feasible. Furthermore, he said that attempts to coordinate between AEC and NEPA had not been particularly successful, which was why Project Lexington was set up. Oppenheimer chimed in by saying that he couldn't get a straight "no" when he asked if nuclear propulsion was worthwhile. Oppy enlarged on his question by pointing out that the style of attack on the problem hadn't yet developed. Between 1942 and 1945, 10 reactors were built and none in the three years since. There were no laboratories that really understood the art of reactor development; reactor talent was much too spread out and one had to ask whether it was worth investing limited manpower in the nuclear-powered airplane.

Jerrold Zacharias asked what the difference was between the nuclear-powered airplane and the nuclear submarine. Oppy jumped on that. He said the sub could use a much more conventional pile than the plane and that both the requirements and the feasibility for the sub are much clearer. Bacher added that the sub had more options for a reactor type, while the plane had to use a gas-cooled reactor.

We plowed on, punctuating our technical discussions with meetings with other high-level groups. We met with Franklin Collbohm, the boss of RAND,[45] a think tank set up by the Air Force, who widened our understanding of alternative ways of delivering payloads at long distances. RAND thought that in the long run rockets were going to have great advantages over turbojets, ramjets, or anything else. They were already exploring the "skip" principle to quadruple a rocket's range. A winged rocket, like a stone skipping across water, would rise in a normal trajectory, descend, then skip again, doing that several times before descending to the earth. The first skip generated too much heat, but the temperatures were OK if the rocket didn't skip but glided. Of course, these alternatives weren't a direct part of the charge to Project Lexington, except to show another technique for delivering long-range payloads besides nuclear propulsion or using normal aircraft supported by aerial refueling and bases at the Soviet perimeter. It was something I would return to later in life.

We encountered severe skepticism on nuclear propulsion. Chauncey Starr of North American Aviation made it clear that we wouldn't be able to solve the problem any better than anyone else who had been studying it—namely, North American and RAND. We were after all depending on the same expertise. And Captain Hyman Rickover, leading the Navy's development of nuclear submarines, visited us in Lexington. I vividly remember sitting at lunch on a pleasant day in front of our bunker in the woods with a few of our people and Captain Rickover. He pointed out that the Bureau of Ships in which his program was lodged was neutral on nuclear-powered aircraft. He added that the Bureau of Aeronautics officially accepted the nuclear aircraft project, would follow its development, but would not put any money into it. Then he cut loose and told us in forceful language that he thought it was nonsense. Later, he did help us probe technical issues, such as closed versus open cycles and shielding, noting that while the AEC had the facilities to probe these issues, it didn't have the motivation. He also asked for help from NEPA in deciding what coolant to use for the nuclear submarine—water, helium, or liquid metal. It was clear to us that his program was well organized, with wide backing from industry, the AEC, and the Research and Development Board. The NEPA project also had its own industrial backing, but, as I remember, for both the nuclear sub and plane projects the industries backing them were mostly potential contractors.

That was followed by a large meeting in Washington with senior Air Force, Navy, and AEC officials in which we asked about the required military characteristics of bombers delivering 10,000 pounds of atomic warhead anywhere and how those requirements would change in 5 to 10 years. They described the planned development of the B-52,[46] plans for 5 to 15 years beyond the B-52 going into the supersonic range, and supersonic air-to-surface missiles as an adjunct to the B-52. We probed hard on aircraft vulnerability, especially since enemy supersonic fighters could in five years give us a hard time. Bombers flying at tran- or supersonic speed afforded some protection, although the real asset for a nuclear-powered aircraft was long flights at very high altitudes.

Air Force officials backed by the Navy again argued for concurrency—that reactor development should not be separated from other components of a nuclear-powered airplane. The AEC continued to disagree. That position was supported by our visit to General Electric, which was heavily into reactor development but whose people felt nevertheless that reactor development in the United States was in sad shape. They argued that reactor development ought to be done in several places, that the AEC concentrating the work in one of its centers was just plain wrong, that they didn't want a monopoly, and that they weren't going to get involved in NEPA until reactors were much further along. When we put this to the NEPA people, they stuck to their guns—it had to be an integral, meaning concurrent, program, and it had to be run by a group that really wanted it, and that was the Air Force. They were direct when we asked what improvements were needed in NEPA: "realistic, active, joint sponsorship and joint interest." They also gave us a schedule: in about 10 years for $500 million to $1 billion, they would be ready to test nuclear propulsion in a complete airplane.

We had more meetings in which we went over the problems raised by various people, such as possible poisoning of the reaction by fission products; the metallurgical problems; whether a chemically fueled aircraft wouldn't serve the purpose; the need for life-cycle testing of components shielding; and not least the weight of the reactor plant forced by the required shielding. We had looked at all caveats. Some were easy to knock down; others such as metals that could stand up to heat and fission products were not. In all, we couldn't provide Robert Oppenheimer's plea for a firm "no." We couldn't find anything pointing to that. We did lean toward concurrency but with a twist: the main effort at the beginning should be on reactor development, then building an airplane. I think we cleared the air. We got people together so that they understood one another. Technical people on both sides saw the problems that needed to be faced and ways to get at them. Administrators realized they had a serious problem in the relationships between the AEC and the Air Force.

Overall, we concluded that a nuclear-powered airplane was technically feasible but that it might take 15 years to fly it. And with my first experience inside the umbrella of AEC security clearance, I began to worry about relatively naked nuclear reactors launched into the air, especially on a series of test aircraft. Therefore, I was not at all disturbed when President Kennedy 13 years and $900 million later killed the program. [46]

I was pleased that I took part but glad it was over for two reasons. I'd tired of spending a lot of the summer in a windowless bunker, not ideal for a New England summer when Bunny and I should have spent more outdoor time together with our firstborn, Guy, Jr., a pleasure left more to her except weekends when I could join. The second reason was that I'd still not settled down to becoming a professor.

A PROFESSOR AGAIN

Jimmy Doolittle,[48] who was later to become a mentor and friend, once told me that as a young lieutenant in the Army Air Force in the 1920s, he sat down to list his assets and debits. He put together many items under each and started crossing them off one by one. He finally got down to one asset: "Brave as hell." And he also wound up with only one debit: "Dumb as hell." So he decided to go to graduate school at MIT in the aeronautical engineering department. He did that, got his doctorate, and then returned to his Air Force career as a better-informed, wiser, and more intelligent air racer, test pilot, and leader of men in World War II.

I did a "doolittle" in late summer 1948, toting up my assets and debits. I had been an assistant professor in the Department of Aeronautical Engineering (later the Department of Aeronautics and Astronautics). My assets were that I had a very good introduction to guided missiles and pilotless aircraft and also some aeronautical engineering, especially from Project Lexington, where we studied aircraft performance at sub-, tran-, and supersonic speeds. Moreover, I was appointed about that time to a committee of the new Air Force Scientific Advisory Board to look at the total wind tunnel picture in the United States, especially its gaps. My debits were that I did not have as much strength in the deeper engineering sciences of aerodynamics and fluid flows, aircraft structures and control, and the like. I needed to teach some courses in aeronautics dealing with the basics, and I needed to get a research program going for graduate students.

On the teaching side I soon got a big break, though it was a setback to the department, when Tsien returned to Cal Tech. Tsien was a brilliant mathematician, engineer, and scientist. He was von Kármán's best student and joined him as a principal contributor to *Toward New Horizons*, setting out the postwar technological blueprint for the U.S. Air Force. He was a member of the SAB for its first four years from 1946 through 1949. He knew as much about the science and technology of air power as anyone in the world. I was an acquaintance, not a friend; he was selective about that. When Tsien returned to Cal Tech in 1949, Mao Tse Tung had just established the Peoples' Republic of China. I did not notice that coincidence but some in government did because when Tsien moved to return to China, his native land, by ship from Canada, he was removed and held in house arrest in Pasadena for three years. There was a rather lame reason given that he had in his possession some classified reports of the National Advisory Committee for Aeronautics, not very strong stuff. The strong stuff was in his head. He disappeared in China, apparently to work on their nuclear weapons and ballistic missiles.

Tsien's story is emblematic of the security environment in which science and technology operated with the onset of the Cold War. The revelation that atomic and hydrogen bomb secrets were given to the USSR by insider scientists affected all of the science and technology community. Security became much tighter, even oppressive—for example the "witch hunts" by the House Un-American Activities Committee, Senator Joseph McCarthy, and others.[49] Some scientists who had been members of communist discussion groups, of which there were several, especially in university surroundings, and even those who rubbed shoulders with them, lost their clearances, sometimes in a very visible trial, as happened to Robert Oppenheimer, who was stripped of his security clearances.[50] That trial[51] was sad for I had friends on both sides.

Mach 7 and Beyond

Hunsaker assigned me two of Tsien's courses, a major one on compressible aerodynamics for transonic and supersonic flight and the other on rockets, which I combined with a guided-missile course that I had originated. I had become fascinated with the problems of hypersonic flight, especially in high Mach numbers—7 and beyond. Wind tunnels in that region are difficult to build because the throat has to be very narrow and both the pressure and temperature of the entering air have to be high so that what comes out has sufficient density, temperature, and pressure. But the air coming out expanded and cooled so rapidly that it was likely to liquefy. So instead of a hypersonic wind tunnel, you'd just have a fancy device for liquefying air. A wind tunnel is a simple device, basically a very large cone that contracts in the middle and that can hold a wing, an airplane, or even a scale model of a building. The heart is where the tube narrows and the entering air picks up velocity, without becoming turbulent. The test section is well instrumented, including cameras using special techniques to visualize the air patterns over a wing at subsonic to super- and hypersonic.

I thought up a scheme for getting high-pressure air and for heating it as it entered the nozzle. My students took on the job designing a Mach 7 throat for the tunnel so that, presumably, it would be a smooth hypersonic flow after it got through. We did get Mach 7 flow, but we also had problems with water vapor condensation. I tried to devise a theory that would enable hypersonic flow in a supersaturated state without condensation. It didn't work, in theory or experiment. So "making lemonade," we wrote a paper on condensation in hypersonic wind tunnels for our supporter, the NACA, and organized several seminars and workshops on the work in both the aeronautical and mechanical engineering departments.

Next we looked at supersonic shock waves with a colleague, Professor Ray Bisplinghoff, who had been asked by the Air Force to study the impact on an aircraft of a large shock from a nuclear explosion. Further, the AEC was planning atomic tests at the Eniwetok Atoll in the Marshall Islands of the Pacific and wanted the data for airplanes that would be flying in the test area. I conceived a shock tube large enough to study the flow following on a supersonic shock wave hitting a plane and the like. A shock tube is conceptually simple. It's a long tube of constant cross-sections with two parts. One is highly pressurized, usually with air, and the other is at low pressure, the two separated by a plastic or metallic diaphragm. When the diaphragm bursts, the high-pressure air rushes into the low-pressure compartment becoming a shock wave, followed, if done right, by a uniform supersonic flow of air. One could then put in the low-pressure zone, say, a model wing and use optical techniques to watch what happened as a supersonic airflow hit the wing.

Other projects joined the work on designing supersonic wind tunnels and shock tubes. One was the measurement of high-speed flows by tracking ions or charged particles. Then I also got involved around 1949 and 1950 in the problems encountered by planes as they moved through the transonic range, from sub- to supersonic speeds. There were horror stories of planes losing control as their aerodynamic surfaces went transonic, the plane becoming unstable, often catastrophically so. No question that transonic aerodynamics—the behavior of a plane at or just below Mach 1—was poorly understood. We created at MIT the Transonic Aircraft Control Project. This project grew a bit more than I expected, went for several years, and was successful both in helping the Air Force better understand transonic phenomena and in training students, many of whom went on to successful careers in academia and elsewhere.

All of these projects related to compressible aerodynamics and the phenomenon of shock waves. In any supersonic wind tunnel the critical design factor is shaping the nozzle entrance so the air can enter the expansion section to ever-higher velocities without creating shock waves. In a shock tube the object is to create a shock wave to convert still air to supersonic flow. In the transonic aircraft case the object is to delay the formation of shock waves to as high a speed as possible. The formation of shock waves requires much energy, and their onset effectively increases the drag of the aircraft enormously, requiring much more powerful engines. Additionally, it changes the pressure distribution on the control surfaces, making control of the aircraft much different.

At this time in the late 1940s and the 1950s the state of development of turbojets for fighter aircraft had not produced enough power to penetrate the high drag of the sonic region, so there was a period where proper aerodynamic design helped. Later there was enough power to push right through Mach 1 into the supersonic area, which also required clever aerodynamic design.

It was a rewarding time to be in education at MIT or in any of our research universities. We were getting extremely well motivated and therefore very good students. Many of them were older, having served in the armed forces, were serious and determined in their work, and had the support of that remarkable legislation, the 1944 GI Bill of Rights. Many of them became enamored of airplanes during the war, not only as fighting machines but also as superb examples of high technology. Winston Churchill once again captured the spirit of the time and the pride of MIT in its wartime work in his talk to the MIT mid-century convocation in March 1949 in the Boston Garden:

> The outstanding feature of the Twentieth Century has been the enormous expansion in the numbers who are given the opportunity to share in the larger and more varied life which in previous periods was reserved for the few and for the very few. This process must continue and we trust at an increasing rate. If we are to bring the broad masses of the people in every land to the table of abundance, it can only be by the tireless improvement of all our means of technical production, and by the diffusion in every form of education of an improved quality to scores of millions of men and women. Even in this darkling hour I have faith that this will go on.

This "darkling hour" was of course the Cold War, which Churchill formalized in his "Iron Curtain" speech three years earlier in Fulton, Missouri. And the Berlin airlift was still on, a magnificent response to the

Soviet attempt to starve West Berlin into submission and force the Allies out.[52] In his peroration Churchill emboldened us:

> Under the impact of Communism all the free nations are being welded together as they never have been before and never could be, but for the harsh external pressure to which they are being subjected. We have no hostility to the Russian people and no desire to deny them their legitimate rights and security. . . . We seek nothing from Russia but goodwill and fair play. If, however, there is to be a war of nerves, let us make sure our nerves are strong and are fortified by the deepest convictions of our hearts. If we persevere steadfastly together, and allow no appeasement of tyranny and wrong-doing in any form, it may not be our nerve or the structure of our civilization which will break, and peace may yet be preserved.[53]

We were indeed in a "war of nerves." About a year after this splendid speech, the Soviets exploded an atomic bomb and in August 1953 the hydrogen bomb; they had an active ICBM program; and in almost every part of the world, the tensions between the two postwar superpowers were rising. The reality was "the uniquely difficult and bipolar world that suddenly arose after World War II: two very different societies and cultures found themselves face-to-face in a world of awesome weaponry."[54] And in our own country we were becoming very concerned about our ability to defend ourselves against a Soviet attack. And at home where Bunny and I now had a full and joyous life with two children, Guy, Jr., and Sarah, I found my attitude quite different from what it had been in London during the war, when I had no family or possessions to worry about. The "war of nerves" was to occupy me as the 1950s began.

4

Defense

1950–1955. The Cold War hardens. Hard pushes on guided missiles and nuclear weapons. I lead the first look at antiballistic missile defense and do my first postwar Washington tour, as chief scientist of the Air Force.

T he nuclear device the Soviet Union exploded in August 1949 also exploded comforting illusions: that our nuclear monopoly made us impregnable, that we could safely and sharply cut our spending on defense, that we were technically superior, and, more subtly, that long-range missiles, because they could not be well aimed, were not worth the investment. This shattering of transient postwar comfort by that single event reshaped the United States in ways that are still with us—and they reshaped my own life. I had embarked on an enormously satisfying and very busy life of family, teaching, research, and government service. I had proven to myself that I could be a university professor, do research, and enjoy it.

In September 1947, de facto become de jure when the Army Air Corps became a separate service—the United States Air Force. The Air Force had come out of the war a fervent believer in the power of engaging first-class science and technology with military purposes. That fervent belief got its "bible" in the report *Toward New Horizons* mentioned in the previous chapter. Late in 1944, the Army Air Forces' commanding general, General Henry "Hap" Arnold, had asked Theodore von Kármán to lay out the postwar role of technology for the Air Force. Von Kármán and his colleagues did that, in 13 volumes delivered to an ill Arnold on December 15, 1945. One outcome was the creation of the Air Force Scientific Advisory Board (SAB),[1] a lineal descendant of the wartime Scientific

Advisory Group, which Major General Curtis E. Lemay said "rendered such signal service to the army air forces during the war that it has made obvious the necessity for continuation of such a service as an essential part of Headquarters staff planning."[2]

DEFENDING THE COUNTRY

On the strength of my guided-missile experiences, I became a member of the SAB and began an ever-closer relationship with its chairman, Theodore von Kármán. At Cal Tech I learned about the unique abilities and character of this genius from some of my friends in the aeronautical department. With his heavy Hungarian accent, they had a lot of trouble understanding him. But his blackboard notes were priceless, and for his research students his close contact night or day created a loving family. When Bunny met him in the receiving line at a reception for my chief scientist job, he grasped her hand, looked her in the eye, and said, "Ze wife of ze Guy is a Doll." He was a charmer to women and friendly to his colleagues. I became an admiring follower.

I remained an SAB member through 1968, except the year plus when I served as chief scientist of the Air Force. It was through the SAB and the implications of the Soviet bomb that I got involved in work on defending the country against bomber and missile attacks. General Muir S. Fairchild, vice chief of staff of the Air Force, called an emergency session in November 1949 requesting assistance in strengthening the nation's air defenses. Until that time the Air Force had expressed concern but not alarm that it could not put up a minimal air defense of our nation's peoples and industries. In 1945, an Air Force officer pointed out:

> [The] importance of the Air Defense of the United States in the future, both in peace and in war, has been thoroughly established by the experience of the past war, by observation of development trends in aircraft and guided missiles toward longer ranges, and in particular by examination of the German plans and projects for long-range offensive guided missiles and for air defense guided missiles. It has become clearly evident that, given a decent interval of time for guided missile development and production, our potential enemies will be able to launch at the United States surprise saturation raids by long range guided missiles, supplemented by very heavy raids by improved models of conventional bombing aircraft.[3]

This assessment carried little weight at the time in part because of the absence of reliable ways for a weapon to find its target, especially at intercontinental distances. Suddenly, with the explosion of the Soviet bomb, that mattered much less. A miss of even a few tens of miles might be good enough. More to the point, if the Soviets could develop a bomb, why not missiles? They also had German rocket scientists. At the same time, our own missile program, especially ballistic[4] missiles, was in disarray. The Air Force ballistic missile program was essentially terminated for budget reasons in 1947. In July 1947, the Air Staff concluded:

> [T]he atomic bomb constituted the nation's main source of power and the subsonic bomber its only means of delivery over the next ten years. As these aircraft would have to penetrate enemy air defenses and return safely, top priority went to air-to-air and air-to-surface missiles. Second priority was assigned to short-range (under 150 miles) surface-to-surface missiles. . . . Air defense missiles and detection and warning systems occupied third place, on the assumption that by 1952 the Russians would have long-range bombers and missile-carrying submarines capable of delivering atomic weapons.[5]

Long-range missiles were a fourth priority, with vertical bombs of the World War II type fifth. The Air Force knew that long-range missiles would become strategically important, but it didn't have the money and figured that development time would be at least 10 years (not a bad estimate, since Atlas, the first intercontinental ballistic missile was declared operational by the Air Force in September 1959).

"Scandalous and Disgraceful"

It was the third priority—air defense—that SAB was now asked to deal with. Our defenses were in bad shape. The Soviets had the bomb, and the United States essentially had no effective defenses against bombers carrying nuclear weapons. "The Battle of Britain had been won through the long-term attrition of Nazi bombers. However, in the atomic age, a near perfect defense against the penetration of bombers was required."[6] Yet in 1949 there was no integrated air defense system, no effective radar system for low-flying aircraft. Airborne interceptor radars failed when looking down at low-flying bombers, and while ground radars worked well at long ranges against high-flying aircraft, they could not detect low flyers.[7] Bob Robertson, a colleague of mine from my wartime days in

Britain and France called the nation's air defense "scandalous and disgraceful."[8] In response, the SAB formed the Air Defense Systems Engineering Committee (ADSEC) but made it an unusual committee in that it gave a lot more than advice. It had an operational role, to set out systems requirements for an effective air defense, identify technical needs to meet the requirements, do demonstration experiments, and provide procurement data and criteria to the Air Force. ADSEC was chaired by my Massachusetts Institute of Technology (MIT) colleague George Valley, who had led the Radiation Laboratory in its development of the H_2X radar bombsight for all-weather bombing of Germany, and before World War II had done research on cosmic rays, going, as I did for my thesis, to high altitudes for his data, in his case to 12,000 feet on Mount Evans in Colorado.[9]

George and I spent time together in Europe during the war when he was on extended visits from the Rad Lab.[10] I played chess with him on a little portable set while we were traveling through the English countryside to visit research laboratories or in London sitting out an air raid. I came to know him well.

ADSEC fell into a pattern of meeting every Friday night at the Air Force Cambridge Research Laboratories, which become the venue for placing contracts and providing funds in response to ADSEC recommendations delivered quite informally. We realized at the outset that there were several things we would have to know to do our job: (1) the state of the Soviets' development of long-range bombers; (2) the capabilities of our long-range radar systems, how they would improve over what time period, and whether and how they would deal with countermeasures used against them; (3) the current state of interceptor aircraft and missiles, both air-to-air and ground-to-air; and, not least, (4) the state of the communications, command, and control (C^3 in today's parlance) to link data on incoming aircraft with our defending missiles and interceptor aircraft.

Whirlwind

Al Donovan, of the Cornell Aeronautical Laboratory, did the calculations showing "that a bomber flying in over the north polar region at high altitude could always detect the ground radar before the radar detected it;

it could thereupon descend under the radar beam and continue undetected at low altitude."[11] For judgments on C^3 systems we would have to gauge the ability of current computer systems to handle the immense amount of data that would come in and be processed at very high speed. And we'd have to be able to project that into the future. We became well acquainted with Project Whirlwind, the very impressive efforts by Jay Forrester (23 years old) and Robert Everett (26) to develop a digital computer reliant on vacuum tubes.[12] Indeed, it was in 1949, when ADSEC began its work, that Forrester and Everett changed their tactics from development of a high-speed analog computer for an aircraft trainer simulator to a high-speed digital general-purpose computer. Two years later came the key inventions necessary to place into operation the first real-time, synchronous, parallel digital computer, including the magnetic core memory that revolutionized computing.[13] ADSEC in its work suggested the creation of a distributed network of small radars to meet the threat of low-flying bombers, with vital presumption of the availability of large digital computers to handle and analyze the data harvest.

I'm jumping ahead. George Valley asked Al Donovan and me to look at the state of guided missiles and of fighter aircraft, including interceptors. We moved quickly by scheduling in early January 1950 a week of visits to the centers of radar and missile development in the Los Angeles area—Cal Tech's Jet Propulsion Laboratory, Northrop, RAND, Hughes Electronics, North American Aviation, and Douglas Aircraft. I was amazed by the progress since I first visited in 1942. Not least, transcontinental travel had changed dramatically. In 1942, with Office of Scientific Research and Development (OSRD) priority, I took my first commercial flight from Dayton to Los Angeles in a TWA DC3, which stopped about every 250 miles, flying low when the weather was bad—and weather or equipment delays were common. By the end of the war, there were longer DC4s and Lockheed commercial flights. We didn't get commercial passenger jets until a decade later.

Now in 1950 we were talking about subsonic interceptor aircraft that with turbojets would go to higher subsonic Mach numbers. And in the works were turbojets with afterburners[14] that would take the aircraft into the supersonic region. Also in 1950 there were experimental prototypes, some in early production, of radar-guided missiles to replace or complement bombers using celestial navigation. Inertial navigation had not yet arrived, but its components such as gyroscopes and accelerometers that Charles Stark Draper and his colleagues were developing at MIT were entering aircraft control and fire control systems. It wasn't until 1953–1954 that the structure of an inertial navigation system became clear, and its operational status was confirmed in 1955, when the Air Force made it the primary guidance system for Thor, an intermediate-range ballistic missile.[15]

ADSEC helped push that technology and made plain the urgency for large-scale digital computing to the nation's air defenses. And ADSEC established a style of working that was to continue well into the 1950s:

> In effect, the Scientific Advisory Board had brought together a group of excellent scientists and engineers who were thoroughly familiar with the problems facing the Air Force but were not inhibited by the bureaucratic inhibitions and restraints characteristic of military organizations. This pattern of applying evolving technology to AF operational needs, largely established by ADSEC, would be repeated again and again as the Scientific Advisory Board continued in the role of major advisor to the secretary and chief of the Air Force.[16]

Charles and Lincoln

The very success of ADSEC—the very large technical territory it traversed and mapped—made it obvious that there was much more than could be done by part-time volunteers: a full-time laboratory was needed. That judgment led to the creation of the MIT Lincoln Laboratory in 1951 and to the mutually agreed upon dissolution of ADSEC in January 1952.[17] The Lincoln Laboratory made real the recommendations of ADSEC; for example, early research at the laboratory focused on the design and prototyping of a network of ground-based radar and aircraft control centers for continental defense.

I'm doing a gross injustice to a much more complicated story, wonderfully told by George Valley.[18] The struggle to create the Lincoln Lab was fierce. After several meetings, punctuated by the usual push and haul of academic politics, MIT accepted the ADSEC recommendation and established a laboratory. MIT stipulated a triservice laboratory, especially having Navy support, which it got rather grudgingly because the Navy was putting all the money it could afford for MIT into the Research Laboratory for Electronics, a lineal descendant of the Rad Lab, and the Army was just as heavily invested in Bell Laboratories. The final score on support by the Air Force, Navy, and Army was proportionately 10:1:1.[19] MIT also demanded that it choose the director. Agreed. Another condition was that the new laboratory, to be called the Lincoln Laboratory, would not be on campus. Agreed. It was put next to Bedford (later Hanscom) Field at the intersection of the towns of Bedford, Lexington, and Lincoln. Finally, MIT stipulated that the ADSEC recommendations for a laboratory focused on air defense were to be vetted and detailed by a larger group of people through a study called Project Charles. Done but not without heat. "Had Margaret Mead attended Project Charles," George Valley wrote, "she might have written a sequel to her well-known book: *Growing Up Among the Physicists.*"[20] Though I agree to some extent with George's sentiments about Project Charles, it did succeed in greatly broadening the knowledge of ADSEC and did entice a number to join the effort at Lincoln Laboratory. Britain's Royal Air Force assigned Air Commodore Sir Geoffrey Tuttle, a much decorated "Battle of Britain" fighter pilot to Project Charles. Geoffrey's wide knowledge and good spirits contributed greatly to the study, which ended in January 1952.

I was pleased with what ADSEC did. The work pushed new technologies, such as real-time processing of radar data, which in turn girded improved detection and tracking of aircraft and ground vehicles. The most dramatic was SAGE—Semi-Automated Ground Environment—conceived by George Valley, in which hundreds of continuous-wave radars were connected via telephone lines and computers to provide detection of low-altitude attacks unseen by conventional radar systems. It became the largest computer project of the 1950s and the training ground for many of the people who led the computer revolution in the following decades and created many of the companies of a new industry. SAGE was the first large control system to use a digital computer and, in fact, because of its central command structure, it was the first computer network.[21]

On June 25, 1950, 90,000 North Korean troops led by 150 T-34 Soviet tanks crossed the 38th Parallel to attack South Korea.[22] The South Korean capital, Seoul, was taken within a few days, and the onslaught pushed south, threatening to drive the South Korean forces into the sea. Within two days President Truman committed U.S. ground troops. Weaknesses in our military capabilities, especially in the Army, quickly became apparent. Some new weapons were introduced but overall very few, which was not surprising given the tight purse strings after World War II. We were again in a shooting war, with pretty much the same weapons, the same tactics. Charles Stark Draper at MIT was delighted with the fire control systems that he helped the Air Force install in some of our aircraft, particularly the F-86 Sabre jet, the country's first swept-wing fighter, which during the Korean War attained a 7:1 kill ratio.[23]

Military budgets went up sharply. The Air Force was still putting together the new Air Research and Development Command, recommended by a panel of the SAB, chaired by Louis Ridenour. There was a strong need, the panel believed, for a command for research and development distinct from logistical functions, such as supply, procurement, and mobilization.[24] But having been established in January 1950, it hadn't advanced far enough to be fully trusted by the top brass to do the work. It just didn't have the capability. So in July, a month after the North Korean attack, the Air Force asked the SAB to review its guided-missile program, in effect mirroring the work of ADSEC looking at air defenses. Specifically, the panel was asked to judge which guided missiles then in development were most promising and when they might become operational. The Air Force also wanted to know if and how these programs could be accelerated if more money became available, as seemed likely. Louis Ridenour,[25] who had chaired the earlier SAB panel that led to the Air Research and Development Command and also served as the first chief scientist of the Air Force, took this one on.[26] It turned out that our ADSEC work fit like a glove with this task, and so we were able within a few weeks to get our briefings and frame recommendations. In fact, the Air Force jumped on the recommendations we discussed at our meeting in July even before we finished our report and accepted a lot of them. The key recommendation we made was to quit thinking of guided missiles as a special case—as an add-on to "conventional" Air Force programs—but rather to integrate them fully as a normal and essential part of its planning. That echoed the systems approach flavoring our ADSEC work. We told the Air Force that if it was to efficiently move guided missiles into deployment to regard them "as being the natural next steps in the cultivation of various aspects of air warfare."[27] This work, paralleling that by other SAB committees, proved over time an excellent kickoff for the new Air Research and Development Command under Major General David M. Schlatter, who commented very favorably on the help given by the SAB and who got the order from the Air Force chief of staff, General Hoyt S. Vandenberg, that the ARDC become by May 1951 "an independent, self-sufficient major Air Force command."[28]

The SAB worked very effectively. There were usually about a dozen panels in existence with titles like "Electronics" or "Armaments" to which each member was assigned, with occasional dual assignments. At each of two annual meetings of the board, following general briefings by leaders, the panels would break out on their own to discuss any topic of interest in their field together with the ad hoc and special committee reports. The panel chairs would then report to the entire board. Thus each board member got complete coverage. That feature broadened my views, stretching me toward the generalist's point of view that helped me in later work.

RETHINKING

The arrival of the Air Research and Development Command to focus and crystallize research and development for the Air Force was one piece of the postwar structure for federal science and technology developing rapidly around 1950. That structure in its fundamental form endures to this day; it drove the exuberance and excellence of American science and technology after World War II. The philosophical underpinning was, and to a considerable degree remains, *Science, the Endless Frontier*, by Vannevar Bush.[29] The organization that Bush created and led during World War II, and for which I worked during the war, had been an incredible success, but Bush knew that it was not sustainable in peacetime and fought against bitter resistance to shut it down. At the same time, he argued fiercely that the central lesson of the OSRD, the essentiality of linking the best science to national goals, had to endure, and that new organizations fitted to peacetime were needed. He argued:

> The Government should accept new responsibilities for promoting the flow of new scientific knowledge and the development of scientific talent in our youth. These responsibilities are the proper concern of the Government, for they vitally affect our health, our jobs, and our national security. It is in keeping also with basic United States policy that the Government should foster the opening of new frontiers and this is the modern way to do it. For many years the Government has wisely supported research in the agricultural colleges and the benefits have been great. The time has come when such support should be extended to other fields.[30]

Bush clearly articulated those "other fields" for peacetime federal research:

> Progress in the war against disease depends upon a flow of new scientific knowledge. New products, new industries, and more jobs require continuous additions to knowledge of the laws of nature, and the application of that knowledge to practical purposes. Similarly, our defense against aggression demands new knowledge so that we can develop new and improved weapons. This essential, new knowledge can be obtained only through basic scientific research.[31]

That Bush report essentially molded and made durable the bond between government and civilian science. The underlying faith that civilian science was essential to national security, meaning in those days military security, set the table for the postwar structure of federal research. The contracting instruments created by the OSRD became standardized in peacetime research supported by the federal government. The decentralization of research entered mainstream belief, as did the faith that, while the government could set general goals, results came by supporting the best. Not least, the enormous success of wartime research through the OSRD model made the military and the American public passionate believers in high-quality science.

Bush was the principal in making that happen. Still he ran into trouble when he tried to impose his ideas for turning the faith in science and scientists into federal structures and budgets. Bush wanted a "National Research Foundation," with sweeping powers vested in civilian scientists. It would be federally funded, but civilian scientists would control its expenditures and choose its director. The foundation would support work in the physical sciences and medicine[32] and would set priorities for long-term military research.

Typical for Bush, this was an audacious, even confrontational plan. Audacious because it would be even more powerful than the OSRD and confrontational because it would weaken if not eliminate military control over its science and technology programs. And it was doomed. The biggest reason was, ironically, the very success of OSRD in convincing the military to love research. The three services fervently believed that if the OSRD went, something else had to rise in its place and that "something else" was going to be controlled by the military, not by a board of civilian scientists setting military research priorities over which they had no control. And Bush's argument for a director appointed by a civilian board didn't sit well with President Truman, his budget director, nor with many in Congress, notably Senator Harley Kilgore (D-W.Va.), who proposed a "National Science Foundation," in a form that he largely shaped. Created in 1950, the NSF was a pale version of what Bush wanted: it had no military division and no medical division, the director was appointed by the president, the board shared its power with the director, and its budget was capped by Congress at $15 million. In fact, it got $250,000 the first year. Bush was understandably not happy and would have little to do with the new agency; for example, he wouldn't let researchers from the Carnegie Institution, which he headed, apply for NSF grants.[33]

What emerged by 1950 was to an outsider a confusing mixture of agencies supporting civilian research, each with different structures and missions. These included not only the NSF but also the National Institutes of Health, the Office of Naval Research, the Atomic Energy Commission, the Air Research and Development Command, and in the fall of 1951, the Air Force Office of Scientific Research. Many people, including me, thought this wasn't bad—that it was a good thing to have many sources of support for scientific research and to have several agencies doing applied research and development. This melange gave the Bureau of the Budget, now the Office of Management and Budget, a strong role in putting together these different pieces.

Clearly, my work for the SAB—I had been a member since 1947—brought me in touch with some of these struggles, although I did not get involved in the hassling about turf, how to organize the various agencies, or budgets. At this time in my life, I began to realize that I didn't relish these kinds of organizational struggles and often took extremes to avoid them. I was much more interested in research itself, in development, and in education. Thus, I wasn't all that upset that I hadn't stayed on in

Washington to work for Vannevar Bush, but rather pushed to return to MIT, first with Project Meteor and then on the faculty. If I'd stayed on, I would have been just another fellow fighting for this or that idea in the complex and confusing organization that is the U.S. government.

I remember a visit I had with Wilbur Goss of the Bumblebee missile program at the Johns Hopkins' Applied Physics Laboratory, someone I respected greatly from the time I first met him in 1944. Wilbur and I philosophically reminisced about how our lives had changed from what we wanted to do. For example, absent the war, I would probably have gone to Stanford in 1941 to start a career in physics rather than to the Rad Lab. Yet we had no regrets in having plunged into wartime work, he into ordnance and proximity fuzes and I into radar and guided missiles. It was that way with many Americans. We couldn't face once again the horrors of war after we just fought the war "to end all wars." Nor did we want to face a world split into two hostile camps, this time with the risks multiplied many times by fission and soon fusion weapons.

I decided that my push into an academic career had been right and that I was going to keep pushing that no matter what. I also realized that with all the investments I had made in certain fields of military importance I should continue to help out, in the main through part-time advising, as I was doing in 1950 with the Air Defense Systems Engineering Committee and with the Ridenour Committee looking at the state of the Air Force's guided-missile program. It worked both ways: my chosen field of aeronautics meant that some of my advising contacts were valuable in deciding my research program.

In the spring of 1951, I was promoted to associate professor and given tenure. It was 10 years after I got my doctorate in physics at Cal Tech and made the decision to go to MIT and the Rad Lab instead of Stanford and physics. Tenure in a university was what I wanted and now I had it. It was in a department in the engineering rather than science school. I certainly still had a strong interest in basic science, particularly physics and more particularly in my own fields of cosmic rays and Geiger counters, and I tried to keep up by talking with friends in those fields and reading the journals. My path at MIT was clearly going to put the finishing touches on making me a professional in aeronautical engineering, or, what it became later, aeronautics and astronautics. Aeronautics of course became one of the great drivers of the postwar American economy, through the development of long-range, wide-bodied aircraft carrying large numbers of people and goods around the world. And certainly not least I now had a family—a wife, two children, a boy and girl, a dog, a lovely home, and a foot in a beautiful place in Randolph, New Hampshire, where Bunny's parents had a home.

PUBLISH OR PERISH?

Getting a promotion and tenure in the spring of 1951 started me thinking even harder about my academic career. I started thinking about that Banquo's ghost of academic life, "publish or perish." I discovered that the topics one could publish on in engineering were different than in science. In academic science, basic research is the ideal of the research that goes with teaching in the universities. In engineering there is wider latitude. Some professors believed that basic research in the engineering sciences was proper for engineering professors, while many others believed that work on major projects not quite basic in their approach would do fine for their own intellectual work and that of their students.

In my three years of graduate work at Cal Tech, I got three nice papers out of two areas of research, on cosmic rays and Geiger counters.[34] When I entered the classified fields of wartime radar and then guided missiles, I found unsurprisingly that I wasn't getting any good publications in the open literature, even though I did an awful lot of writing on reports that were classified. This pattern of hard writing but for a classified audience continued after the war. Still, that MIT gave me tenure signaled that my work, however limited the audience, met the institution's intellectual standards. But I wasn't satisfied. I wanted to change the pattern. I turned to open publications of the work I had started on three projects: (1) the hypersonic wind tunnel with the offshoot technical field of condensation in high-speed flows; (2) the transonic aircraft control project covering high-speed aerodynamics from subsonic through transonic to supersonic; and (3) the shock tube program, which put me into a special area of compressible flow. These all served as a source of publishable work, for me and for my students.

In the fall of 1950, Kenneth Rathbun, a mechanical engineering graduate thesis student of mine and I handed in our report, titled *Theoretical and Experimental Investigation of the Condensation of Air in Hypersonic Wind Tunnels,* to the National Advisory Committee for Aeronautics for publication as an NACA document. That in turn led to me submitting a paper entitled "Condensation of Air in Hypersonic Wind Tunnels" to a meeting in England on heat transfer of the Society of Mechanical Engineers. This occurred at a very busy time for me. I was learning to teach new courses and doing a lot of advising—for ADSEC, Project Charles, and the Ridenour Committee advising the Air Force on its guided-missile program. That meant I had to plan the trip to England very carefully. And I did, not only giving my talk but catching up with what Britain was doing on missile defense, visiting the famous air show of the Society of British Aircraft Constructors at Farnborough, and not least visiting with Geoffrey Tuttle and seeing wartime friends. I returned from three weeks in Britain on a Sunday, for MIT registration day on Monday and to meet a new class on guided missiles on Tuesday. And I had to quickly reconnect to the transonic aircraft control and to the shock tube project, never mind getting our home ready for autumn and winter.

I continued to publish via journal articles and book chapters—on the growth of the boundary layer behind a shock wave, on condensation in high-speed flows, and, in general, on high-speed aerodynamics and jet propulsion. I was engaged in the intellectual network of aerodynamics. In 1954 I published with Ray Bisplinghoff an article in the *Proceedings of the National Academy of Sciences* entitled "The Shock Tube in Aerodynamic and Structural Research." It was used to support Ray's election to the academy and some years later as partial support for my election.

ROCKETS, ICBMS, MIKE, AND TEA POT

It was an extraordinary time for me, but just as the onset of World War II had sharply altered my academic career, so too were events of the early 1950s to change my life again. A signal event was the acceptance of intercontinental missiles mated with nuclear weapons as the country's prime weapon in maintaining its national security—and with that even more focus on how to defend the continental United States against Soviet intercontinental ballistic missiles (ICBMs).

The performance of liquid rocket motors had advanced since World War II. There also were widespread calculations on what could be done with them. In my rocket course I analyzed the performance needed to accomplish intercontinental trajectories, to establish a satellite trajectory, to escape the earth, even to escape the sun. Scientific visionaries like Arthur Clarke foresaw manned space flight; scientists planning the International Geophysical Year, July 1957–December 1958, began to think of flying an instrumented satellite; and military men began planning ballistic rockets of ever-increasing range. The Air Force was assigned the ICBM program. In 1950 Louis Ridenour chaired a committee on which I served on the Air Force's guided-missile program.

In 1952 Theodore von Kármán enlisted another former Hungarian, John von Neumann,[35] to work on the problem of linking the development of ICBMs with that of nuclear weapons, and in 1953 the von Neumann panel was started, and soon pointed out that progress on nuclear weapons was so fast that they would not be the stumbling block to the development of an effective ICBM. Von Neumann and most of his committee were soon co-opted into another venture, the "Tea Pot" Committee, to map out the development and eventual deployment of ICBMs.

The "Mike" thermonuclear tests were set off in October 1952 at Eniwetok in the Marshall Islands. Less than a year later, in August, the Soviets set off their own hydrogen device. Two months later, in October, the "Tea Pot" Committee met, chaired by von Neumann, a fervent believer in lighter and very destructive thermonuclear weapons delivered on rockets. The committee was organized and its members were selected by Trevor Gardner, who in 1953 was Special Assistant for Research and Development to the Secretary of the Air Force and who passionately believed that the nation needed an operational ICBM if it was to avert nuclear disaster. Only a few months later, in February 1954, the committee issued its report, strongly arguing for a crash program on ICBMs, and in 1954 the United States went from a desultory pace to running hard to build its first ICBM. This was Atlas, liquid fueled, first radio and then inertially guided, with a range between 6,400 and 9,000 miles and declared operational late in 1959.

"The Shot Island is Missing"

This is cartoon history, and the reasons for these radical changes are complex and still argued over. Until the early 1950s the Air Force had concentrated on building bombers to deliver nuclear weapons and on fighter aircraft for the Korean War. Resistance to an ICBM program also rested in part on lack of accurate guidance; if V-2s traveling a few hundred miles had an operational error of about 4 miles, ICBMs traveling 3,000 miles would miss their target by 60 miles.[36] But the power displayed in the first U.S. thermonuclear test, Mike, weakened the accuracy argument. As the chairman of the Atomic Energy Commission reported to President Truman: "The shot island Elugelab is missing, and where it was there is now an under-water crater of some 1500 yards in diameter."[37] With "Mike" in principle promising a vastly improved yield-to-weight ratio, the lift needed in an ICBM eased downward. But since Mike weighed 60 tons, that was truly "in principle," and it wasn't until the Bravo test of a much lighter bomb in March 1954 that the yield-to-weight argument became more realistic. Trevor Gardner had taken advantage of a review of missile programs to eliminate unnecessary duplication ordered by President Eisenhower's defense secretary, Charles Wilson, when he created the Tea Pot Committee, formally the Strategic Missile Evaluation Committee, chaired by von Neumann. Gardner deliberately put the committee beyond Air Force control but appointed as its military representative Colonel Bernard Schriever, then a fairly lonely advocate for missiles in the Air Force. This Tea Pot Committee[38] reported in February 1954, urging a strong push toward ICBMs—that is, the Atlas missiles—judging that an ICBM could be operational within six years (i.e., within six years of the Tea Pot report).[39] As to what the Soviets were up to, the committee offered artful language: "While the evidence does not justify a conclusion that the Russians are ahead of us . . . this possibility can certainly not be ruled out."[40] Not least, the committee strongly urged that a new organization be created to develop the Atlas. The Air Force got the point, and moved to establish a facility for accelerating Atlas: the Western Development Division, located in Inglewood, California, under the Air Research and Development Command, and led by Bernard Schriever, now a general officer.[41] General Schriever, faced with the daunting task of rapidly developing a very complex piece of technology, adopted what was then a new management approach called systems engineering. It had many parts, including "change control" to track the effects of changes in a single component on the entire system and "integrated system testing" to look at the whole system under realistic operating conditions. Systems engineering was a management revolution then but is now commonplace.[42]

So in the space of a few years—from about 1950 to 1955—I saw remarkable events, participated in some, and was affected by all. Civilian scientists continued to have a major and to a degree independent role in military issues, not least through the SAB. The need for a separate and strong research and development organization in creating major new weapons systems had been recognized in creating the Air Research and Development Command. A major commitment to command and control systems that were reliant on computers had been made in building SAGE to protect the United States against low-altitude bombers not detectable by other early-warning systems. The United States had committed itself to missiles as the primary delivery system for nuclear and thermonuclear weapons. And, finally, I had just started an SAB study on an anti-intercontinental ballistic missile defense, AICBM.

CHIEF SCIENTIST

In late September 1954 the SAB held its semiannual meeting at Offutt Air Force Base, in Omaha, Nebraska, the home of the Strategic Air Command (SAC). I gave a preliminary discussion of our AICBM study. Our host was General Curtis Lemay, the powerful voice in the Air Force for strategic bombardment and if not hostile at least quite skeptical of the role of missiles to carry a war to the enemy several thousand miles away. SAC was heavily dependent on two subsonic bombers: B-47s and the new, faster, and more powerful B-52s, which first flew that year, 1954, and entered service a year later. For intercontinental-range bombardment, either aerial refueling or a worldwide set of bases was needed, and the Air Force worked on both. General LeMay was especially interested in the SAB's advice on longer-range aircraft that could carry heavier loads (i.e., more and heavier bombs).

At the meeting, several Air Force officers, led by Lieutenant General

Donald Putt, who had recently become deputy chief for development at Air Force headquarters,[43] took me aside to ask if I would become the next chief scientist of the Air Force. It was and is a rotating position. Chief scientists normally came from academia and served for about a year. The term for the current incumbent, Chalmers Sherwin, was ending at the beginning of 1955, and they were beginning to work on a replacement. General Putt made a strong sales pitch, and as I rode home from that meeting I began to think that professors do better if now and then they take a sabbatical. The usual time for a sabbatical is every seven years, and I'd been a professor for eight. And I also felt that it would give me a chance to look again at the work I was doing in the field and maybe take on some new things. Finally, I felt that the results from my research programs were peaking—the transonic control project and the shock tube and atomic blast programs. It would do the projects and me good to have others take those over. I wanted to do it.

I talked this over with Bunny, and she was quite willing; indeed, she had told me she would live happily with me in Boston, San Francisco, or Washington. And it was for supposedly only one year, the nominal tenure for a chief scientist, although it turned out to be closer to 18 months. We had just had our fourth child, Roy, and we would be going to Washington with four children, with Roy six months old and Guy, Jr., seven, bracketing the ages of the two girls, Sarah and Margo—plus one handsome Irish Setter. My mentor Julius "Jay" Stratton, now chancellor of MIT, also thought it was good for me to accept.

Also supportive was Jimmy Doolittle, who was becoming a close friend and who was vice chairman of the SAB, along with Mervyn ("Iron Mike") Kelly of Bell Telephone Laboratories. Kelly became chair of the SAB in January 1955, when the chairman, Theodore von Kármán, decided he could no longer continue.[44] So Kelly and I took office at the same time, although Kelly left before I finished my year, with Jimmy succeeding as SAB chair.

Back to Washington to meet with General Nathan Twining, Air Force chief of staff, and General Putt, and to accept the appointment. Moving from MIT to Washington for a year and half was tougher on Bunny in making all the family arrangements than it was for me. I had secured an attractive rental in Belle Haven, between Alexandria and Mount Vernon,

Virginia, an easy commute to my office at the Pentagon. Early February 1955, one week after I started my job, I returned to Belmont, and Bunny and I drove our two cars to Virginia. Our passengers included four children, one wonderful but aged Irish Setter, and a fine helper, Nanny Blyth, who helped Bunny in many ways, not least the babysitting for a couple of weeks until Bunny got the major chores done.

Washington had the reputation of being a sleepy southern city. From the work standpoint, it was not. We did not get into much of the social and political life of greater Washington, being away both summers and having many friends and colleagues from elsewhere visit us. The local elementary school was a disaster for Guy, Jr., who had dyslexia. Sarah attended a private school, Burgundy Farms Country Day School, run by friends, Kay and Eddie Mayer, and had a long bout of mononucleosis. Bunny had her hands full.

Real Airplanes

Jimmy Doolittle had an office next to General Putt's and mine. Both met me as soon as I came aboard and told me that, while I knew "a lot about guided missiles and rockets and fighter aircraft and all of that, we now want you to get more information about real airplanes." "Real airplanes" meant big ones: bombers.

So off we went, on a plane piloted by Don Putt, flew overnight, and then spent a week flying around to West Coast aircraft manufacturers. It turned out that what these companies wanted to talk with Putt and Doolittle about was the competition to design and/or build the KC-135, which was to be the new jet-propelled refueling tanker for our B-47 and B-52 bombers. I was a bit puzzled and asked Don and Jimmy why these companies were putting so much emphasis on that particular airplane. They laughed, and told me "the company that won the final production contract [would] have a substantial leg-up on the development of the first commercial jet transport."[45]

On the last day of our trip, a Saturday morning, we flew from Los Angeles to Edwards Air Force Base in the desert. When we touched down in our big C-54 transport, we saw extensive black skid marks that rather randomly serpented down the runway for some distance, finally going off the side of the runway and leaving deep gouges in the turf. This was the mark of the first B-52 Stratofortress, about to be delivered from Boeing to the Air Force. Clearly, the landing had been close to a disaster and would have been were it not for rapid action by Captain Magruder, the Air Force officer in charge of the acceptance tests. The jar of the B-52 hitting the runway had released the landing gear, which began to fold up. Magruder, in the copilot's seat, recognized what happened, grabbed the landing gear knob, pushed it back in place, and held it! Nobody knew how he could react that fast and push that hard, but he did. The landing gear unfolded again, but one side locked. Therefore, the wild skidding on and off the runway. But the plane was saved, no one was hurt, and the locking mechanism for the landing gear was promptly redesigned.

THE B-58

I spent some of my time during this tour of the West Coast looking at the status of Convair's B-58 program, since my first real assignment as chief scientist was to a board looking at this plane. The B-58 Hustler was to be the country's first supersonic bomber, going into service in March 1960, about the same time the country's first ICBM went operational. The supersonic B-58 was considered the next step after the subsonic B-52. Smaller turbojet fighter aircraft and even some fighter bombers were reaching the supersonic range, and many in the Air Force now wanted it for bombers. But there were problems, not least that the Strategic Air Command, and especially its leader, General Lemay, didn't much like the plane. Getting to supersonic speed meant slimming the fuselage and wings to reduce the ferocious drag at transonic and supersonic speeds. While more powerful turbojets were then in development to lessen size pressures, the smaller size translated into fewer bombs and not enough space for the electronics and jamming equipment to sufficiently protect a bomber aircraft over hostile ground. The B-52, even though it flew at subsonic speeds, carried the electronics to defend itself, carried more bombs, and with the onset of the KC-135 and aerial refueling, had tremendous range and very long flying times.

My role on the B-58 board was well defined—to judge the aerodynamics of the aircraft. It turned out that aerodynamics were the least of the plane's problems because Convair had worked hard on that, so much so that other elements suffered.[46] For example, there were great difficulties simply cooling the electronics during flight. This was when vacuum tubes hadn't yet been replaced by integrated circuits. Nevertheless, even with its handicaps, we recommended that the program continue. It did, and in December 1955 a contract for 13 B-58s and 30 external weapon pods was let to Convair. The first plane was delivered in August 1960 and by 1964 90 B-58s were deployed.[47] The last plane was retired in 1970. Its problems included relatively high accident rates, that the first ICBMs came on line the same year the plane did, and that the Soviets in the late 1950s built surface-to-air missile capability that forced low-level penetration of enemy defenses, where the plane performed poorly.

THE KILLIAN COMMITTEE

It wasn't until I met with Jim Killian that I began to get some glimmerings of what my term as chief scientist would be like. Jim, then president of MIT, chaired the Technological Capabilities Panel, better known as the Killian Committee, established in July 1954 at the request of President Eisenhower to look at three major national security issues for the country: striking power, continental defense, and intelligence. Many things converged to create the study, but Jim Killian summed it up crisply when he later wrote that "there was a growing realization that thermonuclear weapons in the hands of the Soviets posed a threat of terrible dimensions that required urgent efforts to construct new defenses, to give greater emphasis to the deterrence of war, and to seek arms limitations."[48] The committee was due to report the same month, February 1955, that I was to start my tour as chief scientist. And it did so.

Not only did the Killian Committee report in February, Trevor Gardner was finally confirmed in February 1955 to his appointment as Assistant Secretary of the Air Force for Research and Development, then a new position.[49] Apt timing because Gardner was partly responsible for the creation of the Killian Committee.

The new Eisenhower administration and the Air Force had only moderate interest in missile development, when Gardner became Special Assistant for Research and Development to the Secretary of the Air Force in February 1953. Two and half years later, on September 13, 1955, President Eisenhower designated the ICBM program as "a research program of the highest priority."[50] Gardner got there in good part because of the Killian report, but his methods weren't always appreciated. Jim Killian, in his memoirs about his service in the Eisenhower administration, grudgingly acknowledged Gardner's take-no-prisoners style, describing him as "technologically evangelical."[5a] He learned early of the kind of gadfly pressure that Trevor Gardner could put on his superiors and that Trevor could be abrasive. Jim in a private conversation I had with him used stronger language than in his book in describing Gardner. Others were equally blunt, describing him as "sharp, abrupt, irascible, cold, unpleasant, and a bastard."[53] But he was also "a doer, always direct, skilled at cutting red tape, yet with an amazing breadth of knowledge and interests."

The Killian study[53] was clearly critical. It involved the future role of nuclear and thermonuclear weapons, large rockets, long-range radars to detect missiles and to guide them, and inertial navigation, a technology we were beginning to get familiar with, although as an idea it had been around a long time. It got the strong endorsement of President Eisenhower, who "stressed the high priority he gave to reducing the probability of military surprise."[54] Strategically, the Killian Committee looked at the entire spectrum of military problems facing the United States and the world in the mid-1950s. Its recommendations matched the scope of its charter. It confirmed what Gardner and others already believed, that because "our defense system is inadequate . . . SAC is vulnerable and the U.S. is open to surprise attack." Further, "evidence is accumulating that the Soviets are developing their long-range delivery capability." And most alarmingly, "because of our vulnerability, [the] Soviets might be tempted to try a surprise attack. They might be so tempted in order to attack before we achieve a large multimegaton capability."[55]

There was much more to their findings and recommendations, which they divided into missile programs, continental defense, and intelligence, parsed by progressive time periods, traversing from our present very strong advantage to a future standoff when "an attack by either side would result in mutual destruction."[56] The report of the Killian Committee was presented a couple of weeks after I arrived in Washington to take up the chief scientist job, and the president and his National Security Council moved quickly to give top priority to the development of an ICBM by the Air Force. It also started actions to improve our intelligence, not least starting the development of the U-2 reconnaissance plane. It gave the Army the go-ahead for intermediate-range missiles and encouraged the Navy to develop ballistic missile capability for its warships, particularly its future nuclear-powered submarines. In effect, the report set the table for Atlas, the first ICBM; the Army's intermediate-range Jupiter missile; the Poseidon missile to go into submarines; and the U-2 spy plane.

FASTER THAN A SPEEDING BULLET

And it certainly set the table for me. The report "hailed the establishment by the Air Force Scientific Advisory Board of an Antiballistic Missile panel, and urged that it give early consideration to the formation of a full-time technical group to carry out a rapid but thorough study of defense against ICBMs."[57] The SAB did establish the panel—the Anti-ICBM, or AICBM, panel—and I chaired it. It started before I became chief scientist and was the major item on my plate during my tour. While we were flattered that the White House "hailed" the work of the antiballistic missile panel of the SAB, it also added to our challenges, by asking us to examine the ABM problems facing the Canadians and the British. The Canadians of course also had the same problems with respect to ICBMs that we did, although they were likely to see an incoming one before we did. And the British and NATO faced the problem of IRBMs—intermediate-range ballistic missiles—that we knew the Soviets were developing.

We had a strong membership for the task, including Canadian-British representation.[58] And we had at least one unusual member: Charles A. Lindbergh. Trevor Gardner nominated him. The appointment was unusual in several respects, not least that FDR had banned him from wartime service to the government, angered by Lindbergh's vicious attacks on him and his "America First" campaign before the war. Somehow Gardner got Lindbergh reinstated and on the panel, and he did indeed serve a useful role as a skeptic. Lindbergh later commented about his service on the AICBM committee in a book, saying that antiballistic defense was like trying to stop one rifle bullet with another, and that he felt when he started on the committee some of the technical people were too optimistic that it could be done.[59]

In fact, the problem was worse than hitting a bullet with a bullet, because incoming ballistic missiles are faster than a rifle bullet. That was solvable because recent studies suggested that the incoming missiles didn't have to be hit on the nose. They could be stopped if not destroyed by nearby aerial bursts, by hitting shrapnel, or by atomic explosions generating intense X rays that could possibly burn through the lightweight covering of the warhead. And we looked at other esoteric ways to destroy incoming missiles, studies that continue to this day.

We had growing intelligence information about the Soviet Union's work on long-range missiles. This came in good part from a large radar built on the initiative and leadership of my predecessor as chief scientist, Chalmers Sherwin, on the Turkish coast of the Black Sea at Samsum, close enough to the Soviet border to detect and monitor test firings of intermediate-range rockets flying some 1,500 miles into the central Asian desert. But it was not until the highly secret flights of the U-2[60] over Soviet territory beginning in June 1956 that we could "see" Soviet ICBM facilities deeper in the heartland, in Kazakhstan.[61] Further, the threat of a ballistic missile attack had become much more real now that a thermonuclear warhead could offset bad aim.

The interesting part, of course, was defending against weapons that didn't yet exist and whose specific capabilities and vulnerabilities could only be guessed at. What seemed clear was that a ballistic missile was most vulnerable in its initial phase when the rockets were just beginning to accelerate off the ground. It was tempting to think of antiballistic missile systems attacking at liftoff. That meant the defensive system had to act right over enemy territory. And that in turn suggested very advanced systems in which satellites circled all the time, some to detect and track a launch, others to carry warheads to intercept and kill it. That kind of thinking in the mid-1950s was premature—after all, satellite reconnaissance didn't begin until the 1960s—but people were working on the possibilities of such systems.

The middle range of the flight of a ballistic missile would be very high, far away, and very difficult to intercept. So from the start of the launch we turned to its final stages just before it struck. In that final trajectory, all of the major parts of the launching rockets would have split off, and the incoming warhead would be something like an ice cream cone. It would be well stabilized to descend smoothly through the heavier parts of the atmosphere, the cone pointing forward and protected against the tremendous heat of reentry by shielding. Work was then being done on ablative materials, which would absorb and carry off the intense heat. The first time I heard of this idea was when I was asked to evaluate some Soviet intelligence referring to a mysterious hard wood coating. We were quite puzzled until we realized that hard oak would be a very good ablating coating for the nose cone of a missile: the wood would char, and the char would actually be cooled as carbon particles evaporated into the atmosphere, going instantly from solid to liquid to gas and in the process carrying a lot of heat away. Also, as the charring continued and deepened it would become a very good heat insulator. In fact, that's the kind of coating that eventually came about, not wood but embodying the same principles of char simultaneously carrying away heat and insulating.

So those of us in the ABM business at that time considered these incoming cones with a great deal of interest, moving at great speed, not maneuverable since they were in the later stages ballistic, and very hard to hit. That the nose cone couldn't duck gave some potential advantage to a defense that could fire at the expected trajectory of the missile. The first need was for radar to detect a very small nose cone with very low radar reflectivity. Second was a computer system that would integrate and analyze the radar information in time to launch a defending missile.[62] Another part of the puzzle was what kind of destruction was needed, whether one could use just shrapnel and let the incoming nose cone destroy itself when it hit shrapnel or whether one could depend on a blast some distance out from the nose cone to destroy it. Of course, talk about damage at a distance immediately invoked nuclear weapons.

One of the largest technical problems then—and still today—was an ABM system differentiating the real thing from decoys. Of all the various schemes we thought about, the best one rested on weight differences. The decoys were likely to be lighter than the actual weapon and hence would decelerate faster from atmospheric drag on entry into the atmosphere and then be detectable. We spent a lot of time calculating how heavy decoys could be and how many were needed and still have an operational missile. Of course, if one could send decoys, you could with a little more push in the rockets also send multiple warheads. That would make defense even more difficult.

We also looked at reshaping the nose cone so that it slowed down just as the decoys did. That stratagem was unlikely to work because, if the nose cone was shaped differently than the decoys, it would have a different and likely detectable radar signal. So the game of measure and countermeasure was played but all on paper, since at the time there were neither ICBMs nor the kind of technologies available that would be embodied later in, say, the Nike series.

We had our final meeting in late March 1955, concurrently with a spring meeting of the SAB and also with the SAB nuclear weapons committee established in 1953 and chaired by John von Neumann. Johnny, who died in 1957, led the panel in drawing for the Air Force the future role of nuclear and thermonuclear weapons. We had a joint session with Johnny's panel to discuss how nuclear weapons fit into the whole ballistic missile defense program, specifically the size and type of nuclear weapons likely to be available by the end of the 1950s. In it we emphasized the urgency of ICBM development.

Even before our report was finished in May, I had to begin briefing it, first to the full SAB. While we felt that it might be possible to intercept a missile as it was coming down, intercepting it in space was then beyond the state of the art. Systems had been looked at to do it, but the technology just wasn't there. But ground interception fit into the aircraft defense role given the Army, which was pursuing it very strongly through its Nike program. I was a little disappointed that we weren't able to come up with stronger and more brilliant recommendations on ballistic missile defense. The ones we did offer were good as far as our technology went, and several would come into play as new technology developed.

JIMMY

At this time in my life, I established a great friendship with Jimmy Doolittle, a friendship that lasted until his death in 1993. Jimmy had retired to the business world after his exploits in World War II but came in from time to time as a special consultant to the Air Force chief of staff.

Jimmy was a superb mentor—he was so good with people, thinking of their interests, receptive to their ideas, and throwing himself fully into anything he did with them. He had a strong role in brokering relationships between the Air Force and the White House at a time of tectonic changes in thinking about national security, most sharply expressed in the charge to the Killian Committee, of which he was a member. Indeed, he was a close associate of Jim Killian, a member of the MIT Corporation, and a distinguished graduate, earning a doctorate in aeronautics in the mid-1920s—indeed, he was one of the first in the United States to earn this degree.

My personal relationship with Jimmy Doolittle did not end when he finally finished his tour with the SAB in 1958. He helped me get established with a small club of avid landlocked salmon fishermen in northern Maine. The club opened only when the lake was ice-free, the smelt were numerous, and the salmon would rise to feed on them. Fly fishing with smelt-like lures was done from a canoe motoring just fast enough to make our imitation smelt attractive to a salmon. It proved at times very effective. One fisherman sat in the middle, another in the bow, and the guide in the stern, handling a paddle or a small outboard. Jimmy and I also spent summers climbing in the low coastal mountains of California, doing so until he was close to death. When he was in town during my time in Washington as chief scientist, he would come to our home for dinner. Our children remember him for his wonderful Doctor Doolittle stories. Jimmy taught me a lot, not the least of which was to keep my speeches short! I did.

Washington was justly famed for its summer heat and humidity. In those days, the mid-1950s, it was also known for the fact that along with most everybody else it had no air conditioning. The children could take it better than their parents could, but we also had gotten used to living in Boston, where in the summertime a cooling east wind often prevailed. What made the 1955 summer terrifying was a serious polio outbreak the summer before. Despite the arrival of the Salk vaccine, people were frantic that in the cities polio would again become a serious threat. Bunny's mother, Lillian Newell Risley, herself not too well, thought this situation too frightening and invited Bunny and the children to come to their place in Randolph, New Hampshire. It turned an ominous summer into a glorious one, but it ended sadly. The day we returned to Washington for the start of school we got a telegram that Nana Risley was dead, her heart gimpy for all the dozen years of our marriage had plain given out. We thanked her in our prayers.

That sad change in our lives was accompanied by many others. My term as chief scientist would nominally be done by the end of the year, and I was to face new challenges on my return to MIT.[63] More broadly, I was now enmeshed as one cog in the deterrence and defense posture the country had adopted. For deterrence the Eisenhower administration had adopted the policy of massive retaliation, the threat of using nuclear weapons delivered on an aggressor by bombers, cruise missiles, and soon ICBMs. For defense the country was to depend on early-warning radars, fighter aircraft, and long-range antiaircraft missiles, such as the Army's Nike series. And just to keep things interesting, another card was soon to be added: entry into space, to detect what the enemy was doing and to attack him.[64]

5

Into Space

1956–1964. Back to an academic life but hardly a peaceful one as missile programs driven by "mutual deterrence" intensify and as a little metal "basketball" puts me into the politics of civilian space.

I returned to the Massachusetts Institute of Technology (MIT) in the summer of 1956 as a full professor of aeronautics and astronautics and, more tellingly, as associate dean of engineering. "More tellingly" because until then I had put off academic administration in favor of mounting a strong program of research and education. People I admired enormously—Jim Killian, president of MIT; Julius Stratton, chancellor; and C. Richard Soderberg, the new dean of engineering—had pressed me hard to take the job after finishing my tour as chief scientist of the Air Force. What I did turn down was an offer by Don Quarles, Air Force secretary, to be Assistant Secretary of the Air Force for Research and Development after Trevor Gardner[1] resigned over what he considered inadequate funding for the missile programs. I wanted to be back at the university. I did accept Jimmy Doolittle's request that I become vice chairman of the Air Force Scientific Advisory Board (SAB), to help Jimmy, its chair, thus broadening my coverage of Air Force science and technology.

PHASE CHANGE

In 1956 many forces were carving the adversarial terrain between the two superpowers for a quarter century. The intercontinental ballistic missile (ICBM) program was now a national priority, hard thinking was being done about how to defend the country against missile attacks, and the United States and the Soviet Union had each committed to orbiting a space satellite. The Cold War was being hardened by new weapons and technologies, including development of a suite of intercontinental and intermediate-range ballistic missiles (IRBMs), the onset of small and lighter nuclear weapons to arm those missiles, confirmation that inertial navigation was both possible and reliable, and the very secret emplacement of a national program for reconnaissance—"spy satellites"—from the upper atmosphere and space.

The notion of mutual deterrence was now in play. "Mutual deterrence" was an evolved concept. The first major stage was "massive retaliation," publicly articulated in a speech in January 1954 by John Foster Dulles, Eisenhower's secretary of state. The nation's security, he argued, should "depend primarily upon a great capacity to retaliate instantly, by means of and [in] places of our own choosing."[2] There was at the time good reason for the policy. While we could never hope to match the Soviet's land forces, we were far ahead of them in long-range bombers. And Eisenhower from the onset of his administration was determined to control the federal budget, not least military spending. He didn't see military spending and economic strength as a zero-sum game, but rather saw disproportionate military spending as posing as great a threat to economic security as too little spending did to national security. He stated his feelings bluntly in a talk he gave to the American Society of Newspaper Editors in 1953 after just three months in office:

> Every gun that is made, every warship launched, every rocket fired signifies, in the final sense, a theft from those who hunger and are not fed, those who are cold and are not fed. . . . We pay for a single fighter plane with a half million bushels of wheat. We pay for a single destroyer with new homes that could have housed more than 8,000 people. . . . This is not a way of life at all, in any true sense. Under the cloud of threatening war, it is humanity hanging from a cross of iron.[3]

Indeed, Dulles cited the policy of massive retaliation as providing "more basic security at less cost."[4] Yet "massive retaliation" had the proverbial problem of using a flamethrower to kill a fly—that is, applying it to what on the scale of things are minor security threats to the United States, such as the Soviet suppression in 1956 of the Hungarian uprising. The policy invited "brinkmanship."

And "massive retaliation" was credible as long as the other side couldn't respond. But the Soviets in the same year, 1957, tested an ICBM and launched *Sputnik I*, the first earth satellite. That and the increasing sophistication of weaponry—especially better guidance systems—led to a variant of massive retaliation called "first strike." It was a counterforce strategy, destroying the other guy's ability to retaliate. That policy was destabilizing, since both sides were now paranoid about a surprise attack, making imperative the urge to strike first. By the early 1960s, the concept of mutually assured destruction—MAD—emerged. This meant that even if the other guy attacked first, you still had more than enough left to destroy him. The United States was acquiring MAD capacity in the form of ICBMs moving into full production in the early 1960s, in missiles launched from submarines, and in missiles loaded into hardened underground silos.

As chief scientist I had worked hard on both sides of the missile business. First, how to defend against a missile attack, through my chairing the Anti-ICBM Panel of the SAB, and, second, on the development of long-range missiles. It's a complicated story, made even more chaotic in the late 1940s and early 1950s by severe cuts in the federal budget, which decimated missile programs, and by the Air Force's skepticism about the efficacy of long-range missiles compared to its very large bomber fleet. That changed sharply around 1954 when Eisenhower made the development of the ICBM a national priority and the Air Force created the Western Development Division[5] to build the *Atlas*, the first ICBM. That missile—and *Titan I*, begun after the *Atlas*, using a different liquid propellant—had a special structure and a liquid propellant engine.[6] But it was difficult logistically to handle a large liquid propellant rocket in the field, never mind the hazards and complexity of liquid propellant pumps. The same sort of problems afflicted *Thor*, an intermediate-range missile. The Army also had an IRBM[7] in development, first called *Redstone* and then transmuting into the *Jupiter*. In any case, considerable attention shifted to solid propellants, led by the Navy's development beginning in July 1956 of the solid-fueled *Polaris* fired from submarines. In early 1956 the Pentagon agreed to let the Air Force develop its own solid-fueled missile that could be launched within 60 seconds—the *Minuteman*. (See Table 5-1 for a listing of the various programs.)

TABLE 5-1 U.S. Ballistic Missiles, 1960–1970

Atlas	First U.S. ICBM, deployed in September 1959; liquid fueled (kerosene and liquid oxygen).	Air Force
Titan	ICBM, alternative to the Atlas in case the latter failed; a two-stage, liquid-fueled (hypergolic) rocket.	Air Force; *Titan I* successfully launched in 1960
Thor	IRBM.	Air Force
Polaris	Solid propellant.	Navy
Jupiter	IRBM.	Initially Army-Navy, then only Army
Minuteman	ICBM, launchable within 60 seconds, solid propellant.	Air Force; in service in early 1960s

Note: Each missile type had several successive variations (e.g., Titan I, Titan II, or Atlas A, B, C, D). An ICBM (intercontinental ballistic missile) has a range of 4,000 or more nautical miles. An IRBM (intermediate-range ballistic missile) has a range between 1,500 and 4,000 nautical miles.

FISHING, CHARADES, AND DEANING

We settled into our house and our old social life in Belmont. Young Guy was 9, Sarah 7, Margo 4, and Roy 2, all established in local schools. We came back in August and spent a good part of that month in Randolph, New Hampshire, and I spent some time fishing. One of our neighbors was James B. Conant, in wartime the deputy director of the Office of Scientific Research and Development, OSRD,[8] and founder of its London mission, former president of Harvard, High Commissioner to Germany and then our first ambassador there. I told Jim, like me a fisherman, that one of my favorite places was the Moose River gorge, where my father-in-law had introduced me to fly-fishing. It's not a big river, but it has a beautiful wild gorge, well away from civilization, with an old, almost abandoned, railroad that had only one train go through each day. It's a good hike to get into the gorge. I told Jim about this and he said, "Gee, I'd like to go there some time." I said, "How about tomorrow?"

But he had a meeting in New York. The next day while I was fishing in my favorite hole in the gorge I saw a fisherman coming upstream. It was Jim Conant. Turned out his meeting had been cancelled. I invited him to fish at my favorite spot. He did that and soon netted five very nice fish, before going upstream. I resumed fishing, but there were no more fish to be caught. He used wet flies, I dry ones.

One more story about Jim Conant, this at the annual picnic in August of the Randolph Mountain Club, where we played charades. We did a scene from *Don Giovanni*. I played Don Giovanni and Mrs. Conant the statue of Il Commandatore murdered by Don Giovanni. Jim Conant was handling the stage effects, two children played a flute and violin duet, I sang "Là ci darem," and then danced with a young lady, scheming to seduce her. The statue came alive, descended from its pedestal to drag me off to Hades, portrayed by "smoke" created with dry ice and alcohol by Jim Conant, a famous chemist. Mrs. Conant, to be a proper statue, had been wrapped from head to toe like a mummy with just two little openings for her eyes and nose. She immediately began to choke, not able to breathe because of the dense fog. Dr. Conant rushed on to the stage to rescue her. Everybody said it was a great charade. I don't remember the word we were miming.

When I wasn't fishing or trying to seduce young ladies in charades, I was associate deaning. When I started in the fall of 1956, I was asked right off the bat by Dick Soderberg to meet with all the department chairs in the engineering school to talk about their plans for promotion, for granting tenure, and for hiring new faculty. It was a great pleasure for me to do that, for the quality of this group of department heads in engineering was high. In aeronautical engineering, Charles Stark Draper had succeeded the longtime head of the department, Jerome Clark Hunsaker. In chemical engineering there was Walter Gordon Whitman, whom I knew well from our work together on nuclear-powered aircraft studies. For civil and sanitary engineering[9] there was John Benson Wilbur, a strong civil engineer who was trying to adapt civil engineering to new powerful uses of computers. In electrical engineering there was Gordon Stanley Brown, whom I had met way back in World War II, and I worked with him quite a bit. In mechanical engineering there was Jacob Peter Den Hartog, a brilliant teacher and engineer who had come to MIT from a position on the Harvard faculty after World War II (I met Den Hartog during World War II in Paris a few days after its liberation where he was serving as a captain in the U.S. Navy and as head of a group in technical intelligence work). In metallurgy[10] there was John Chipman, a senior leader in the metallurgy field. In naval architecture and marine engineering there was Laurens Troost, Jr., a brilliant man from the Netherlands, which had a history of high-quality naval architecture and marine engineering.

These powerful leaders in their respective fields could be trusted to assure the future of their departments by recruiting and promoting the best. The problem was too few slots. I worked with Jay Stratton and Dick Soderberg in zeroing in on the strong cases and also the marginal ones where the department head was forced, sometimes several times, to make a stronger case—and sometimes lost. That often took months. There were some rough tussles. Dick Soderberg had strong opinions and the discussions would at times become acrimonious.

As with all great research universities, MIT had substantial research support from the outside, principally the federal government, which since the end of World War II had invested heavily in graduate research in the universities. Principal funding agencies included those focusing on fundamental research, such as the National Science Foundation and the National Institutes of Health, and more strongly applied agencies, such as those of the armed services[11] and the Atomic Energy Commission. These federal infusions of substantial funds weren't controlled top down by the administration. Every "cask was on its own bottom," meaning each professor had to find the support for his research, graduate students, and, at times, undergraduate students. Managing a professorate endowed with tenure and independent funds made for a complex management challenge (and still does). There were also some large laboratories, some controlled by one or two departments, others more autonomous. For example, the Research Laboratory of Electronics, which Jay Stratton, its first director, had helped pull together out of the remnants of the Radiation Laboratory, but with an accent more on the research side rather than the development of radar, was bridged between the departments of physics and electrical engineering. The Instrumentation Laboratory in Aeronautical Engineering, founded and led by Charles

Stark Draper, fit almost entirely into one department. The Lincoln Laboratory, set up to do highly classified work in air defense, was outside any department but nevertheless offered pertinent research to several professors and their students. The faculty members who did their research in one of these laboratories had their problems. The departmental structure was often so strong that people working in one of these laboratories had a harder time gaining departmental promotion and tenure, not least because many if not all of their publications, however excellent, were classified.

MIT not only led in securing federal funds but was also very aggressive in obtaining private support from companies, foundations, and individuals. I was directly involved in securing a large grant for the School of Engineering from the Ford Foundation, at the time a supporter of universities. Dick Soderberg put together a committee of very good and hence very busy people. To get us together he promised to hold meetings at the end of the working day and, since that was normally cocktail hour, to serve martinis. Sure enough, at one of the early meetings, out of a small office refrigerator came a jug of martinis. We never got to the second martini. Soderberg had mixed a large amount of the necessary components, gin and vermouth, and cooled them in the refrigerator. He forgot that normally the mixture is shaken with cracked ice to cool and dilute it. Without the cracked ice we relaxed too much and accomplished little. We stopped that routine. And we got the Ford grant.

The program supported by the grant had several parts:

- revision of the engineering curricula and teaching materials, including new texts ($3 million),
- development of instructional laboratories integrated with classroom theory ($1.5 million),
- endowment of seven additional professorships in newly emerging fields of engineering ($3.5 million),
- establishment of postdoctoral teaching internships and research fellowships to encourage young people to enter the field of engineering education ($1 million),
- fellowships and loans to graduate students anticipating teaching careers ($150,000), and

• faculty exchanges with industry and other colleges and education conferences as the program developed ($125,000).

The program was very successful. But most of the things high on our list for support in the late 1950s are still high today on people's list. I have always believed that the important things in life never change quite as fast as people think they do.

"A Mouse Learning to Be a Rat"

Beginning my second year as associate dean in the fall of 1957, I realized what many others had found before me: academic administration is very repetitive. Obviously, course content, teaching methods, and the like change, but to administer it all, you repeat the formula: do the work to ensure a high-quality faculty and outside support because education is never fully supported by tuition. In any case, the problem of being bored by repetition was solved for me when Dick Soderberg retired at the end of the 1958–1959 academic year. While I was some people's candidate to succeed him, I didn't take it too seriously because I felt the top bosses at MIT had a better candidate. Gordon Brown, from electrical engineering, was selected, and I returned with some delight, to being a full-time professor in the Department of Aeronautical Engineering.[12] Some of the people who had wanted me to get the job would come up to me and say how sorry they were and I said, "Thank you very much for your sentiments." But the best approach at that time was from General Jim McCormack, whom I had known in Washington when he was on active duty in the Air Force. He said simply, "Guy, into every life a little rain must fall." I remembered that at other times of my life when, in fact, a little rain fell. I also remembered the comment by Theodore von Kármán when he learned I was going to become associate dean. He asked me if I knew what an associate dean was and when I said "No," he said, "An associate dean is a mouse learning to be a rat." I felt there was some truth in that too.

I remembered one other comment about deans. At an earlier time I was on a trip with George Kistiakowsky, a great Harvard chemistry professor. We roomed together, and one evening we were discussing the debate on the formation of the National Science Foundation and how its grants would be given. I innocently asked, "Why do they fuss about all of this business of individual investigators and peer reviewers and so on? Why don't they just pick good universities and give them money and let them select their best professors to support?" When I said that, George's face got a little redder than normal and he raised his fists and shook them and pounded on the table, and he said, "By damn, I don't want any [deleted] dean telling me what research to do."

WAKE-UP CALL

In the wee hours of Friday morning, October 4, 1957, Bunny was awakened at our home in Belmont by the insistent ringing of the telephone. The voice said that he was a reporter for *The New York Times* and that he wanted to speak to Dr. Stever. When Bunny told him I was out of town, the reporter got rather excited, and said he had to speak with me because the "Soviets have just launched a satellite of the earth by rockets." It was a 184-pound satellite about basketball size that emitted a continuous beep as it circled the earth every 98 minutes.[13]

Bunny wasn't too surprised by the news. I had talked often about the satellites. And both the Soviet Union and the United States two years earlier had announced intentions to launch a satellite into space as part of the International Geophysical Year. Bunny told the reporter that reaching me was well nigh impossible "because he's in the Maine woods with a friend, hunting deer with bow and arrow." A long silence on the telephone followed, which Bunny and I later interpreted as the reporter thinking, "No wonder we're behind the Russians."

Although I wasn't surprised that the Soviets had launched *Sputnik*, I was surprised by the immense reaction in the United States and indeed around the world. A Japanese paper, perhaps ironically but I doubt it, called *Sputnik* "a Pearl Harbor for American science." The prime minister of Great Britain declared it "a real turning point in history. Never has the threat of Soviet communism been so great."[14] That apocalyptic reaction echoed what happened in the United States, where newspapers, Congress, and indeed much of the population saw *Sputnik* as a beeping symbol that our country was suddenly vulnerable to a frightening enemy, that our vaunted technological superiority was delusional, and that our political leaders—notably, President Eisenhower—not only failed to anticipate the threat but even after *Sputnik* still didn't grasp the magnitude of what had happened. That kind of reaction was capsulated in a little verse by the governor of Michigan, G. Mennen Williams:

Oh Little Sputnik, flying high
With made-in-Moscow beep,
You tell the world it's a Commie sky,
And Uncle Sam's asleep.[15]

It's easy to pooh-pooh this as overreaction. It was, but even very savvy people, such as Jim Killian, whose life was transformed by *Sputnik* as was mine, confessed that the Soviet achievement found him "psychologically vulnerable and technically surprised."[16] In retrospect, all of us were a bit naive in underestimating public reaction and too complacent in our technological strength. The warning signs that the Soviets were capable— or at least felt themselves capable—of putting up a satellite were apparent since the idea emerged at the start of the 1950s. In 1950, several people met one evening in suburban Washington at the home of James Van Allen, then a physicist with the Applied Physics Laboratory of Johns Hopkins University, to plan for what would become the International Geophysical Year (IGY). They had a common purpose—global coordination of high-altitude research. One of the people at the informal get-together, Lloyd Berkner,[17] suggested another International Polar Year, such as the ones held in 1882 and 1932. The timing they hit on was an 18-month interval from 1957 to 1958 when the 11-year sunspot cycle was at its maximum. Working through the International Council of Scientific Unions, their idea was accepted by 67 countries. The National Academy of Sciences, the national ICSU member, appointed a U.S. National Committee for the IGY in March 1953 and secured federal funding to prepare for the IGY. Most tellingly, a few months earlier, in 1952, Berkner and several of his colleagues in Rome waiting for a meeting to start had suggested that a space satellite to do science be part of the IGY.[18] ICSU formally endorsed the idea in October 1954, and a few months later, in July 1955, both the United States and the Soviets announced on successive days their intent to launch satellites.[19] In September 1955 the Naval Research Laboratory's Vanguard program was selected to launch the satellite using a *Viking* rocket. That decision didn't go over well with all, especially the Army, which felt, rightly it turned out, that it had in its *Redstone* missile a reliable and virtually ready booster for launching a satellite. Then, the countdown came. In June 1957 a Soviet rocket scientist reported that a satellite was being readied for launch in a few months. In mid-September Radio Moscow announced that the Soviets were ready to launch a satellite. On October 1 the Soviets released the satellite frequencies. Three days later *Sputnik* was in space.

"Like a Shuttlecock"

I returned from Maine to MIT on Monday, October 7, when fortuitously the MIT Corporation was meeting. Jimmy Doolittle, a member of the corporation, came over to my office to ask how the SAB (Jimmy was chair and I was vice chair) should adjust to *Sputnik*. First, we decided to accelerate the report of the study I chaired on military uses of space, focusing on cislunar space, the region between the earth and the moon. That study started in the summer of 1957, a few months before *Sputnik*, and we submitted our report[20] to the Air Force chief of staff, Thomas White, on October 9, five days after *Sputnik*. Our key message was that the successes so far of the missile programs would yield rocket components and other elements for military uses in space. We also strongly urged Air Force support for scientific research in space, arguing that what was learned would also serve whatever military functions in space the Air Force took on. And not least we noted the urgency of the Air Force working much more strongly on space flight.

Others joined these reactions of the Air Force. General Bernard A. Schriever offered a good sense of the climate and pressure in a 1995 talk:

> In February 1957, in San Diego, I made a speech concerning military space and indicated that space would play an important role for national security in the future. The next day I received a wire from the Secretary of Defense's office: "Do not use the word 'space' in any of your speeches in the future." The launch of *Sputnik* I later that year . . . changed that. Suddenly, everyone got space-minded. I was flying back and forth from the West Coast to Washington, like a shuttlecock in a badminton game, making presentations to

people in the Pentagon and Washington. "Why can't we go faster?" they demanded. "Why can't we do something?" They were thinking mainly about international prestige, because we had been outmaneuvered by the Soviets. . . . We could easily have been first ourselves; we had the capability to do it.[21]

Secretary of the Air Force James Douglas convened a committee chaired by Edward Teller[22] to look at the future uses of military space. The committee met on October 21 and 22 and submitted its report six days later. The resulting report of the Teller Committee was a strong attack on the complacency and overorganization of the military bureaucracy, including that of the Air Force, in stifling technical advances critical to the Air Force in the space era. From the secretary of defense on down, the management and organization of missile and space programs needed to be consolidated and simplified. And, not least, it warned that there was no quick fix. An "attempt to counter the sobering effects of *Sputnik* . . . by a spectacular, but technically superficial demonstration would be to seriously and perhaps fatally deceive ourselves as to the gravity of the present technical position of the country."[23]

Jimmy and I also decided to modify the agenda for the fall meeting of the SAB, December 4–6 at the San Marcos Hotel in Chandler, Arizona, from a think session on limited war to a session on space. Coincident was a strong push to reassure the country that there was strength on the U.S. side as well, and the Air Force seized on the fall meeting of the SAB to make such a statement. Jimmy Doolittle, working closely with General White, arranged for presentations by the senior officers for the three services—General White, General Lyman Lemnitzer, and Admiral Arleigh Burke—and senior officials from the White House and the State Department. For the first time, the SAB got major publicity. A big spread in *Life* had pictures of the board and explained its work and composition.

The work of the Teller Committee and my cislunar group overlapped, and since members of both groups were at the meeting, the board quite sensibly directed us to come up with a unified report and recommendations. An ad hoc committee[24] on space technology that I chaired sent its report to General White a few days after the meeting. We recommended prompt and vigorous action:

1. Obtain a massive IRBM and ICBM capability as soon as possible.

2. Establish a vigorous program to develop second-generation IRBMs and ICBMs having certain and fast reaction to Russian attack.

3. Accelerate the development of reconnaissance satellites.

4. Establish a vigorous space program with the immediate goal of landings on the moon.

5. Obtain as soon as possible an ICBM early-warning system.

6. Pursue an active research program on anti-ICBM problems. These were decoys, discrimination, and radar tracking. When these problems are solved, a strong anti-ICBM system should be started.

All of these recommendations moved forward in the Air Force, although the problems of an anti-ICBM system proved much more difficult than we thought at the time.

The aftershocks of *Sputnik* continued. The day after we finished our fall meeting, Soviet leader Nikita Khrushchev boasted that the Soviet Union would surpass the United States in heavy industry and consumer goods, piling on to the earlier claim that the Soviets had an operational ICBM.[25] And the same day of Khrushchev's boast, President Eisenhower gave a long television and radio speech to the nation on science and national security in which he sought to reassure Americans that we were not ignoring space, that we were strong in many ways,[26] and that he was appointing Jim Killian to be his special assistant to the president for science and technology. In concert, the Office of Defense Mobilization's Science Advisory Committee was reconstituted as the President's Science Advisory Committee, or PSAC. Those events formalized at the presidential level the strong postwar role of science and technology in the U.S. government.[27] At the same time, the tremendous influence that Dr. Killian had because of the nature of the crisis—the country reaching out almost in desperation to him—has, in my mind, never been equaled since, although all presidents since Eisenhower have had a science advisor in some form.

The launch of *Sputnik*, and with it the debut of the Space Age, gave my expertise a bright new polish. Soon, instead of being just an aeronautical engineer or guided-missile expert, I became an expert on space flight. Just a few small changes in my curriculum vitae did that, since I had effectively been in that business since World War II, albeit on the military side. And we changed the name of our department at MIT from aeronautics to aeronautics and astronautics. This was done in the usual academic fashion—awkwardly. After a long and sumptuous dinner at Lock Ober's, then the place for seafood in Boston, the professorate from the department, augmented by humanists from Harvard and MIT, hit on "Department of Archophorics," incorporating supposedly the Greek words for propulsion and direction. The next day I told people who weren't at the dinner what the name was to be. I got blank looks until one of the secretaries looked up the word in a dictionary, and the closest definition she could find was a reference to diseases of the rectum. It became the Department of Aeronautics and Astronautics.

THE BIRTH OF NASA

Once *Sputnik* forced the issue of how the country's space programs were to be organized, an above-average state of confusion hit Washington. There was most obviously a dichotomy between military and civilian interests, echoing the debates during the creation of the Atomic Energy Commission that led to placing nuclear energy and weapons under civilian control. Space posed some of the same issues. There would be intense use of space by military and national security agencies. For example, the secret satellite reconnaissance program—spy satellites—was well under way, something the country was told about but paid little attention to in a *New York Times* report 10 days after *Sputnik*.[28] But important civil and commercial applications were also emerging. Bell Telephone Laboratories was far along in building *Telstar*, the first communications satellite.

Another facet of the civilian versus military management of space was that many of the scientists who wanted to do research in space and eventually explore the planets understandably wanted to do it in the open—as unclassified nonmilitary programs. They worried that excessive classification and military control would severely hamper their efforts. There were other confounding factors. What would be the future role of the military in space and, more particularly, the respective roles of the three services? Where were the boundaries of civilian and military space programs, especially since the large boosters were military? What was to be the driving philosophy for the nation's venture in space: national and secret or international and open? And what were the prospects for the National Advisory Committee for Aeronautics (NACA)? Politics, of course, was in this stew. The Senate and House, both controlled by the Democrats, each established their own committees to deal with space. In the Senate it was the Special Committee on Science and Astronautics, chaired by Lyndon Johnson with the rest of the members serving as chairs of other committees, and in the House the Select Committee on Astronautics and Space Exploration, chaired by John McCormack. These were politically powerful committees—especially so in the Senate given Johnson's ambition—meaning that whatever new space organizations emerged would have strong political backing. And they were also counters to the powerful armed services committees, meaning that civilian goals would dominate the new organization.

But what were the goals? Pursued by what organization? The goals were first defined by the PSAC in a report issued publicly on March 16, 1958, entitled *Introduction to Outer Space*,[29] endorsed by President Eisenhower, including the goals[30] it set for the nation's venture into space. These goals proved durable, set the tone for the debate that followed, and are worth quoting:

- The first [goal] is the compelling urge of man to explore and to discover, the thrust of curiosity that leads men to try to go where no one has gone before. Most of the surface of the earth has now been explored and men now turn to the exploration of outer space as their next objective.
- Second, there is the defense objective for the development of space technology. We wish to be sure that space is not used to endanger our security. If space is to be used for military purposes, we must be prepared to use [it] to defend ourselves.
- Third, there is the factor of national prestige. To be strong and bold in space technology will enhance the prestige of the United States among the peoples of the world and create added confidence in our scientific, technological, industrial, and military strength.
- Fourth, space technology affords new opportunities for scientific observation and experiment that will add to our knowledge and understanding of the earth, the solar system, and the universe.

As to the organization, many variants were flying around—for example, NACA proposed a space program under joint control of the Department of Defense, NACA, the National Academy of Sciences, and the National Science Foundation. But what was key to the new organization was a bill drafted by PSAC and submitted to the Congress on April 2, 1958. The bill proposed the establishment of a National Aeronautics and Space Agency into which the NACA would be absorbed. This agency was to have responsibility for civilian space science and aeronautical research. Civilian space programs under the new Advanced Research Projects Agency in the Department of Defense would be transferred to the new agency.[31] The bill underwent some changes but not major ones. Separation of civilian from military space programs bedeviled much of the debate, even if there was tacit understanding that in many instances the separation was artificial. The "agency" became an "administration," then thought a step up in the Washington pecking order. After a final negotiation between Eisenhower and Johnson, the bill creating the National Aeronautics and Space Administration became law in July 30, 1958, and NASA was legally born on October 1, 1958.[32]

Noise and Hot Argument

I entered into the arguments about the future organization of space because I was working so closely with Jimmy Doolittle on the military space programs of the Air Force. Jimmy, in addition to chairing the SAB, was NACA chairman. The NACA director was Hugh Dryden,[33] a fine aeronautical scientist and soon to be a fine space scientist. Hugh called me about a week after the SAB meeting in Chandler, Arizona, and after the announcement of Killian's appointment, to say that NACA was going to try to become the space agency as well as the aeronautical research agency. He wanted me to chair a committee to advise how to do that— what technology programs, what research and development, etc., would be needed. I was pretty busy with SAB work, but after meetings with Doolittle, Dryden, and Killian, I agreed.

There was a lot of noise to do something in a hurry. This worsened when a *Vanguard* test failed the day before Eisenhower spoke to the country to reassure it and to announce Killian's appointment. The Army, led by Major General John Bruce Medaris, head of the Army Ballistic Missile Agency, renewed its demand that its *Jupiter* missile be used to launch a scientific satellite. That was approved, and on the night of January 31, 1958, the first U.S. satellite, *Explorer I*, was launched using a *Jupiter* rocket. More U.S successes—and failures—followed, as did Soviet ones, summarized in Table 5-2. The Soviets didn't tell us about their failures, but we eventually learned that there were at least four Soviet launch failures during the period covered by Table 5-2, both in their Sputnik and Luna series.

The Special Committee on Space Technology[34] I agreed to chair on defining the future of NACA met for the first time on February 13, 1958. It was heated and tense. While there was convergence among the scientists and engineers involved in space as to objectives, the political battle on the form the new agency was to take was boiling. NACA conducted its political battles quietly, trying to prove it was strong enough to take on the job. The Space Technology Committee was certainly part of the campaign. But there were tough opponents with a lot of political clout, with the Air Force and Army pushing hard to control the total space program, including basic science, arguing that the most important applications of space were military. In February 1958 Eisenhower had approved deployment of four *Thor*, four *Jupiter*, and four *Titan* squadrons (see Table 5-1), that is, Air Force and Army intercontinental and intermediate-range missiles.[35] In a novel maneuver, Secretary of Defense Neil H. McElroy, in mid-January 1958, just before the Space Technology Committee met, created the Advanced Research Projects Agency (ARPA) to be "responsible to the Secretary of Defense for the unified direction and management of the antimissile missile program and for outer space projects." Most presidential historians and scientists involved in defense matters credit Eisenhower with the creation of ARPA. He was determined to bring some rationality to the rival and redundant missile programs of three military services, and that was the initial ARPA mission.

Rough Treatment

I got some early warnings of what a brutal business politics could be. I was asked by Hugh Dryden to help him with the two committees established in each House to deal with space. I went with Dryden to the House hearing, chaired by John McCormack. It was a very good session. I was even invited to report on what we were finding in the Space Technology Committee. A little later I went with Dryden to the Senate com-

TABLE 5-2 Soviet Union and United States into Space, 1957–1958[a]

1957

Sputnik I	October 4	184 lbs.; reentered the atmosphere and disintegrated on January 4, 1958.
Sputnik II	November 3	Carried dog; reentered the atmosphere on April 13.
Vanguard TV-3 (Vanguard 1)	December 6	Rocket malfunctioned almost immediately after ignition and crashed on the launch pad.

1958

Explorer I	January 31	10.5 lbs.; a joint program of the Army Ballistic Missile Agency and the Jet Propulsion Laboratory using a *Jupiter-C* launcher; discovered Van Allen radiation belts.
Vanguard Test Firing (TV-3BU)	February 5	Failed 57 seconds after launch.
Explorer II	March 5	No orbit; last stage did not ignite.
Vanguard I	March 17	Confirmed Van Allen belts; geodetic data confirmed that earth is indeed pear shaped.
Explorer III	March 26	A joint project of the Army Ballistic Missile Agency and the Jet Propulsion Laboratory, launched by the Army's *Juno II*, the antecedent to the *Saturn* rocket; data on radiation belt, micrometeorite impacts, and temperature; returned to earth on June 28.
Sputnik[b]	April 27	Launch failed.
Vanguard (TV-5)	April 28	Failed to orbit.
Sputnik III	May 15	Carried 1.5-ton satellite into orbit, including a geophysical laboratory.
Vanguard (SLV-1)	May 27	Flawed orbit, due to problems with second stage.

TABLE 5-2 Continued

Vanguard (SLV-2)	June 26	Launch failure due to premature second-stage cutoff.
Explorer IV	July 26	Fourth U.S.-IGY satellite.
Pioneer 0	August 17	*Thor-Able* launcher with IGY lunar payload exploded 77 seconds after launch.
Explorer V	August 24	Failed to orbit.
Luna	September 23	Third-stage failure.
Vanguard (SLV-3)	September 26	Failed to orbit; destroyed on reentry.
Luna	October 11	Third-stage failure.
Pioneer I	October 11	NASA probe, using an Air Force *Thor-Able* rocket, traveled 70,700 miles before returning to Earth; measurements of the radiation belt; observations of the earth's magnetic field and the interplanetary magnetic field and of micrometeorite density in space.
Jupiter-C	October 23	Failure owing to premature separation of second stage.
Pioneer II	November 8	Second NASA probe; second and third stages did not separate.
Luna	December 4	Third-stage failure.
Pioneer III	December 6	Third NASA IGY probe; failed to place a scientific payload near the moon but did discover that the radiation belt has two bands.
Project Score	December 17	U.S. Air Force *Atlas* orbits first communications satellite (150 lbs.); Eisenhower's Christmas message beamed from *Score* two days later, the first voice from space; stopped transmissions on December 31.

[a]Based in part on information provided in http://www.hq.nasa.gov/office/pao/History/timeline.html and McDougall (various). Local rather than Greenwich Mean Time used.

[b]The information on Soviet failures was obtained by the late Charles Sheldon, of the Congressional Research Service. I'm grateful to Marcia Smith of the CRS for providing the information.

mittee chaired by Lyndon Johnson. Dryden was treated badly. He was raked over the coals by Johnson, who implied that Dryden didn't know what he was doing and asked why NACA hadn't led the Germans during World War II in transonic and supersonic aircraft, jet engines, and rockets. I was shaking in my boots when my turn came, but I was treated politely and positively. Afterwards, I asked Dryden why the difference. He simply said, "Guy, you are a private citizen and I am a civil servant." I remembered that years later when I was director of the National Science Foundation and had my turn at rough treatment by some members of Congress.[36]

As our NACA Space Technology Committee proceeded through the summer and fall of 1958, it became obvious that NACA would become NASA and that some of the lifting capabilities of the Army and the Jet Propulsion Laboratory would be transferred to NASA. The committee became a much happier group, and we could now turn to recommending specific programs for the new agency. Von Braun took a strong lead on this because he was already thinking about getting to the moon, even though the next steps for NASA were clearly to do what the Soviets were doing: get satellites up and start a manned program. And the scientists who wanted to explore deep space were also demanding their place.

Who would lead NASA? Hugh Dryden asked me to be associate director, with him as director of NASA. Hugh felt strongly that he could lead NACA reincarnated as NASA. However, Hugh was not in favor among some in Congress—witness his brutal treatment by Lyndon Johnson—and the politicos in the White House knew it. Why? After all, NACA had run civilian programs, such as they were, since 1915. Yet NACA had no strong allies in the fierce battle that was under way. Few of its funds went to private contractors, in contrast to the heavy private investments by the military. It had a limited role in aeronautical research, although within that it did some effective things, particularly building and operating wind tunnels. But it was seen by some as unimaginative and narrow in its vision. Fairly or not, Theodore von Kármán labeled it "skeptical, conservative, and reticent."[37]

When I talked with Jim Killian about Hugh Dryden's invitation, he pointed out that there was no chance Hugh would get the call. He wasn't going to lead NASA. I felt for Hugh but was glad to have an easy decision. I had the same feelings about that job as when I had to choose between being Assistant Secretary of the Air Force and going back to MIT. I still felt that my life was in New England—Boston, Cambridge, Belmont, and MIT. Until I talked with Killian I hadn't turned Hugh down. I did then. Soon Hugh himself knew that he would not get the top job. Since Jimmy Doolittle had been the chair of NACA, he was also offered the post, but he turned it down, feeling that as a military man he was not the right person to lead a civilian agency. Then T. Keith Glennan, president of the Case Institute of Technology, was selected, accepted, and turned out to be a very good choice.[38] Dryden became second in command. Right after Keith took the job in August 1958, he came to MIT and asked if I could come down from Randolph, New Hampshire, to talk with him. We had a good discussion about the Space Technology Committee, and he urged me to keep that going. He asked if I would join him at the new NASA. I told him that I would help as much as I could but that I was going to stay at MIT. That also had some benefit to NASA. Many of my students went on to work for NASA, two as astronauts: Buzz Aldrin, the second man to step on the moon, and Rusty Schweikert.

I helped NASA more directly through the report of the NACA cum NASA Special Committee on Space Technology, entitled *Recommendations to the NASA Regarding a National Civil Program*. Issued in October 1958, the same month NASA opened for business, the report defined the scope and priorities for the nation's endeavor in space. We argued, in considerable detail, that "the major objectives of a civil space research program are scientific research in the physical and life sciences, advancement of space flight technology, development of manned space flight capability, and exploitation of space flight for human benefit. Inherent in the achievement of these objectives is the development and unification of new scientific concepts of unforeseeably broad import."[39] Perhaps the most salient of our many recommendations, most of which bore fruit in the following two decades, was our firmness on then a contentious issue—manned space flight. While "instruments for the collection and transmission of data on the space environment have been designed and put into orbit about the earth . . . man has the capability of correlating unlike events and unexpected observations, a capacity for

overall evaluation of situations, and the background knowledge and experience to apply judgment that cannot be provided by instruments; and
in many other ways the intellectual functions of man are a necessary
complement to the observing and recording functions of complicated
instrument systems."[40] This was, again, a recommendation that turned
out to have legs.

The manned space program became a prestige race between the
USSR and the United States. Both countries started astronaut training
programs, and the USSR scored first with Yuri Gagarin, first man in
space, followed by Alan Shepard and John Glenn in orbit. Then the
manned space program got in effect unlimited political support for *Apollo*
and landed several times on the moon. It cost $25 billion; was, I think,
the greatest engineering program in history; and was shut down when
the "unlimited" political support disappeared. With *Apollo*, there were
other space thrusts, including the manned U.S. Space Lab, and many
civilian and military satellites for communications and intelligence. By
the end of the *Apollo* program, several distinct space communities had
emerged to battle for dollars. The strong military and intelligence community quietly but effectively got the most money, often spinning out
technology improvements to the civilian space programs. The Space
Shuttle has its pros and cons. It is good for ferrying into space very large
items such as the Hubble Space Telescope and the pieces for the Space
Station. For space science we now have much smaller, lighter-weight,
unmanned launch rockets that have proven valuable. But manned and
unmanned all compete for space dollars in the annual budget, which
itself is shrinking. I expect the government will keep a tight lid on any
new ventures. The legs of manned flight are tired but still useful. If plans
to build a manned station on the moon or for a manned visit to Mars
ever develop strong public and political support, they will require much
more effective propulsion devices than the chemically fueled rockets we
have now—more on this in Chapter 10.

In short, the Space Technology Committee strongly supported a civilian space agency. One could ask "why not?"; but again the military
fought hard for control of the nation's space program. And the military
after NASA still had a very powerful presence in space.[37] But for all my
strong career in military space and astronautics, I felt that space science,
exploration of the solar system and of deep space, would not be well

served by the military. Although the military in fact created many of the technologies on which NASA would have to build, NASA soon developed its own industrial sources and laboratories. And it was a lot easier for us in the academic world to get research results from a civilian space agency than a military one.

Ranch Work

In a way that belief in a duality of military and civilian space roles reflected what I was doing in the *Sputnik* era. I was trying via the NACA/NASA Space Technology Committee to help a new civilian agency think about its goals. At about the same time, I agreed in the summer of 1957 to chair a new committee of the SAB to examine the entire organization for research and development in the Air Force. So in parallel with my work on missiles, reacting to *Sputnik*, and teaching and research at MIT, I spent a good part of my time in 1958 and 1959 chairing the Air Force's ad hoc committee on research and development.[42] We met for the first time on November 21, 1957, when Air Force Chief of Staff General White asked for an "impartial and searching review of the organization, functions, policies, and procedures of the Air Force and the Air Research and Development Command in relation to the accomplishments in research and development over the past seven years" (i.e., since the Ridenour Committee). And he asked us to recommend "how we can do the job better in the future."[43]

Like the Ridenour Committee, we came to our task during a budget crunch, with another president determined to stay under the debt ceiling of $275 billion. But this time, rather than the newly born Air Force looking at bombers versus missiles, it had to examine its mission in space. What did the term "military space" mean in this new era? And what was the role of research and development in whatever the space mission was? What would the Soviets do, and how should the Air Force respond? On a different plane, many in the Air Force were embittered when Secretary of Defense Neil McElroy in effect blocked the Air Force's plans to accelerate ballistic missile plans in the wake of *Sputnik* (and in the process favored the Army's IRBM program, the *Jupiter*, over the Air Force's *Thor* missile).[44]

When we were about to write our report, it was decided we needed a quiet, remote place to do our work. We wound up north of the Salton Sea in Indio, California, at the desert ranch of Floyd Odlum and his wife, Jacqueline Cochran. Both were Horatio Alger stories: Jackie, an orphan, succeeded in the cosmetics business, learned to fly, raced airplanes, set speed records, ferried Lockheed Hudson transports to Britain during World War II, and founded and led the Women's Air Force Service Pilots. Floyd, who had to work to support himself at a young age, put himself through college and became a major financier and head of the Atlas Corporation. Not least, he cashed the corporation out of the stock market before the October 1929 crash, putting himself in a powerful financial position. Their ranch was well set up for our needs, with pleasant cottages for our members and an immense meeting center with conference rooms and an unbelievably large sitting room. We made good use of the swimming pool, from which Floyd Odlum often conducted his business floating about on a rubber raft, with a drink, his papers, and telephone on a tray in front of him.[45]

Making the General Smile

We submitted the report to Jimmy Doolittle on June 20, 1958, and then briefed it to senior Air Force officers, General Sam Anderson, head of the Air Research and Development Command, and General Curtis LeMay, who had left the Strategic Air Command to become vice chief of staff. It turned out that Sam Anderson had given a speech the day before at the Air War College in Montgomery, Alabama, which to Sam's great embarrassment had made it into the *Washington Post* because of its strong criticism of the administration and Congress for cutting the Air Force budget and for micromanagement. Because my friendship with Sam started way back in London during World War II, I thought I could start the briefing with a flip remark to LeMay: "Curt, we've written this report to help Sam Anderson make better speeches." Anderson turned brick red in the face. LeMay broke into a big grin that he quickly wiped off. As far as I was concerned, I'm the only one who ever made him smile.

Our report wasn't universally pleasing to the Air Force. We expected resistance, especially by Sam Anderson and his colleagues in the Air Research and Development Command, ARDC. While we praised the progress made since the Ridenour report that had looked some seven years earlier at Air Force research and development, we also wrote that budget problems and "excessive administrative controls" had eroded these gains. Decision making on R&D had become too ponderous and slow for an effective and responsive program. Our principal recommendation was that the heavy hands of the higher echelons of the Air Force should be lifted from its R&D centers, which had both very good scientists and engineers and very good managers. We redeemed ourselves with the Air Force headquarters a bit by telling it to work to get the Department of Defense and even Congress off its back. And we pointed out that the budget constraints imposed by Secretary McElroy were seriously handicapping progress on ballistic missiles, aerodynamic research and development on new aircraft, and new electronics. That recommendation sat well with the Air Force. What sat less well were our recommendations on organization and management—to correct duplication, micromanagement, flawed procurements, and poor support of basic research. The Air Force rejected our suggestion for reorganizing Air Force laboratories around military systems and that the ARDC adopt a systems strategy for its research and development. But like bad weather, you sometimes need to wait a while. Sure enough, when Bernard Schriever took over the ARDC in April 1959, he immediately set about implementing our recommendations. And in April 1961 the ARDC was reorganized into the Air Force Systems Command and the Office of Aerospace Research.

I learned something from that experience. Most people in government work hard to improve the system, but when they hear a new idea they may resist it for a while. With time they see the good parts and begin to accept it, especially if you can convince them that it was their idea in the first place. However, General Schriever gave our committee full credit.

I continued with the Air Force SAB—as vice chairman until December 1961 and as chairman from January 1962 until 1968. It was an extraordinary time. Ballistic missiles moved from development to production to deployment. The space race intensified. The decision to go to the moon in May 1961 and the Cuban Missile Crisis in October 1962 loomed. And of course there was a sharp change in administrations, from

Eisenhower to Kennedy, with the sharpest change for the Air Force (and the other services) being the arrival of Robert S. McNamara as secretary of defense. McNamara had established his reputation at the Ford Motor Company, where he and his colleagues—the "whiz kids"—had put into place "system analysis" methods to manage programs, measure results, and control costs. McNamara was the first "outsider" to become president of Ford, taking the job the day after Kennedy was elected and five weeks before he accepted Kennedy's invitation to become secretary of defense.

McNamara immediately questioned the quality of the huge number of Department of Defense laboratories, including those of the Air Force. The Air Force had in retrospect anticipated McNamara by having the redoubtable George Valley with several well-qualified colleagues[46] look into how the Air Force could build stronger ties in basic research with the universities. The panel affirmed that the Air Force laboratories had many very good scientists but that in many cases they had tasks in which they were neither particularly competent nor interested. Bureaucratic obstacles to research contracts with universities amplified these failures in research management. The Air Force, despite its obvious need to maintain technological supremacy, was doing a poor job at being an active player in the larger research community.

But McNamara wanted more, not only strengthening basic and applied research in the military but also technological development. General LeMay, who became Air Force chief of staff five months before Kennedy's election and McNamara's nomination, embraced the charge, building on the work of the Valley Committee and a second committee, chaired by Leonard Sheingold, just starting his term in July 1961 as the ninth chief scientist of the Air Force. The Sheingold Committee added one strong recommendation to that of the Valley Committee: that the research and development agencies in the Air Force get direct control over sizeable research funds now in the military construction budget.

Things moved rapidly. LeMay approved a new research and technology division within the Air Force Systems Command[47] to respond to the Valley Committee's criticism and giving control over funds directly to the new division in response to the Sheingold recommendations. Thirty-seven laboratories across the country were merged into seven within the research and technology division. And each laboratory was given its own line item in the budget, showing, said the director of the research and technology division, Major General Marvin C. Demler, that "someone really trusted the Lab Directors to do a good job with these unfettered funds."[48]

These moves to strengthen research and technology development in the Air Force were remarkably well timed. Already in 1960 the role of information technology for communications, command, and control of military operations was emerging. And while liquid-fueled ballistic missiles were coming online, they were to be relatively short lived, removed from operational status by 1966 in favor of solid-fueled missiles, including the *Minuteman* ICBM and the submarine-launched *Polaris*. The concept of multiple independently targetable reentry vehicles emerged in the 1960s. Investments in reconnaissance from space were ramping sharply upward as its technological frontiers continued to be pushed hard. And, finally, the art and science of space communication systems was gathering speed.

In short, postwar investments in military science and technology were in the early 1960s reaping a revolutionary transformation in military strategy and tactics. It was a heady time. One had a confluence of a new generation of computers; the realization of their deep application to military communications, command, and control systems; and a very dangerous world of nuclear-armed missiles and airplanes. We had created a world vulnerable to horrific destruction through "accidental" war. Each side could now destroy the other—and a lot of the human race with it. It was now extremely urgent for critical information to be accurate, usable, and available for military and political leaders to make quick and effective decisions. The SAB through various panels worked on these problems, notably on reducing the danger of unauthorized attacks, of false alarms, misinterpretation, and communication failures.

The SAB hit problems in the early 1960s, coincident with the start of the Kennedy administration. One was size. The other was conflict of interest, now chronically familiar but then a more muted issue. The SAB was extraordinarily busy and productive. In 1961, for example, the board produced 21 formal reports and 29 special memos and held 86 panel or ad hoc meetings.[49] Sustaining that required a lot of people, all of whom had "daytime jobs." Thus, the board in 1962 had 88 members, a reality that got the attention of Secretary McNamara who in turn "suggested" to Secretary of the Air Force Eugene Zuckert that the number go from 88 to 20. After some brouhaha over the matter, an agreement was negotiated that capped the SAB at 70 members and also gave the secretary of the Air Force approval over new appointments along with the Air Force's chief of staff.

The conflict of interest issue hit the fan with several articles in December 1961 in the *New York Times* on the propriety of General Donald L. Putt serving as chair of the board while also a senior executive with the United Aircraft Corporation, a major Air Force contractor. The board of course had been aware since its creation after World War II that given its tasks conflicts were sure to arise. It had handled them by biasing its appointments wherever possible toward people from universities and nonprofits and by trusting to the integrity of its members. Those days were over. The *Times*, for example, questioned why the Atomic Energy Commission declared Don Putt ineligible but not the Air Force.[50] Don Putt, citing the press of business but more likely because of the conflict issue, declined reappointment as SAB chair. I was appointed chair, although my formal appointment was held up until the attorney general issued a report on the matter. On February 26, a presidential executive order set forth new procedures for using advisors in the government, requiring, among other things, federal agencies to update semiannually financial holdings for each advisor and their immediate family.[51] And I became the SAB chair in name as well as in fact in April, after my financial and related information had been vetted.

HAUGHTY ADVICE

You can't be a respectable professional in science or engineering without belonging to the principal society in your field. When I was a young scientist in graduate school and then at the Radiation Laboratory, my field was physics and I considered radar and radiation and all of that a branch of physics, so my principal society was the American Physical Society, and my key publications were *Physical Reviews* and the *Review of Scientific Instruments*. Then, one day, after some time in the MIT Department of Aeronautics and Astronautics, I mentioned to the department head, Jerome Hunsaker, a publication I had just submitted to the *Review of Scientific Instruments.* Jerry said quietly and almost haughtily: "Guy, the society for aeronautical engineers is the Institute of Aeronautical Sciences." I got the message and became active in the institute. The institute was founded in 1932, and its first president in fact had been Hunsaker. Another of my mentors, Jimmy Doolittle, also served as president. So I was greatly honored when I was elected institute president in 1961. It was for a one-year term but a very busy one, not least that I got the momentum moving to reconcile the increasingly clumsy institutional division between travel in air and in space, including the large overlap between the aircraft and space industries, which were asked to support both the Institute of Aeronautical Sciences and its counterpart for space, the American Rocket Society. The division no longer made sense, and after a considerable amount of work the two groups merged, becoming the American Institute of Aeronautics and Astronautics.

On yet another track, I had since I finished my tour as chief scientist of the Air Force in 1956 served as a consultant to the United Aircraft Corporation.[52] I did that for 18 years, helping initially on propulsion systems for missiles and rockets, a new area for the company and one to which they were latecomers.[53] Nevertheless, the corporation, especially when it acquired new leadership at the top, made solid gains in propulsion systems. It started on a guided-missile program and established a development and test center in Florida for liquid-propelled rockets and a similar facility albeit for solid propellants in Palo Alto, California. Don Putt directed the latter after he retired from the Air Force in 1958. It was this job combined with his chairing the SAB that got him front and center in the *New York Times.* I also worked hard to involve first-class scientists in helping the corporation in other areas, including Frederick Seitz, a solid-state physicist of world stature; I. I. Rabi, the Nobelist who had for a time been at the Rad Lab; and George Kistiakowsky, the Harvard chemist I mentioned earlier who devised the explosive mechanism for triggering the atomic bomb and succeeded Jim Killian as presidential science advisor.

Amidst all that, the Stever family in June 1960 embarked on a 12,000-mile trip to the West, a first for our four children—Guy, 13;

Sarah, 11; Margo, 8; and Roy, 6. The trip took two months and had its moments. Roy refused to eat or drink anything other than peanut butter sandwiches, ice cream, milk, and Coca-Cola, until he discovered that he also liked Dinty Moore beef stew. Bunny taught Sarah French. Margo and Bunny had a scary nose-to-nose run-in with a moose. The moose retreated. A bear pillaged our cooler at one campsite. And we saw the West at its most spectacular. For example, we went through the Wind River range in Wyoming through Lander just across the Continental Divide. I returned to that spot some 17 years later when Roy, just out of college, laid out a 17-day trip above 10,000 feet for me in my first summer after I retired from service in the White House.

We returned to our home in Belmont about three weeks before the start of the MIT fall 1960 term. My return signaled another change in my professional life. Dean Gordon Brown and now President Jay Stratton had for some time been trying to persuade me to head up two departments, the excellent mechanical engineering one and the much smaller but very distinguished naval architecture and marine engineering. This was a rather unusual assignment. It came about in part because Dean Brown wanted to fold the naval architecture and marine engineering department into the very large mechanical engineering department. Also, he needed a new head of mechanical engineering, following rough times with the two previous department heads. I told him that the department of naval architecture and marine engineering was a jewel. Why should he bury it in a much larger department?

CAIR PARAVEL

The day I accepted the dual assignment I called the president of Yale to decline for the second time an offer to become its dean of engineering. There were many reasons for turning it down, including our love of Boston, of MIT, and of Randolph, New Hampshire, where we now purchased a place to relieve the pressure on the Risleys, a place we called Cair Paravel, the magical castle in C. S. Lewis's *The Lion, the Witch, and the Wardrobe*.[54] Four English children discover a magic land that lies beyond an ordinary wardrobe closet. In this land, Narnia, one of the children, Edmund, betrays his siblings to the wicked White Witch, who has been holding all Narnia in thrall to winter. Only when the lion Aslan agrees to die at the witch's hand can the betrayal be forgiven and spring come to Narnia.[55]

There was no betrayal but Randolph became our Cair Paravel. Bunny and I when we were married in 1946 went to Randolph sparingly—it wasn't ours but that of Bunny's parents and so we went as guests. And it wasn't winterized. Fifteen years after we married, we took possession of our own place at Randolph, now with four children and a firmer financial base with my increased MIT salary and consulting. We started winterizing as soon as we could, finishing up shortly after Christmas 1961.

The original land was seven acres of mostly woodland, a large house, and a small garage. Over the years we added a large barn-garage, which I had the pleasure of designing myself and that has become a major part of our existence. We helped Guy, Jr. and Debra buy land across the road, and they built there. Later, when he and his wife and child moved to a larger house nearby, we bought their original property for a spillover for our other children's families as they grew up. Also, since Bunny and I grew more and more allergic to dogs and cats, we named that spillover home *Casa di Cani*, the dog house. Visiting dogs and cats were always put over there, along with their owners, and we were dog-free and cat-free at our house.

By several purchases the woodland grew to about 20 acres, with several healthy soft wood stands of white pine, spruce, balsam, hemlock, tamarack, and white birch, as well as hard wood of maple and oak. Bunny had five flower gardens that she loved, and I had a couple acres of lawn that I mowed. The whole family helped cutting trails and firewood with an assortment of chain saws. We all kept healthy.

We made many friends, many having also made Randolph into a summer retreat for their families. There were constant movable feasts with different groupings centered on one or more of the activities. For example, there was a group that just loved to square dance and we had a lot of fun doing that. Another group liked golf. Others were particularly interested in climbing, but we all shared and had a wonderful set of relationships. Tops on many of our families' list was fly-fishing for trout in the wild streams of the area, Bunny having taught me and I willing to teach the children. Both Roy and Margo became expert. Lots of skiing in the winter, downhill and cross-country.

We unified around the Randolph Mountain Club, which kept open a hundred miles of trails, using, usually, the younger members of our families; our children served as hut boys at a couple of the cabins. The Randolph Mountain Club also had a tea to open the summer season. And an annual picnic that was the high point of the summer, with elaborate charades competitions and group singing. There was a regularly scheduled Tuesday climb and a Thursday climb, one more difficult than the other. You could sign up for those, and the attendees would vary from 3 years old to 90.

Cair Paravel was the vital balance to the pressures of my professional life in the 1960s: presidency of the Institute of Aerospace Sciences, vice chairman and then chairman of the SAB, growing consulting responsibilities with the United Aircraft Corporation, teaching and research, and now head of two engineering departments at MIT. Both my professional life and my personal life ahead seemed clear and even predictable. Little did I know.

6

Going Public

1964–1972. I left MIT for the presidency of the Carnegie Institute of Technology, which merged a year later with the Mellon Institute to become Carnegie Mellon University. And the 1960s hit: civil rights, Vietnam, campus protests, and money squeezes.

In the summer of 1964 during a consulting trip to the West Coast, Don Putt spoke to me about the presidency of the Carnegie Institute of Technology in Pittsburgh. Don was an alumnus of Carnegie Tech, studied engineering there, had been a member of its ROTC (Reserve Officers' Training Corps), going on after graduation to flying school and a long career in the Air Force. The presidency was open and Don wanted to know if I was interested. It wasn't the first presidency I was asked about. I had already turned down offers from Yale and Cornell to become the dean of their engineering schools, and now my alma mater, Colgate University, had offered its presidency to me. I loved Colgate, and turning down that offer was wrenching. It was a fine liberal arts college but not an institute of technology where I had spent much of my professional life, first at Cal Tech and now at the Massachusetts Insitute of Technology (MIT). Colgate's community and its undergraduate mission were far removed from mine, and I could see the strong professional bonds I had built and treasured weakening if I went to Colgate. Carnegie was different. It was an institute of technology, and while it had no program in my fields of aeronautics and astronautics, the technological milieu was a familiar one.

The first toe-dipping trip was in September 1964. I was met by Jim Bovard, president of the Carnegie Tech Board, and John "Jake" C. Warner,[1] president of Carnegie Tech[2] since 1950 and retiring in 1965,

149

and Don Putt, an alumni trustee and later a life trustee. We talked a bit and then went to the top floor of the Gulf Oil Building to meet Richard K. Mellon,[3] Roy Hunt, a past CEO of Alcoa, and other Carnegie Tech trustees. Then a quick visit to the campus to meet and dine with Carnegie Tech vice presidents.

Carnegie Tech[4] had impressive strengths, not least Pittsburgh itself, with its industries in iron and steel, aluminum, heavy electrical machinery, and the like—and the corresponding wealth. Tech had created very strong engineering departments mirroring Pittsburgh's industry—in metallurgy and materials science, mechanical, electrical, and chemical. And because big business was beginning to see the value in business schools, it now had the Graduate School of Industrial Administration, funded by the Mellons. But the picture wasn't monochromatic. There was a very fine college of fine arts with an exceptionally strong drama department. I was introduced to the Margaret Morrison College for women, a favorite of tech founder Andrew Carnegie, named for his mother, and specializing in the "practical arts for women" such as nutrition, clothing design, and the like.[5] No law school. No medical school. No college (then) of humanities and social sciences, but there were strong departments in those fields, justifying Jake Warner's claim that it was a "liberal professional" institution.

I returned to MIT leaning toward taking the job. But first I had to talk to my mentor, Jay Stratton, president of MIT. Jay, as always, was helpful, telling me in some detail about the infighting among the various Pittsburgh institutions, especially with the University of Pittsburgh and its strong leader, Edward Litchfield, about whom later. But at the end of the conversation, Jay said with a grin, "You've got one priceless asset at Carnegie Tech." I immediately thought of some great endowment, but he said, "That asset is Herb Simon." Herb Simon,[6] he pointed out, was among the broadest and deepest thinkers in the social sciences, and that he, with Allan Newell and Alan J. Perlis, gave Carnegie Tech enormous strength in the emergent field of computer science.[7]

I decided to accept in September. That simple declarative hides a lot of agony: leaving MIT and the strong community of friends and colleagues we had in Boston, Cambridge, and Belmont. But there were things I wouldn't give up, especially Randolph, our spiritual home, with my family, with friends, and the camaraderie and solitude that came with my love of fly-fishing.

Carnegie Tech announced my appointment in October 1964, on Columbus Day, and with that I became for the first time in my life a public figure in contrast to being well known in professional circles and in Washington. I was inundated by publicity. The first go-rounds in the press were flattering, but after the first wave was a second, in the form of thoughtful pieces, especially from the two Pittsburgh papers, the *Post-Gazette* and the *Press*, that hinted at a strong desire to become a distinguished national university. The *Press* editorial, in particular, set some markers for me:

> Although some Carnegie Tech faculty members and alumni may have preferred a president with more background in liberal arts, the appointment of a leading scientist seems appropriate in this age of technological revolution. Pittsburgh's industrial research community, as well as Carnegie Tech, will benefit from the choice.
>
> Dr. Stever will inherit from Dr. Warner an ambitious expansion program that should help to attract space age industries to this district. He is well qualified to supervise the program, under which Tech's graduate enrollment is expected to double to 1300 students in ten years while undergraduates increase by 25% to 2500.
>
> Dr. Warner, who has been president for 15 years and a faculty member since 1926, has had a distinguished career at Carnegie Tech. Pittsburgh will wish Dr. Stever equal success in his new position.

"WHERE'S THE TONNAGE?"

I had to continue the drive begun by Jake Warner to strengthen the position of Carnegie Tech in the total university galaxy. And I certainly had no argument with introducing Pittsburgh to the modern Space Age. The city had a ways to go. At one of my early meetings with Pittsburgh-based trustees of Carnegie Tech, they reported on a visit by a Pittsburgh delegation to meet with the head of the National Aeronautics and Space Administration (NASA) to find out how Pittsburgh could help. They heard what space rocketry was all about, but a member of the delegation, bred in producing steel and aluminum, asked, "Where's the tonnage?" There would have to be a little bit of an adjustment in the center of gravity if they really wanted to get into space-age industry.

That little anecdote and the editorials put in front of me the gap between my technical experience and connections and what was facing me at Carnegie Tech. I decided that it was important to continue the work I had been doing with the government, especially the Air Force, and my technical consulting for the United Aircraft Company. But I also decided that I'd better look into new areas where Carnegie Tech could gain a national reputation. I went back to Jay Stratton's statement that Carnegie Tech had one tremendous asset in Herb Simon and his pioneering use of computers to probe human behavior in organizations. That thinking about computers proved extremely apt in what was billed as "just a pleasant dinner" with leaders of major Pittsburgh foundations: the A. W. Mellon, Scaife, and the Richard K. Mellon foundations. As we talked, it became obvious that they wanted to know what really good ideas I had for Carnegie Tech. I said it's very clear that the computer is still in its infancy and will have a tremendous effect on our future.[8] All relaxed when I said that because they were hoping that is what I would say. That session led to a $10 million gift to endow an R. K. Mellon chair in computer science (to which, unsurprisingly, Herb Simon was appointed) to help us buy a new computer—they cost a lot in those days, and of course the very large supercomputers still do—and to set up new research programs. This money enabled us to establish a computer science department with Alan Perlis as chairman.

Bunny and I made two important decisions before I actually started my Carnegie job. One was not to live in the house intended for the new president. Bunny felt, rightly so, that the traffic in front of the house on Morewood Avenue was too heavy for our children, especially the two youngest just moving into their teens. Carnegie sold the house within six months and bought a new home for us on Devon Road, a quiet residential street a block away from campus. The second decision was that I would go alone on February 1 to Carnegie Tech and the new job. Bunny would stay in Belmont with the four children, so that Guy and Sarah wouldn't have to split their junior year in high school. The wisdom of that decision was amplified when two very close friends, Helen and Gordon Scannell, asked us if we would like to leave Guy and Sarah with them for their senior year in school. That was enormously helpful for I had discovered, whenever I tried to hire someone who had children in high school, they did not want to move at the critical last year when the children have to face the trauma of college entry.

THINKING OTHERWISE

On February 1, 1965, my pay started at Carnegie Tech. I was 48. I left my family, my home in Belmont, and the faculty and students of MIT to fly to Pittsburgh, carrying a couple of suitcases. I moved into the president's house on Morewood Avenue, complete with maid, cook, and driver. I didn't need much attention at home, but the driver I used all the time. I reported in at my office and met Miss Priscilla Brown, who had worked for my predecessor and would be my secretary throughout my tenure; Edward R. Schatz, a talented, typically tough-minded western Pennsylvanian, vice president for academic affairs; and Robert Kibbee, who became my vice president for planning. Another ally in my office was Professor Raymond Parshall, who helped me with correspondence, especially the tough kind. He could cut through the chaos that at times besets a campus and its president. One time when we were having a little set-to between faculty and administration, Ray calmed me down by pointing out that "a professor is someone who thinks otherwise."

And I met the deans, including Norman Rice, dean of the College of Fine Arts. As we stood in front of the fine arts building, he waved his hand toward the science and engineering buildings and then at the fine arts building and said: "Now, President Stever, I want you to know that this is the principal axis of the institution." He was most convincing. Dean Rice was not the only salesperson for fine arts. There was the outstanding success record of the graduates of the college.[9] And even more pleasant for the campus life, there were the regularly scheduled student plays, frequent symphony and chamber concerts, individual musical performances, and art exhibits. These were enjoyed not only by the campus community but also by the surrounding city community.

The Graduate School of Industrial Administration began in the late 1940s as the School of Industrial Administration with $6 million from William Larimer Mellon, who wasn't happy with the "caliber of managers he had been able to hire at Gulf Oil, where he was chairman."[10] It became the Graduate School of Industrial Administration, coming into its own as a strong institution during Jake Warner's tenure as president, and when I arrived ranked with the top business schools in the country. We had just lost its first dean, Gordon Leland Bach, to Stanford. He had laid out a unique approach to management education and assembled a powerful faculty. The new dean of the school was Richard Cyert, and his first point to me was that I should understand that a portion of Carnegie Tech's endowment was given for the Graduate School of Industrial Administration and that therefore it had more say on its budget than other schools. True, but the school also needed general funds from the university. We always had an interesting session with Cyert and his colleagues at budget time. He also gave me two other messages. One was that I should understand that the president's salary didn't have to be biggest in the institution because we had some very important and first-rate professors. His final message was that I shouldn't waste a lot of our ability on raising money for new buildings. The money should go instead to support our fine and growing faculty.

I ignored his first point even though I knew we had "some very important and first-rate professors." I partly agreed with him on the second point. I didn't see any great building needs. And I knew from the comparative salary data I had looked at from the American Association of University Professors that Carnegie Tech needed to boost faculty salaries if it was going to be competitive.

We had a gem as leader of our social sciences in John Coleman. Later as president of Haverford College,[11] he did a weekly tour with garbage collectors in the area surrounding Haverford saying that brought him down to some very earthy matters and served as a reality check on the application of the social sciences to worldly problems. Some students indeed did follow his example, not simply in collecting garbage but in trying to link their work and studies closely to real problems. You practiced what you preached.

I had one serious problem: Margaret Morrison College. Jake Warner urged me to close it and told me why: competition from state-supported public schools, such as Penn State, offering the same courses at lower cost; a low research effort; many tenured faculty nearing retirement and difficulties in attracting quality replacements; and programs that were no longer sustainable—for example, the degree program in social work had to be phased out.[12] After one year on the job, I learned that the "Maggie Murphs" also had many distinguished alumnae, were leading campus citizens, and added quality and pizzazz to the campus atmosphere, making that recommendation hard to take.

In my inaugural talks with students and faculty, I emphasized what I saw as the great strengths of Carnegie. It balanced strong programs in science and engineering with the fine arts. It was gaining national leadership, notably in the cognitive sciences, the first real efforts in artificial intelligence, efforts to model human behavior in computers through symbolic manipulation, and the study of organizational behavior from different perspectives—all driven by people who became giants in these fields, including Herb Simon, Alan Perlis, and Allan Newell.

One of Jake Warner's points in a farewell speech to the faculty was that they should not let their professional activities cut out close contact with the undergraduates' social life and extracurricular life. I worked to get close to the undergraduates, athletes, housing units, fraternities and sororities and social clubs, and Bunny and I dined with them often.

Soon after I arrived, I learned a great deal about Carnegie's relationship with the University of Pittsburgh, in particular its chancellor, Edward Litchfield. He had come into office in 1955, when Pitt was a "middling American university, perhaps somewhat above the median but not by much."[13] Chancellor Litchfield pushed hard, and spent hard, to change that. By 1965 when I arrived, he had increased graduate and professional school enrollments, increased SAT scores of entering freshmen, increased faculty salaries, increased the proportion of full-time students, and increased the university's physical facilities—even acquiring the Pittsburgh Pirates baseball stadium.[14]

While pleasant and smooth in his personal relationships, Ed was hard driving with strong ideas on what the university needed and how to do it—his academic specialty was, after all, administration, somewhat ironic given what was to happen to him. He was very aggressive toward his academic neighbors. Litchfield, in trying to recruit a colleague of mine to be Pitt's dean of engineering, which he earlier had tried on me, took him to the top of the Tower of Learning, a skyscraper on the campus, waved his arms toward the Carnegie Tech campus and declared: "This is your college of engineering of the future." Jake Warner told me that when Lichtfield complained to him that "you don't like me," Jake said: "Ed, I like you very much. I just don't trust you." When Bishop, later Cardinal, Wright was invited to receive an award as the "Man of the Year in Religion," he said that he'd "like to think about it." He then checked that Litchfield was going to be there, too. He came and joked that he was delighted that Chancellor Litchfield was also there because he was a little afraid to leave his cathedral lest it be taken over by Pitt.

The Pittsburgh research and higher education institutions did try to work together, primarily through the Oakland Corporation. Ed Litchfield had one big idea for the corporation: putting a big building in Panther Hollow, the deep cut with a railroad running through it between the University of Pittsburgh, Carnegie Tech, and the Mellon Institute. It was to be the equivalent of Route 128 around Boston and Stanford Park in Palo Alto, both entrepreneurial hot beds. It was an expansive view of what the Oakland Corporation could do over and above what the Mellon Institute did with industry. I had my suspicions about that because I knew that large industries—for example, TRW, for which I had consulted—were discovering that they would rather have their own research parks with campus-like settings. Big companies were simply not willing to get into Panther Hollow-type operations. Nevertheless, I was a good citizen, attended meetings, and tried to be constructive in Oakland Corporation activities. But I had a feeling that Ed's plan was grandiose and would fail. It did.

Ed Litchfield's end at Pittsburgh was a sad caution for his fellow presidents. By 1965, Pitt had a deficit of almost $5 million with current accounts $25 million in the red. Litchfield had a mild heart attack and resigned two months later. Private funding dried up, and the state had to rescue it with a $5 million dollar bailout. In 1966, Pitt became a public university, and its quest to become a leading private university ended.[15] It had become a very good public university. Tragically, Ed Litchfield, then a corporate leader, his wife and two children were later killed in a plane crash.

While Pitt was going through its agony, I was trying to learn the ropes of being a president and spent spring 1965 in an unrelenting round of meetings with faculty, students, community leaders of Pittsburgh, especially contributors—in fact and potential. And by prearrangement with Carnegie Tech, I continued to consult with United Aircraft, to finish up my chairmanship of the Air Force Scientific Advisory Board, and to take on other governmental tasks, notably President Johnson's Commission on the Patent System. Among the pleasanter additions to my portfolio was serving on the board of the Pittsburgh Symphony. For that board I served on a committee that analyzed, at Jack Heinz's request, a large grand old movie house in downtown Pittsburgh for possible conversion to a symphony and drama facility. We supported that after looking at a similar conversion in St. Louis. It became Heinz Hall. Not least, I joined the committees of several local development groups, such as the Allegheny County Development Corporation.

The high point of the spring came at Easter time when the Carnegie Tech Community Orchestra presented Pablo Casals's "Easter Oratorio," and we held a special convocation to award Casals an honorary degree—my first. Sidney Harth, head of our music department, who played with Casals in some of his European events, arranged this event, which took some doing because Casals had for some time refused to visit the United States because of a contretemps with our government.[14] He spent several days with us giving master lessons to students. Bunny had brought our four children to Pittsburgh. They were in awe with front row seats at the concert and convocation. In turn Bunny and I were awed the next day at a lunch for Casals with members of the music faculty. At our small table were the Casals, famed musicologist Fritz Dorian and his wife, Sadie, and we listened to Pablo Casals and Dorian reminisce about the Salzburg festivals before World War II.

June graduation—my first as president—went very well, followed by my family moving full-time to Pittsburgh, followed almost immediately by an about-face for summering at Randolph. In October came my inauguration. It was lovely, a gathering of many college and university presidents, other dignitaries and personal friends from around the country, including two people who greatly influenced my career—Lee Dubridge, president of the California Institute of Technology, and Jay Stratton, president of MIT. John Coleman, who was still with us, had assembled a two-day symposium on higher education with Vince Barnett, then president of my undergraduate college, Colgate, giving a keynote speech on undergraduate education. Lee Dubridge gave the keynote address on graduate education, and Jay talked about the broader side of professional education.

But these wonderful reunions aside, my real object in the inaugural was to assert my views of how Carnegie Tech could go in the future. I felt strongly that the liberal arts and professional education the institution was providing was very good, but I wanted to set new standards and new goals. Other institutes of technology, notably MIT and Cal Tech, had quite deliberately broadened and strengthened their programs so that they could justly call themselves universities. Carnegie Tech wasn't there yet, but I was determined to take it there. So I entitled my talk "A Great University" and said in part:

> The highest standards for research and scholarship we inherit from our past. Our problem is to continue the evolution, recognizing the importance of the newly emergent, and discontinuing the areas that have served their purpose. The past has shown we can. Carnegie Tech has led or been amongst the leaders in the establishment of fields recognized as important throughout the land—industrial psychology, drama, industrial management, computer science, metallurgy, solid state physics and a number of others. For this important portion of our evolution, we depend upon our strong faculty and its professional leadership at the department and college level. They are the well-heads for such change. No president or administrator can have all these ideas, but for illustrations of my point, let me mention my own professional field, engineering, where there is needed a further maturing to face the large scale technological problems of our times, such as cleaning our air, land, and water; designing transportation systems; combining architecture, city planning and engineering to solve our urban building problems . . .
>
> A great university is a balanced one. This institution can achieve its greatest potential when all of its professional work is first class so that its graduates can walk in the front ranks of professionals throughout the world; when it recognizes that it's not here to stamp out a standardized product, but to enable each of its students to search for and determine his own path for service to society—when each professor and professional field is enhanced by the presence of quality work in other fields on the campus and when students can benefit from quite different types of students working with them; and when the impact of the leading minds on the campus is felt across the entire spectrum of education.

ADDING AND SUBTRACTING

Following up on my inaugural talk, I managed in my first year to get two new and major things going, with a lot of help on and off campus, the latter including substantial funding. We first established the Transportation Research Institute, focusing academic work on an area of keen interest to Pittsburgh because of the role of local industries in many transportation modes—aluminum in aircraft and steel in railroads and automobiles. The institute was proposed as an interdisciplinary program by the heads of civil, mechanical, and electrical engineering—Tom Stelson, Milt Shaw, and Everard Williams, respectively. Jim Romaldi became professor in charge. It was successful for a while but ran into problems after I left Carnegie for various reasons, one being that the automobile was going to dominate transportation no matter what, and hence there was little incentive for private funders to pay to look at other systems. The second was that people in the institute had supported Pittsburgh building a metropolitan rail system. That came a cropper, denting its reputation.

A more lasting addition was the School for Urban and Public Affairs. There was tremendous postwar interest in Pittsburgh in urban renewal, and the R. K. Mellon Foundation supported fellowships in urban affairs at several universities. However, they didn't go very far in attacking the real problems of cities, and they lacked a rigorous quantitative method to examining the problems and interdisciplinary approaches that knitted together engineering, management, and other perspectives on problems that didn't fall neatly into this or that disciplinary box. At a meeting of university presidents to look at urban issues, I gave my thoughts on an educational institution that might not have these weaknesses. Nathan Pusey, president of Harvard, didn't think much of the idea. But the R. K. Mellon Foundation did. It asked Carnegie to establish the School for Urban and Public Affairs and gave us $10 million to get it started. The foundation's reasons were well put in a letter written to me in December 1968 by a Mellon Foundation trustee: "We see a great opportunity for this School to provide assistance to the city, the state and the nation in the massive effort which is needed to educate, train and motivate managers in the field of urban affairs."[17] The school got a good start with Dean William Cooper and Associate Dean Otto Davis leading, and after some early difficulties went on to do extremely well, emphasizing quantitative and problem-solving approaches to urban issues. In the 1990s it was renamed the H. John Heinz III School of Public

Policy and Management[18] for Senator John Heinz, who was killed in a plane crash in 1991. The John Heinz III Foundation liberally endowed the college.

Those were the additions. The subtraction was Margaret Morrison College. In 1968 the faculty senate seconded the recommendation of a visiting committee to the college that four programs within the college be shut down. In November 1969 the board of trustees approved my recommendation to close the college. Many of the students at Margaret Morrison transferred to the new school, later to become the College of Humanities and Social Sciences. Morrison's dean, Edwin Steinberg, became the dean of the new college. All that sounds a bit glib, but it wasn't easy, and it was emotionally taxing on all involved. But the reality was we had no choice. The college was no longer sustainable.

Rumblings

What happened to the University of Pittsburgh and Edward Litchfield, as unfortunate as it was, proved to be one factor in the biggest event during my seven years at Carnegie: the merger with the Mellon Institute. Jake Warner had told me that R. K. Mellon wanted the institute and Carnegie Tech to come together somehow, but it didn't happen in part because Jake had other priorities at the time.

Like most successful marriages, the merger made more sense after it happened and even more so as the years passed. At the time it wasn't a sure thing. Carnegie Tech and the Mellon Institute were quite different. Carnegie Tech was a place to teach students, while the principal raison d'être of the Mellon Institute was contract research for industry, with particular strengths in chemistry. Not surprising, since the idea for the institute came from a chemistry professor at the University of Kansas, Robert Kennedy Duncan, who in the early 1900s wrote books and articles lobbying for closer ties between academia and industry. The Mellons were entranced and built Duncan a research institute on the University of Pittsburgh campus. It thrived and moved in 1937 from the Pitt campus as the Mellon Institute for Industrial Research to its own quite magnificent building. However, it ran into trouble after World War II. It didn't build connections with federal funding. It didn't add strength in fields becoming important to industry, such as microelectronics. And while it built up strength in fundamental research, it didn't build up the funding, including industrial interest, to maintain it.[19]

My first serious thinking on the subject came in May 1966 when I was invited to lunch with the Mellon trustees, who asked me point blank what we would do if the Mellon Institute offered to join with us. I was well aware of what the institute brought to the table: a quite magnificent building next to our campus, a sizeable endowment with a book value of $40 million but more than that at market value, a field laboratory outside of Pittsburgh with excellent facilities and equipment, basic researchers who were good, at least some of them, and those applied researchers still earning some money. After I recovered from their surprising question, I asked a lot of questions, told them that I favored the idea and that "I would like to put together a strong working group with some of our vice presidents, deans, and best professors to do a proposal." They drew back in horror. "No! We don't want any of them knowing about this. It has to be kept close and confidential."

They did agree that I could discuss it with a small group of colleagues—Aiken Fisher,[20] chairman of the Carnegie Tech Board; Edward Schatz and Robert Kibbee, my vice presidents; and Paul Cross, president of the Mellon Institute. That was the working group. We totted up the minuses and pluses. In the minus column went an oversized chemistry department and long-term commitments to the current employees of the Mellon Institute. Our first proposal was more or less shoehorning Mellon Institute scientists and applied researchers into Tech's organizations with a residue of people to be decided later. When I tried it on trustees of the A. W. Mellon Foundation in New York, I got little enthusiasm and, when I went on to our Randolph summer home, I got even worse from Jim Killian, the leader of the Mellon trustees working on this merger. Liz and Jim had driven over from their vacation spot in Sugar Hill for lunch with Bunny and me. After lunch Jim and I sat on our front porch looking up at the peaks of Mount Madison and Mount Adams, relaxing I thought until he opened the conversation with "It isn't bold enough." He pointed out that the new entity wouldn't be that different.

I flew back to Pittsburgh the next day, determined to come up with something "bold," starting with the principle that we would also include a number of moves that I wanted to do anyway. I called Schatz, Kibbee, and Cross to a confidential meeting in the President's House. I pointed out that this opened the door for us to become a university in name as well as in fact—the Carnegie Mellon University. And I pointed out that while, yes, Andrew Carnegie was the founder, in fact the Mellons had made major contributions to transforming Carnegie Tech, including the new School of Urban and Public Affairs, the Graduate School of Industrial Administration, Scaife Hall for Engineering, Warner Hall for Administration, and several endowments.

It went well. But there was one problem: we didn't involve the faculty. And I knew when the faculty was told, there would be trouble. I often wonder what would have happened if we had done that. We'd probably still be arguing today about the merger. And late that evening in the President's House, after some beers and after Kibbee had stretched himself flat on his back on the floor, I said to Schatz and Kibbee: "I'm going to have an awful lot of headaches when this is announced." "Guy," Kibbee said, "you can afford a lot of headaches for a $25 million plant and a $40 million endowment." After another sip of beer I asked: "What will the colors be for the new university?" Bob immediately said: "Green and gold." When I asked about a motto, Bob, son of humorist and actor Guy Kibbee, said: "Lux and Bux." Thus was born Carnegie Mellon University. Because one of the Mellons objected to their name being included in the new university, we first called it Carnegie University. And we replaced the timid transformation of the Mellon Institute into a graduate department of Carnegie University with a bolder arrangement, a Mellon Institute of Science as the umbrella for our science departments. And that meant splitting into two our College of Engineering and Science—a combination that didn't make sense to me anyway. And most important for me, we proposed to establish the College of Humanities and Social Sciences. The executive committee of the Carnegie Board enthusiastically backed the merger, especially when I gave them the numbers to back my point that the merger translated into one of the largest increases in the resources of a university in the history of higher education.

I told the faculty at the start of the fall semester in 1966 that we had agreed to discuss a merger (although it was a fait accompli). About a half dozen committees, including both Carnegie Tech faculty and deans and Mellon Institute staff, were created to look at various details of the merger. Each of the different proposed changes had to be treated rather differently, and there were different time scales. The College of Humanities and Social Sciences took the longest, and we worked hard to get the remnants of Margaret Morrison transferred to the college. I felt badly about that for Morrison had contributed to Carnegie Tech's character. On April 19, 1967, the full Carnegie Tech board approved the merger.

I became briefly president of Carnegie University, which became Carnegie Mellon University when, in a splendid talk Paul Mellon gave at our June 1967 commencement, he added "Mellon" to the name of the university. He said: "I'm very happy to express the pride and pleasure of all of us in the realization that the [Mellon] Institute will add its dedicated scientists, its physical and financial resources . . . toward the formation of what we all hope, and what we all know, will be an outstanding American university." Paul Mellon's blessing was seconded by Andrew Carnegie's daughter, who wrote to me: "Your founder would share our pleasure today. . . . I can hear my father say as he did so often, 'All is well since all grows better.'"[21,22]

Interlude

Yes, the 1967 merger was a big deal for me. But the year was also a big deal for me as a fly fisherman. From the beginning of our marriage two decades ago, Bunny, who had fished for trout since she was a little girl, had whipped me into shape as a fly fisherman. And fly-fishing had become an important part of my life at Randolph. And now it became important in Pittsburgh, through my membership in the Rolling Rock Club and a wonderful trout stream the club managed in the hills east of Pittsburgh. The Rolling Rock Club stocked Loch Leven Brown Trout,[23] a wonderful breed of brown trout, that is justly regarded by "many anglers as the perfect trout, both for its graceful form and its sporting qualities." My first day on the Rolling Rock stream was perfect. I caught a goodly number of trout and took some back home to cook, both for dinner and for breakfast, trout being a breakfast delicacy, with Bunny being an absolute expert preparer of trout.

Roy and Margo were the two of our four children staying with us in Pittsburgh, and Roy would fish with me in the Rolling Rock stream. Margo joined in eventually, so with Bunny we had four enthusiastic fly fishermen in the house. Sarah wasn't as interested, focusing rather on academics and going to Europe for her humanities studies, particularly history. Guy had a terrible fish allergy and just touching a fish led to swollen eyes and difficult breathing.

In the year and a quarter between first being asked by the Mellon trustees about a merger and its realization, Bunny and I celebrated our twentieth wedding anniversary and I turned 50. I began to do some very careful thinking. Was I going in the right direction in my life? How was our family doing? The happiest time in my life was at MIT when I finally got to teach and do research. And I had added variety to that life that fit very well, in my government and industrial relationships, mainly the Air Force Scientific Advisory Board and the United Aircraft Corporation. Now at Carnegie, I had given up teaching, research, and the day-to-day intellectual charge that came with being on the MIT faculty. But I slowly came to accept that change, emotionally and intellectually. I evolved into full engagement in making the brand new university work.

Making the university work meant money. In the press conference that Paul Mellon, Aiken Fisher, and I had announcing the merger, Paul Mellon denied that including "Mellon" in the university's title would guarantee the continuance of gifts from the Mellon family, foundations, or corporations. That surprised neither Jake Warner nor me. We'd gotten the same message from the Carnegie Corporation of New York, the main conduit of Carnegie money. We were told gently but firmly that just because "Carnegie" was in our name didn't guarantee money from the corporation. That proved true with Carnegie—and with Mellon. However, when we had a good proposal, both were willing to consider it.

A RED INK TIDE

But we needed more money! How could that be when the new university had through the merger gained enormous assets and endowment money? A substantial part of the answer was the commitments we made as part of the merger: to keep the Mellon Institute programs, including the people who weren't moving to teaching and research. No mass firings.

Just a few weeks before the merger became official, on July 1, 1967, we closed a money-raising campaign started 10 years earlier by Jake Warner. We had collected $30 million, we had more income coming in through federal research grants, and our budget was in the black. But we had closed down our capability to do fund drives, and anyone familiar with universities knows they always need more money. As soon as one fund drive starts, the wheels for the next one must start turning.

But, still, why should Carnegie Mellon University and other universities run into financial trouble when student enrollments and research support were booming? After all, the 1960s marked for public and private universities and colleges "the largest enrollment increment for any decade in enumerated history."[24] Two hundred thousand more students enrolled in private undergraduate colleges and universities between 1965 and 1970; the numbers were more dramatic in the public colleges and universities, with enrollments increasing by almost 2.5 million.[25] Those students needed more housing, classrooms, and faculty to teach them, all of which cost money that the private universities had to raise, with little of that coming from government, federal or state.

The research boom of the 1960s lured many universities and colleges into a financial swamp in at least two ways. The first was that more research capacity, meaning buildings and laboratories and graduate education and students, was needed on campus to attract federal funds and quality faculty. "A direct correspondence existed between average faculty salaries and overall institutional ratings."[26] Private institutions had to find the money to do this for themselves. That meant unrelenting hunts for money and that a university president was judged by how much he found—a reality still true today. The second way that universities became mired in a financial swamp after the boom deflated was when they used "soft money"—federal research funds—to pay faculty salaries. And faculty salaries were rising sharply, far ahead of inflation. Soft money worked as long as the federal funds flowed more or less unabated. But in the goodness of time it became a very costly shell game as research funding declined in the late 1960s, and the universities suddenly found themselves with a large cadre of tenured faculty having lost research support but still harboring the quaint notion that they should be paid their full salary—if not by federal research funds, then by the university directly. By the end of the 1960s, it was obvious that "the insatiable appetites of research universities had outstripped their financial resources."[27] A Carnegie Commission study done in the late 1960s projected that by 1975 private research universities would fall short of needed income by 30 to 40 percent.[28] Not only Carnegie Mellon, but other major private research universities—Princeton, Columbia, Harvard, Cornell—all ran into severe financial problems. The intensifying competition for research dollars and hence prestige encouraged by the postwar boom in federal research translated into major construction and staffing commitments—commitments responsive, for one, to the rise in information technologies as a major need of a campus and new programs to support minority and low-income populations. These were all good causes but none were cheap, especially when anticipated income didn't arrive. As Roger Geiger aptly put it:

> In the late 1960s, the financial environment became decidedly less congenial for the endowed private universities. The stock market made a secular peak in 1966 that was not surpassed in real terms for two decades. Giving to higher education after 1965 went into a ten-year slump. . . . [I]t became harder for private institutions to increase revenues, even as the costs faced by research universities continued to escalate.[29]

Carnegie Mellon didn't escape that "decidedly less congenial environment"during my time as president. By the 1970–1971 academic year we faced a substantial and climbing deficit. Even though I knew in my mind that "in large part the accumulated deficit had been caused by factors beyond the university's control,"[30] it was for me personally, and certainly for the university, very painful. Desperately needed renovations and repairs were put off. We didn't launch a department of biological sciences, delaying the university's establishment of a standing in an exploding field. We didn't turn into reality our dream of building a center for the dramatic arts. We started losing good faculty. We urgently had to reinvigorate our fund-raising effort, and we did, although by the time I left for Washington we were considerably short of what we needed, much to my disappointment. Could I have done more? Even from the vantage of several decades looking back, I doubt that I could have escaped the tidal wave of red ink that beginning in the late 1960s ripped through the major research universities. At the same time, I inherited what was in retrospect a budgeting mistake that bit us hard: operating and capital funds were commingled in the same budget, so that money intended for, say, maintaining our physical plant went instead to covering rising construction costs. Doing that means that "financial oversight can be muddied. Accurate forecasting of total annual expenditures becomes difficult, and money slated to balance general operating expenses can be more facilely transferred to cover the costs of a project whose funding is based on restricted sources."[31] When income fell off, these "convenient" arrangements turned first into cash flow problems and then real deficits. Capital and operating budgets need firm separation; that wasn't true for Carnegie Mellon at the time but is now.

Blinded

The fall in research income, a falling stock market, and sharply higher faculty and related costs all contributed to the financial problems of research universities. More deeply and more harmful was the most dramatic event for American higher education in the 1960s—the student unrest if not downright rebellion on many campuses, fueled by the convergence of Vietnam, the civil rights movement, vague dissatisfaction with a "consumerist" culture, and growing use of illicit drugs.

Drugs—or rather the fear of what they could do—hit us in January 1968, just when we were about to launch a major new funding push, the Fund for Distinction. An official of the Pennsylvania commonwealth told reporters that at a university campus in western Pennsylvania students on LSD were blinded when they propped their eyes open and stared at the sun. It was a stunner. I immediately called together my senior administrators, and they were adamant that Carnegie Mellon was not the campus where this happened. Moreover, whatever the campus, they didn't believe the story. George Brown, our highly respected dean of students, insisted that simple logic dictated that the story was false; that you couldn't have five or six blinded students in hospitals without nurses or doctors talking about it, never mind screams of rage from the families.

The national media thought otherwise. Not just western Pennsylvania but the entire country was treated to this horrible result of drugs on campus. We were flooded with reporters, but that was nothing compared to what hit us when one of the leading TV news shows reported that the culprit campus was Carnegie Mellon University. We were in for it. Media flooded us. One TV crew set up on campus to stay and tried to get students on camera to talk about it. Most refused. But one of our acting students, a young and very attractive woman, did go on camera—after all free TV time is gold to an aspiring actress—to tell the viewers that "It couldn't happen on our campus. We're all too bright for that."

The end of the story came when reporters finally went back to the official who made the original announcement. He turned out to be blind himself and was understandably horrified by stories that people could blind themselves with LSD and drugs. So as a warning I suppose, he had concocted the story. TV cameras and reporters and the headlines melted away. We looked in vain for some sort of apology from the media or even a correction. I respect the media, but they could have done better on this one.

Even for those who lived through the 1960s—and more especially those like me who were hit by the 1960s—the pace and drama of what happened seem in hindsight almost unimaginable. Coming home on April 4, 1968, after a day with visiting committees and dinner with some students, I learned that Martin Luther King, Jr. had been murdered. The next day we closed down operations for a memorial service. The terrible sadness that enveloped all of us was soon joined by shock as riots and fires occurred in the major cities, including Pittsburgh. We made sure that students were protected. Many of the students living off campus moved in with dormitory friends.

Even before King's murder, we had tried to attack at least one of the root problems in American race relations: the participation of African Americans in higher education. Some 10 years earlier, Carnegie Tech had started the School College Orientation Program, which held summer institutes to help minority students prepare for college.[32] Dean Norman Johnson, an African American, urged us to go beyond that, to help minority students not only prepare for college but stay there. We created in 1968 the Carnegie Mellon Action Program, CMAP, to do that, with Norman Johnson as director. The program started well, but Johnson pointed out that none of the CMAP students were going into engineering and the sciences, and he and others worked on changing that. "By 1976," long after I had left, "Carnegie Mellon was the third largest producer of minority engineers in the country."[33] It wasn't easy to get there.

We tried to no avail to get individual and institutional support for the program. And it wasn't until the early 1990s that the university was able to provide operating funds out of its resources. At no time when I was at Carnegie Mellon do I remember an African American participating in activism. They were too interested in their educational opportunities. Our activists were white students who had seized on the construction union's refusal to accept African Americans.

These efforts—to prepare minority students for higher education and to keep them there—were the university using its special strengths effectively and pointedly to confront problems forced on it by the "outside world." But more was to come, a time of what Herb Simon called "The Troubles."[34] Perhaps the worst, for the university and for me, came in 1969. Two realities collided violently: (1) United States Steel and Carnegie Mellon University both had major construction under way that year and (2) the local building craft unions had either none or only a few black members. Minority workers were effectively shut out of construction jobs.

A Terrible Mistake

In September 1969 I came back from a fund-raising trip to New York to be met at the airport not only by my driver but also by two of my vice presidents. That was trouble. They briefed me on the emerging crisis—a direct confrontation between protesters and the construction workers and union officials at the construction site for our new science building. We spent the weekend planning what to do. We had already laid the groundwork in my letter on student protests on campus published earlier in September in *The Tartan*, the student newspaper. I emphasized that campus protests were entirely apt but that:

1.   The protest or demonstration must be of an orderly nature so that no acts of violence shall occur and the normal orderly operation of the University shall not be impeded.

2.   The protest or demonstration shall not infringe upon the rights or privileges of students not in sympathy with it . . . .

3.   Finally, the freedom to demonstrate on the campus shall be limited to members of the campus community only.[35]

We had already earlier shut down construction for three days, at the request of Mayor Joseph Barr. But the week that letter was published it was also clear that next week some students were going to demonstrate on campus, demanding that construction be shut down again. We shut down construction again on Monday, September 22, and cancelled classes in favor of an all-campus convocation where I would be the sole speaker. We put in a lot of work over the weekend getting ready for that, but I was skeptical that we would change any minds. We held to our basic positions: that the university community, students and faculty, should be free to voice their opinions but peaceably and that while individuals could certainly do as they wished, we were not going to put the university between the black coalition in Pittsburgh and the construction unions.

I was nervous when I entered the gymnasium for the convocation. I was the only act, and there were to be questions afterwards. I started reading my speech but felt that I was coming off as rather cold. In trying to ad lib to soften the tone and to underline my sympathy with the black students, I made a terrible, terrible mistake by saying "why I had lunch with some of them yesterday and found that they're real people." I was so shocked by expressing myself so badly that I lost my usual aplomb and couldn't figure out how to correct the unfortunate remark. I went on with my written talk and then took questions. I recognized a long-haired young person, very slight and short: "Here's a young woman in the front row." He rose and said: "Young man, sir." The place cracked up, but it didn't save the day.

I was so upset that I went home to lunch. Bunny, bless her, suggested we go to the zoo, one of our favorite places to go with our kids. It was a weekday and would be uncrowded and quiet. We went to our two favorite exhibits. One was the lion—old, shuffling, with sore feet and aching limbs. I had at that moment a lot of sympathy for him and spoke to him a lot. The otters were our second favorite. The young otters played wonderful games, plunging down the water slides and chasing each other. They always refreshed us when we went to the zoo, and they did that day.

We laid out a communications center in my conference room that night and assigned deans and other academic officials equipped with walkie-talkies to key points around the campus. We were very glad we did, for the next day was a near disaster for the university. Police helicopters flew over our campus. Police massed in Schenley Park just over the hill from Carnegie Mellon. I had a front-row view from my office of the crowd gathering for a demonstration at 10 a.m. at the construction site for our new science hall. Three hundred workers at the construction site were about to pour concrete for the basement. One of the large cement mixer trucks, instead of dumping its load at the site and leaving, circled the site just off campus, a way I suppose for the construction unions to taunt the demonstrators.

It heated up. The Pittsburgh chief of police called to tell me that he had police stationed nearby, that we were on the verge of a serious riot, and that the police should move onto the campus. I told him that we knew police units were nearby but asked him not to send them on campus unless we requested it. I thanked him for being so well prepared. The police chief appealed my request to Mayor Barr, a member of our board. I again pleaded that police not be sent in. "If we have a confrontation of our students, faculty, and police, we won't get this campus settled down for years. Please hold this police chief back unless we call." He reluctantly agreed.

The demonstrators divided into three groups with each marching from a different direction to the construction site. They faced off with union officials, calling them names and understandably angering them even more. An explosion was near. We had one report that a young faculty member was encouraging students to lie down in front of a truck. One student just managed to crawl to safety as a truck went by. I was terrified. And angry with the faculty member who did this. Then our network reported that the construction workers were getting ready to charge with two-by-fours. I still refused to call the police but told my staff that if it worsened I'd have no choice. My heart sank.

Then I got a better call: the students were backing away. The prospect of being charged by construction workers using two-by-fours was too much. We arranged for discussions of what had happened and what to do next by students in their own colleges led by deans. The next day was to start with a meeting of our policy advisory board, which included students. But that morning I got a call at home quite early and was told: "Your office is occupied by some students. Why don't you stay away?" The sit-in, thanks in good part to some thoughtful faculty, turned out to be a minor one. I eventually got to my office. Everything was neat as a pin, and the only thing missing was a telephone charge card for my Defense Department business. Somebody did eventually use it for a while, but I have no idea if they were caught.

The close call with the construction union was for me the scariest event of the student activism of the 1960s. We had more events, with of course the Vietnam War a major provocation. Young people cared about the war for many reasons, not least that young people did the fighting. The war had a long history, and U.S. involvement began innocently enough when, first, President Eisenhower and then President Kennedy sent "military advisors" to help South Vietnam in fighting North Vietnam. That changed quickly: From 17,000 military advisors at the end of 1963 to rapidly climbing draft calls in 1964 to 500,000 and 80,000 U.S. casualties by 1968.[36] Vietnam became the "longest and ultimately the most unpopular war in American history."[37]

In my first year at Carnegie Tech, students and faculties at campuses across the country had begun organizing "teach-ins" on the war. That also came to the Carnegie Mellon campus but a bit later than other places and I'm glad to say without the disruption of academic life or the violence (and deaths at Kent State and Jackson State) that marked some of the protests. "Compared to Columbia, Berkeley, Harvard or Wisconsin, Carnegie Mellon was distinctly peaceful. No students were killed, no faculty members injured, no buildings destroyed, no police called in."[38] Part of the reason is that what was in some ways a motley group called the Pittsburgh Council of Higher Education[39] agreed on the principle of maintaining the rights of free speech on campus and issued a strong statement affirming that right and setting out guidelines. While there were rough moments, that action turned out to be very effective. There was one visit to our campus by a noncampus group on its way to a large rally in Washington. They settled on our campus for a rally. While many students had polar views, they behaved well. No violence. No damage. We didn't appreciate the invasion by the media with TV trucks ripping up our lawns and getting a lot of attention in the press. We all breathed a sigh of relief when the protesters continued on to Washington.

The government tried to disengage from the war even before Richard Nixon took office in 1969. The Paris peace talks had begun, and on

November 1, 1968, President Johnson suspended bombing of North Vietnam to pressure the North Vietnamese to settle. Nixon came into office in 1969 with what proved to be contradictory goals: reduce American involvement and win the war with the South Vietnamese army. The failure was almost absolute. The policy of withdrawing destroyed the morale of American troops wondering why they should die when the United States was pulling out. The South Vietnamese army was simply never up to the task. And Nixon in other ways escalated the war. Cambodia was bombed in March 1969 in an attempt to destroy North Vietnamese sanctuaries. A year later, American and South Vietnamese troops invaded Cambodia, a campaign that failed in its purpose but left the country badly destabilized and prey to the coming horrors of its "killing fields."[40]

Of direct concern to me was that in May 1969 Nixon changed the selective service system so that now the youngest were called up first rather than the oldest. While, in fact, the actual numbers being called up declined, there was suspicion that the change was aimed at the campus activists. That wasn't proven, but the activism became enormous and then superheated with the killing of 4 Kent State University students by national guardsmen on May 4, 1970, followed by the killing of 2 and wounding of 11 students at Jackson State College. The terrible news from Kent State and then Jackson State reverberated at Carnegie Mellon as it did on campuses across the country. On the evening of May 4, students had a bonfire on campus. I met with students and agreed to shut down the campus for a day to allow attendance at a memorial service to be held at the Heinz Chapel at the University of Pittsburgh.

I joined 79 college and university presidents in October 1970 in petitioning President Nixon to step up the Vietnam pullout. We emphasized our common concern for the young people whose education was being seriously and perhaps irreparably damaged by the war. Our petition got good coverage in the press but no detectable action. Some alumni supported my signing; others didn't, saying Carnegie Mellon shouldn't take a position, even though the petition said that we were signing as individuals and not for our institutions. I did not ask the university's board for approval, not wanting to get the board members involved in the issue. I would take the heat.

LOSING A THOUSAND GRANTS

Like many other campuses, the protests damaged our efforts to raise funds for what was still a new university. I think it wasn't so much the causes—civil rights, Vietnam, environmental damage, the murders of Martin Luther King, Robert Kennedy, and students at Kent State and Jackson State—but more the tone and the targets of some of the protests. A bad moment came in a fund-raising session with the Sun Coast Clan of alumni in St. Petersburg, Florida. One older alumnus sitting in the front row asked some hostile questions about the students. He made it clear he didn't like anything about them—long hair, beards, their activism, and so on. I tried to answer gently, but he broke in with a snarl and an expletive. That set off other alumni on an antistudent tirade. Fortunately, a forceful graduate of Margaret Morrison College told them to shut up and listen, which they did. Still I was insulted and upset and wrote in my appointment book "the last time." It was a low point.

The activists had also in effect taken over the student newspaper, *The Tartan*, and the senior yearbook, *The Thistle*. They used these and other platforms to launch mean-spirited attacks on the Mellon enterprises and other businesses, such as the Gulf Oil Company, that had so strongly supported us.[41] Then there were the marshmallows thrown on the stage when Senator Strom Thurmond was speaking, a break-in of ROTC offices, and other incidents, all of which got us publicity and angry alumni around the country. Other campuses also had their problems, some much worse, but this was happening on my watch and absolutely crippled my efforts to raise the funds the university badly needed. Both Richard Scaife and Richard Mellon, major supporters of the university directly or through their families and foundations, resigned from the Carnegie board, pleading other commitments but leaving us wondering whether the turbulence on campus and the attacks on them were the real reasons.

The year 1971 brightened with a palpable wearying by students of the turmoil of the past two years; stronger support by the trustees, especially Aiken Fisher, who was a stalwart throughout the combative years; and more active engagement by the faculty. We organized sessions for students, administration, and faculties to discuss issues, including an all-night vigil, and maintained academics in the face of traumatic events outside the campus that could have overwhelmed us. I especially wel-

Plate 9

My first supersonic flight, in a TF 131 Trainer, 1964. *(U.S. Air Force)*

General Donald S. Putt, who recruited me to be Chief Scientist of the Air Force and later President of Carnegie Tech. *(U.S. Air Force)*

Jimmy Doolittle, who was my mentor, great friend, and splendid fishing companion. As a young lieutenant in the Army Air Force in the 1920s, he sat down to list his assets and debits, and by his own accounting wound up with one asset, "brave as hell," and one debit, "dumb as hell." So, he decided to go to graduate school at MIT to earn a doctorate in aeronautical engineering.

Plate 10

A Boston beauty: Bunny before I met her. Easy to see why it was love at first sight. *(Alfred Brown)*

Bunny and I with our rods off to some serious fishing.

Our wonderful children: Clockwise from left, Guy Jr., Sarah, Margo, and Roy. *(Samuel Cooper)*

Plate 11

"A great university is a balanced one." My inaugural talk as the new president of Carnegie Tech, 1965. *(Carnegie Mellon)*

At least five were paying attention: Bunny and our four children listening to my inaugural speech. *(Carnegie Mellon)*

The deed is done. Aiken Fisher, Paul Mellon, and I at a press conference to announce the merger of Carnegie Tech and the Mellon Institute. *(Carnegie Mellon)*

Plate 12

Keeping 'em scared: Telling ghost stories at Crag Camp on Mount Adams, Randolph, New Hampshire.

Flyfishing, TV, and energy policy mixed in a wet interview in a Colorado trout stream. I didn't want to do it but it sure doesn't look that way. *(KMGH TV)*

Plate 13

Five ambitious guys – including Aiken Fisher. chair of the CMU Board, and Fred Foy, head of Koppers, second and third from left, respectively, planning new campus construction. The U Pitt Tower of Learning is in the background. *(Carnegie Mellon)*

The B-52 Stratofortress went into service in 1952 and is still flying almost 50 years later. I saw it, albeit indirectly, at Edwards Air Force Base in California through the skid marks and gouging left when the first B-52 being delivered to the Air Force had landing gear problems. *(Air Force Flight Test Center/History Office)*

Plate 14

Confirmation in the mid-1950s that missiles tipped by nuclear weapons were technologically plausible and awareness that the Soviet Union was embarked on a vigorous effort to build guided missiles led the United States to commit itself to building intercontinental ballistic missiles, ICBMs. The first was Atlas, test launched August 1958 at Cape Canaveral (*left*). Then came Titan, also liquid-fueled like the Atlas, and the construction of Titan launch sites in the United States, such as this one near Denver, Colorado (*below*). Both were succeeded in the early 1960s by the solid-fueled Minuteman. (*U.S. Air Force*)

Plate 15

Jim Killian taking office in 1957 as the nation's first science advisor – "science czar" was the hyperbolic phrase then, as Jim was lionized as the nation's guide to responding to the shock of Sputnik. *(NASA)*

President Kennedy awarding the first National Medal of Science in 1961 to Theodore von Kármán, the great leader in aeronautics and astronautics. Von Kármán was quite rightly the only recipient that year. *(Courtesy of the Archives, California Institute of Technology)*

Plate 16

The National Medal of Science returns to the White House in 1973, as President Nixon receives that year's recipients. *(The White House)*

The original presidential science advisers. From left to right: James Killian, George Kistiakowsky, Jerome Wiesner, I, Donald Hornig, Lee Dubridge, and Edward David. *(Franklin Institute)*

comed the new mood of the faculty evident in the start of the new academic year in September 1970. Prior to that, with some notable exceptions, the faculty had been rather passive. One reason may have been that a "great many university faculty members . . . were dependent on the love of their students"[42] and were unwilling to anger them. And certainly some faculty members shared at least some of the students' views or at least believed that while their "values were nearly all correct, their ideas for reforming the world [were] nearly all wrong."[43] It was about this time when I realized that we had a positive effect from keeping the merger discussions from the faculty because that prompted them to organize a faculty senate with Sergio di Benedetti as chair. Jim Langer, who proved to be an outstanding helper in handling the encounters with activists, followed him.

Still, it wasn't all tea and crumpets. As I mentioned, student activists—Maoists citing the "poverty of student life," a slogan used in 1968 by Daniel Cohn-Bendit, the French revolutionary student leader—took over *The Tartan* and used it to push their views and to attack the administration, trustees, and others. "Being annoyed especially by the growing incivility of our students," Herb Simon "decided to get into the act." Herb wrote a letter to *The Tartan* noting that the nonvoluntary student contributions to support the newspaper "subsidizes the propaganda of a little band of self-appointed 'radicals' who use it to preach a muddle of tea-party anarchism." There ensued some to-and-fro between Herb and the editors, ending with Herb writing eight columns for the newspaper, under the title (of course) of "Simon Says." His penultimate column, a New Year's message written from a mountain town in Peru, where he and his wife, Dorothea, were retracing Inca history, wonderfully expressed how Herb saw the world:

> The past decade was a Pandora's box of unsuspected marvels; that which commences today promises to be, without fear of disappointment, even more marvelous in the field of science. But it ought to be equally the bearer of something grander, more sublime, of which both science and man have lost the vision: tranquility of spirit, peace of heart, the source and germ of human happiness—happiness, the eternal and at the same time, unreachable dream of man.[44]

I didn't know it at the time, but in 1968 in the midst of all the turmoil a call from Lee Dubridge signaled another phase change in my life. Nixon was elected, transition teams were forming, and on Thanksgiving Day while I was watching a football game Lee called. He had just been announced as the president's science advisor, an appointment that delighted the scientific community and of course me. I had known Lee since 1941 as the wartime director of MIT's Rad Lab. Nixon was putting together several task forces to aid in the transition, and Lee wanted me to chair the one on science. I accepted and quickly began to put together the rest of the task force.[45] It was quickly formed, met the following Saturday, and then weekly through December. I reported to the president-elect and his major advisors on January 11, 1969, at the Hotel Pierre in New York. It was an intense and long session, something a little different from what I had done before.

"WOW"

One of our recommendations for appointments included several for directorship of the National Science Foundation. Lee Dubridge looked at the recommendations and then asked me if I would take the NSF directorship. "Wow!" I was excited and certainly interested. But I thought it over and said to myself and then to Lee: "Gee, I've been at Carnegie Mellon almost four years, but we're in the middle of a fund drive and I ought to continue that." And I felt that four years was too short a stay, even though I had balked inwardly when my predecessor, Jake Warner, in introducing me to faculty and students, said that I was young enough that I could serve a full 20 years before retiring, which would make me the longest-serving Carnegie president. I wasn't going to stay 20 years, but four was too short.

  With some reluctance, I said "no" to Lee, who immediately asked me to join the National Science Board, the governing body for NSF. I said "yes" to that. But that turned out to be the proverbial camel's nose. My work on the National Science Board, particularly in strengthening the foundation's programs in engineering research and applied science, brought me in much closer contact with broader and also higher aspects of national policy for science and technology. Heretofore, my governmental involvement had focused on military research and development, especially radar, guided missiles and defenses against them, and civilian space. Now my view was much more cosmopolitan. For example, after I joined the National Science Board, I had extensive contacts with the White House Office of Science and Technology, led by Lee Dubridge, who was succeeded in 1970 by Edward E. David, Jr., from Bell Telephone Laboratories.

In June 1971 Ed David asked me to come see him in Washington, which I did. He then asked if I would consider succeeding William McElroy as NSF director. Bill had since 1969 directed the foundation effectively and ably but had now accepted an offer to become chancellor of the University of California at San Diego. Ed gave me the summer to think about it. He kept asking. I finally said "yes." I saw a chance to tackle on a national stage many of the science and technology issues I had experienced firsthand at MIT and then at Carnegie Mellon, dominantly maintaining and strengthening the "compact" between universities and government that had led the United States to excellence in fundamental research.

President Nixon announced my nomination on November 15, 1971. I was immediately flooded with publicity and congratulations, this time on a national scale for every research university in the country had and sought more grants from NSF. I regretted leaving Carnegie Mellon while there was still unfinished business, especially our fund-raising drive. Fund-raising in the aftermath of the 1960s was then a severe problem for us as it was for most private campuses. We raised enough funds to build a research-computer building but fell short of funds for a fine arts building, an expansion of the student center, Skibo Hall, and other projects.[46] I took comfort, however, in what had been accomplished in my seven years at Carnegie. There was the merger, which could have been disastrous but in fact came off very well. We were now a university. New colleges were in place and starting to do well, including the College of Humanities and Social Sciences, the Mellon College of Science, the Carnegie Institute of Technology, the Transportation Research Institute, and the School of Urban and Public Affairs. Our endowment had gone from $55 million to $112 million.[47] We now had a very strong position in computer science and technology and the facilities to back that up. We became "one of the first universities in the country—in the world—to train students in computer science at the doctoral level."[48]

And we did more than survive the 1960s; we prevailed. We had our problems, but nothing like the experiences of our brethren campuses, not only at Kent State, Columbia, and the University of California at Berkeley but even some of our immediate neighbors such as the University of Pittsburgh. We didn't censure publications even when they were really beastly to the administration. Rather, faculty, notably Herb Simon, took them on directly. We never called in police, even under intense pressure to do so. We set out what I believed to be fair guidelines for on-campus protests and made them stick. We spent a lot of time on campus dialogues, vigils, meetings—anything to enable students and faculty to express their opinions, however disagreeable. I'm very proud of what we did.

In the two and a half months I had left at Carnegie Mellon, I was torn between making all the connections in Washington to get confirmed, which I was unanimously, and saying goodbye to all our Pittsburgh friends.[49] For example, we had a dinner for Acting President Ed and Virginia Schatz, Vice President Jack and Jane Johnson, and Dean Erle Swank. At the end, the Cameron Choir appeared in front of the house and sang Christmas carols, came in for cider and doughnuts, and sat on the floor to sing some more. Especially after all the tumultuous events all of us had lived through, I was moved when the leaders of the student government gave me a plaque that read:

> In Recognition for Seven Years of Outstanding Leadership and Service to Carnegie Mellon University, the Student Body Presents this Certificate and Their Appreciation to H. Guyford Stever.

On the evening of February 17, 1972, Bunny and I crossed the Potomac River in our packed station wagon to enter Washington and a new life.

7

To Washington

1972–1973. My first year as the fourth director of the National Science Foundation. "Uneventful" against what came, but a wonderful time for science and the portfolio of the foundation as we learned what moves mountains, probed the stunning violence of the universe, and explored the coldest, driest, windiest, and highest continent.

At six in the morning on May 10, 1950, from the back of his train stopping in Pocatello, Idaho, President Truman signed into law S. 247, creating the National Science Foundation (NSF).[1] It was a second birth, the first effort to create a new agency aborted by President Truman unhappy with a structure that made the government the paymaster but ceded governance to scientists by allowing the National Science Board to select the director. Truman rejected an agency "divorced from control by the people to an extent that it implies a distinct lack of faith in democratic processes."[2] The new structure met Truman's complaints but also left something for the scientific community. The director was nominated by the president and confirmed by the Senate but served a six-year term to obviate presidential elections and, in principle, politics. The governing body, the National Science Board, while appointed by the president was given a fair degree of autonomy to guide not only the affairs of the new agency but also—naively, it turned out—the growing science programs across the government.

MODESTY AND TENSIONS

The agency's purpose was laudatory and critical: " . . . to promote the progress of science; to advance the national health, prosperity, and welfare; to secure the national defense; and for other purposes."[3] Its beginnings blended modesty and tensions. "Modesty" was in its first budget of $225,000 for fiscal year 1951 to pay for administrative costs. FY 1952 was the first real "science budget," and that was only $3 million against a congressional cap of $15 million and Vannevar Bush's call for $33.4 million out of the starting gate. While the foundation came into life a month before the Korean War, it got little of the new monies for military research. *Sputnik* changed that with a vengeance, as it did so much else; the foundation's budget tripled between 1957 and 1959. In FY 1973, my first as director, the budget was $610 million.[4]

There was also modesty in the new agency's portfolio. It didn't include the social sciences, carrying on Vannevar Bush's distaste for them.[5] And it did not include military research. Further, and a feature of the NSF particularly attractive to me, was that it was not permitted to do research on its own. Its principal role was to judge proposals made by external units. From the beginning, these external units were primarily in the academic institutions of the country, more particularly grants to individual researchers, then and now the foundation's principal raison d'être.

This was not quite what Vannevar Bush envisioned. Bush saw the NSF[6] as a much more powerful, all-embracing patron of science than actually resulted. His proposal was to have *all* basic research that was supported by the government overseen by the NSF with much of the research itself done by the universities. This was an idea that clearly stemmed from the success of the Office of Scientific Research and Development (OSRD) during World War II in mobilizing academic research.

The "tensions" were largely reducible to the foundation's "split personality,"[7] between the agency itself and the National Science Board, the latter created to govern the foundation and to guide government-wide policy for the support of research. When I joined the National Science Board in 1969, this comment was apt:

> The National Science Foundation has been split structurally by an ambiguous authority relationship between the Director and the National Science Board. It has been split in functional orientation between its own program of grants to support basic research and its wider role in developing national science policy and evaluating federal research programs across the board. As a policy-making and evaluating body, it has simply failed to do the job assigned in its statutory charter; for its own support of basic research, it has, however, generally been accorded very high marks.[8]

The notion that a small agency could somehow carry on the work of the wartime OSRD in coordinating government-wide research was naive. The political scientist Don Price pointed out that in the five-year interval between the end of the war including the OSRD and the birth of the National Science Foundation, "the National Institutes of Health had gotten rolling, the Office of Naval Research had gotten rolling, the Atomic Energy Commission had gotten rolling, and by the time the Science Foundation was really set up with about $3 million in appropriations, it was not the great new post-war overall research program doing military research for the government and so forth. It was just the smallest and youngest and weakest of the scientific research programs."[9]

The understandable resistance of agencies to having their research programs taken over by a new agency was in fact seconded by the first director of the foundation, Alan Waterman, chief scientist of the Office of Naval Research before he came to NSF. He had no interest in empire building but wanted to focus on the foundation becoming first rate. He had the right priorities. He was also politically astute and knew that the way to get traction on research programs of other agencies was through the president's Office of Management and Budget (OMB). The NSF, quoting Don Price again, "could be the provider of dispassionate advice to a place which really has got some muscle and leverage, and that's the Budget Bureau. Give them the obligation to look at the total program and then if they've got any influence it's bound to be through the budgetary process, which won't be their action at all, but advice to the Budget Bureau."[10]

There was yet another tension facing me as the new director, affecting all of science and with it the federal agencies supporting it. This was the impact of the 1960s, one that I certainly felt during my tenure at Carnegie Mellon University and that followed me to Washington. Science came out of World War II a hero, the tool that gave us radar, the bomb, penicillin, and so much else vital to winning the war. Now the halo was askew, and science and indeed almost all institutions of our society were being questioned, often hard and sometimes with little regard for the civilities of discourse.

> Between 1965 and 1975, protests revolving around the war in Vietnam, civil rights, environmental pollution, and the development of civilian nuclear

power plants sparked a reexamination of conventional policy structures for science and created a demand for expanded scientific advice. There was a general increase in the public scrutiny of science, and scientists were increasingly viewed as an interest group susceptible to the same politics of open debate as everyone else. . . . Scientific research was also competing with new demands for funds. For the first time since the close of World War II, Federal support for basic research (when adjusted for inflation) actually began to decline in 1967.[11]

That decline wasn't reversed until the mid-1970s, and federal funding didn't return to the 1967 level until 1982.

Those very real tensions aside, the NSF quickly earned high marks for the quality of its research support. That was due in part to its delayed creation. The Office of Naval Research (ONR) had stepped into the vacuum for support of basic research caused by the delay in launching the foundation. As the de facto NSF between 1946 and 1951, ONR established peer review by knowledgeable colleagues as the principal guide to where money went.[12] That principle of money being awarded on scientific merit traveled to the NSF when Alan T. Waterman moved from the ONR to NSF.[13] That principle has prevailed, albeit with occasional dents, twists, and sometimes outright assaults. It was in place when I was sworn in on February 1, 1972, and has remained to this day the bedrock on which the foundation has grown.[14]

Ticketed

Bunny and I arrived in Washington on the evening of February 17, 1972, in a snowstorm, which meant little traffic and a beautiful snowy scene as we drove to Georgetown and our new home, the "Innkeeper's House" on 33rd Street. It also meant a parking ticket: the snow covered a No Parking sign where I left the car. That aside, the movers arrived, and by the weekend with a lot of hard work we were well settled, thanks to Bunny's careful division of the Pittsburgh furniture between Washington and Randolph.

On the drive from Pittsburgh where I left the president's house and the presidency of Carnegie Mellon University to come to Washington to be the director of the National Science Foundation, Bunny and I had time to talk over what this move meant to us, to think of the happiness and joys of the past and the challenges and changes of the future. We were excited about the change, although it was an abrupt ending to my university life—at Cal Tech, where I earned my doctorate; at the Massachusetts Institute of Technology, where I achieved tenure; and at Carnegie Tech cum Carnegie Mellon, where I learned something—at times the hard way—about running an institution.

Bunny and I found Washington transformed from the city we had known 20 years earlier when we lived in Alexandria. The metropolitan area had grown. Officials and staff in the large numbers of embassies and the international financial banks had made the nation's capital truly cosmopolitan. There was a lively cultural scene, including the newly opened John F. Kennedy Center for the Performing Arts. And there was a growing restaurant trade. It was a different city than the sleepy provincial capital with remnants of wartime temporary buildings dotting the Mall that we had known when I was chief scientist of the Air Force.

I had spent lots of time in Washington off and on during World War II, on my trips either from the Radiation Laboratory to Washington in 1941–1942, or in going to and from the London Mission of the OSRD in 1943–1945. We had also lived there for about two years while I served, on leave from MIT, as chief scientist of the Air Force. And I had often come to Washington to consult with the various branches of government and to serve on committees, boards, and panels of government agencies, the National Research Council, and so on.

President Nixon announced my nomination as director of the National Science Foundation on November 15, 1971 at his morning press conference. It was a relief to no longer have to dodge reporters tracking down rumors of my nomination and in making the news public to the Carnegie Mellon community. Two days later I flew to Washington to prepare for the confirmation gauntlet, one considerably less onerous and lengthy than what often happens nowadays. Edward E. David, Jr. was the president's science advisor, and between him, his colleagues at the White House Office of Science and Technology, and the congressional liaison people at the NSF, a very smooth set of meetings were arranged with Republicans and Democrats particularly involved in NSF affairs.[15] And I was already well known to those members of the House interested in science and technology. All the work on my confirmation got an A when I learned while at a meeting in California that the Senate had confirmed me without dissent.

I learned a great deal about NSF in the previous three years after declining in January 1969 my first invitation to be director but becoming a member of the National Science Board. Turning down the 1969 offer inadvertently triggered some problems, since the next candidate on the list was Franklin Long, a distinguished chemist at Cornell University. Long had served on the President's Science Advisory Committee, and while on the committee but speaking as a private citizen he had opposed the administration's push to build a supersonic transport.[16] In obvious retaliation, Nixon blocked Long's nomination. There was a huge outcry from the science community. Nixon relented and withdrew his objections, but Long declined because of the controversy. Nixon then somewhat mollified the science community by appointing William McElroy, a biochemist[17] and professor of biology and director of the McCollum-Pratt Institute at Johns Hopkins and a registered Democrat.[18]

THE NIXON LOOK

I went to my first meeting as a member of the National Science Board on May 21, 1970, and the next day we went to the White House to meet with President Nixon. We traipsed into the Cabinet Room in our usual haphazard way, and Lee Dubridge, who had arranged the meeting, was feeling good because the president had agreed to his request to add $10 million to the foundation's budget for research programs.[19] The foundation's overall budget had in fact declined in the latter 1960s, and the FY 1970 budget was less than that for FY 1966, even less so with inflation.[20] Funding for research was "boom or bust." The boom was the very sharp spike in money after *Sputnik*. The bust was the decline in federal funding for nonmilitary programs, including science, forced by the Vietnam War hitting at the same time as the scientific and technical doctorates from the boom years hit the labor market.[21]

The president came in and after friendly greetings and supportive remarks about Lee said, "I'm very pleased to announce to you today that we've just added $10 million to the budget of the National Science Foundation for basic research." He then looked straight at Philip Handler,[22] who as chair of the science board, was seated directly opposite. Phil didn't smile, didn't say thank you. Nixon's face changed almost as if a hood had dropped over his eyes. He smiled and said, "I know, Dr. Handler, it isn't enough, is it?" Phil said, "No, Mr. President, it isn't." That settled— *ended* is maybe a better word—the good relationship of the National Science Board with the Nixon White House.

I learned a lot about NSF in the year and a half I was on the board before I became director. I learned for example that any hope that the foundation's research program would focus entirely on fundamental work was short lived. NSF's role in applied research became a hot topic in my two and a half months of "suspended animation" between my nomination and being sworn in. That issue became the scalpel for dissecting me, because it was the dominant issue for the scientific community dependent on NSF funding, a community that tends in any case "to be mildly manic-depressive or at least subject to sudden swings from supreme self-confidence to self-pitying gloom."[23] But it also focused on me because I was a basic scientist gone astray, having traversed from a fundamental physicist in earning my doctorate at Cal Tech to an engineering professor at MIT doing work that ranged from what I considered very basic to clearly applied.

Another issue that served to set the stage for my confirmation and tenure as director was that the foundation was low on the political totem pole. I can't recall any director of the NSF ever being asked to make a political speech. For the most part, both parties in the Executive Office and the Congress agreed on the importance of supporting basic research and, hence, the foundation. That support has continued unabated to this day, if not without arguments at the margins—pushes for particular programs; disputes about how to move public money to the researchers and the relative weight given to peer review, geographic distribution, equity for minorities, and helping the have-not states bootstrap their research quality so they could compete with the haves; and the value of the social sciences and indeed whether the foundation should support them at all. Of course, there were some special features of my tenure, notably the Vietnam War and the corrosive effect it had on all nonmilitary agencies. And then came the Watergate scandal, and the way that hit me. That's a story for the next chapter.

Some of my friends asked me when my nomination was announced why I did it, given that I would take office in President's Nixon fourth year in office. I thought the risk low, since the director nominally had a six-year term and to date there hadn't been any leadership changes for the NSF when a new president came in; again, remember the foundation was a minor character in Washington politics. And a second reason I wasn't worried was that I could defend the programs that the administration wanted the NSF to carry out, although I knew I was going to push the administration in some new directions.

The biggest issue in my in-box when I arrived as director was the foundation's role in applied research. NSF's programs had always had an applied side. An OMB official, Carlyle E. Hystad, pointed out that early in its life the foundation had to choose between purity and financial growth; and when it sensibly opted for the latter, it also made a Faustian bargain when by 1970 "the Foundation, of its own volition, had become a diversified agency. Were its leaders sincere about touching nothing but pure research, they should long since have settled for a budget in the neighborhood of $200 million a year."[24]

The "bargain" got a tougher bite to it in the early 1970s as the Nixon administration confronted a troubled economy: slowing productivity, recession signs, and severe squeezes on the federal budget imposed by the Vietnam War and by the growth of federal entitlement programs launched by the Johnson administration. The initial response had been to whack at the federal budget, including the sciences and therefore the NSF budgets. However, the Nixon administration reversed course in the FY 1972 budget and asked federal agencies to spend more money to "prime the economic pump." That government-wide instruction was particularized with the NSF, strongly abetted by senior OMB officials who asked that the foundation spend more of its funds on making sure jobs were there for all those new Ph.D.'s produced by the flush times of the early 1970s. That pressure was joined by the demand—still with us today—that we find faster ways to move all that wonderful research to market, to transform it into commercial innovation, into new products and processes.[25]

The budget tension cast into relief a "central ambiguity"[26] in the nation's postwar research system. The premise of federal support of research was that it benefited agency goals and hence the government's goals. The issue was how sharply to define the benefit before the money flowed. For research at universities the support was rarely tied to specific outcomes; rather, the belief was that a rising tide of knowledge would benefit all. That was fine, when money was plentiful. When it drained away, the uncertain links of academic research to specific administration or congressional goals became a political target.

The result was a dog's breakfast of successive programs that in one form or another was intended to stimulate the application of science and technology to "practical needs." There were several such programs for NSF, each swallowed up by a bigger successor. Easily the largest was Research Applied to National Needs, RANN, a clever title since it didn't call for "applied research" as such but only that the programs seek to apply the research supported by the foundation. RANN was born out of a "bribe." In 1969 the foundation had, at the behest of the Congress (which had changed the foundation's organic act to make it possible), created a program grandly if clumsily called Interdisciplinary Research Relevant to Problems of Our Society, IRRPOS. With an annual budget of about $13 million, IRRPOS supported interdisciplinary research projects—that is, research across the usual disciplinary boxes of science—in universities, national laboratories, and the like.[27]

On December 13, 1970, in the OMB director's office in the Old Executive Office Building, that marvelous pile next to the White House, NSF Director Bill McElroy was given a startling "gift": $100 million more for the foundation's budget—almost a 20 percent increase—in return for the foundation phasing out its support to institutions,[28] cutting back its educational programs, and substantially increasing its efforts in applied research.[29] That was a remarkable offer, especially with the budget cuts of the past years. McElroy accepted the offer. [30] I would have done the same thing.

IRRPOS disappeared and RANN arrived exactly one year before I became director, on February 1, 1971. Ray Bisplinghoff, my former MIT colleague and deputy director of the foundation, became acting director of the Research Applications Directorate, essentially RANN.[31] On February 23, RANN got over $11 million of the director's special reserve fund to pay for seven projects already under way. Projects included the role of deterrence in criminal behavior, a study of the quality of American life as a guide to developing "social indicators," and an examination of "critical environmental problems affecting modern society, a study that led to changes in California's air pollution regulations."[32]

It wasn't all hosannas and rose petals for the new program. Some influential members of Congress—in particular, members sitting on the House Committee on Science and Space, the foundation's authorizing committee—were dubious. They were unhappy about the cuts in institutional support, in basic science programs, and "increased emphasis on research in the social sciences and very heavy emphasis on applied research." To them the foundation appeared to be "almost moving into a mission-type agency."[33] The unhappiness got teeth when the FY 1972 budget emerged. The administration's request for RANN was cut by $30 million to $54.1 million, out of a total of $622 million for the foundation. And the foundation's funds for institutional and educational support were restored—only in theory since the Nixon administration impounded those funds.[34]

The next big event after those shock waves hit July 19 when Bill McElroy announced his departure for the University of California at San Diego. I was called on to succeed Bill as NSF director in November, and on February 1, 1972, I was in the hot seat. That first day began appropriately enough with a presidential prayer breakfast at 8 a.m. I didn't feel a need for prayer at the time, but later I felt I ought to begin most of my days with one. I was sworn in after the prayer breakfast by Ed David at the NSF offices at 1800 G Street, N.W., virtually across the street from the Old Executive Office Building and its next-door neighbor, the White House.[35]

The next several days were largely for getting acquainted—with the senior staff of the foundation, especially the people running its major directorates, and with White House staff. Ed David invited me to lunch in the White House mess two days after confirmation, where I met John Erlichman, who greeted me enthusiastically while also bragging loudly about all the things that he had to get done. I wasn't impressed. A few days later at a White House reception I met Bob Haldeman, another of the senior White House staff whose reputation was destroyed by Watergate.[36] I was even less impressed.

A much pleasanter encounter came a few days later when I had a reunion with George Shultz whom I had known since our postwar MIT days when we were both young assistant professors trying to scramble up the academic ladder.[37] George was now director of the OMB and special assistant to the president. As OMB director, George was tremendously important to me because OMB had in recent times had a very strong say in the foundation's programs—for example, the $100 million "bribe" mentioned earlier that, among other things, launched RANN.

Until I became NSF director, I didn't have much contact with the people at OMB. I came to admire them for they really did work very hard to make the government run effectively, efficiently, and economically. They were tough but underneath pleasant people. We arranged a luncheon meeting in my office with Jack Young, who was in charge at OMB of the science and technology agencies, and Ray Bisplinghoff, my deputy. Jack had many questions about how the programs were going and where one might expect them to go in the future and it was a good exchange. Jack was a graduate of Colgate, as I was, so the people at NSF who discovered this thought it would be just great for our dealings with OMB. They couldn't have been more wrong, for Jack treated me and the NSF just exactly like he treated all of the agencies—with skepticism and sympathy and hard work to get them into shape.

Reading a Script

I soon faced my first real task as director: to present and argue for the foundation's FY 1973 budget, first to the House Science and Aeronautics Committee and then to several other committees. Given the budget calendar under which Washington operated (and still does), I was in reality presenting, and defending, Bill McElroy's budget.[38] My main task was to convince the Congress by convincing the committee to raise the foundation's budget for FY 1973 by about 12 percent, from $600 million to $675 million. In 1972, $675 million was considered a lot of money (and it still is). But that $675 million had to be divided among some 15 different line items, which could be grouped into three major areas: the relatively new applied programs, especially RANN; science education; and, of course, research, the core of the foundation's raison d'être and then and now easily the largest part of its budget. And the research part—in FY 1972 about 60 percent of the foundation's budget, or $374.2 million—consisted of support for individuals, decided by peer review, such as very active work on molecular evolution; national and special research programs, such as the Global Atmospheric Research Program; and national research centers, such as the Kitt Peak National Observatory.

We proposed about a 10 percent increase for science education, to $70 million. Aims included improving the quality of science training in order to expand career options and developing curricula and other materials to provide students with a better understanding of science so that they could participate more effectively in our technologically oriented society. More specifically, we wanted to address some widely recognized problems: the need to find ways to improve the cost effectiveness of science education through the application of improved curricula, modern educational technology, and new instructional strategies and methodologies; the need to assure the nation of the appropriate quantity and variety of scientific and technical manpower; and the need to improve science education for a broader range of students and to foster science training and awareness for nonscientists so that they could function effectively as both workers and citizens. All fine goals, but there seemed to be always a great argument as to whether the National Science Foundation should spend its money on these programs.

As I presented testimony on the science education program, I couldn't help but notice that we didn't have the same zip and excitement in them as we did in the research programs, where we had more specific examples and a more attentive audience. I well remembered what my predecessor, Bill McElroy, had said—that the tensions and dissension we incurred in the government on the science education program were so great that they were hardly worth a candle. Of course, it was worth it, even if it was—and still is—frustratingly hard to move a system as enormous and ponderous as education and to deal with the political infighting that meets the merest hint of change.

Our real meat was the science for which the foundation served as steward. It's impossible to report on everything, but I'll offer a sampling to show why so many others and I were so excited by the exuberance of the national research effort. There was enormous ferment in the fundamental fields—in physics, astronomy, chemistry, and geology. And the promise and style of the International Geophysical Year that led to *Sputnik* and the exuberant growth of U.S. science, and the foundation's budget, were echoed by more programs knitting together research efforts of many countries: the Global Atmospheric Research Program, the International Biological Program, the International Decade of Ocean Exploration, and, not least, research programs at the poles of the planet, the Arctic and Antarctic. And major new facilities for doing fundamental science were being built or coming on line, facilities still serving us today. Not only Kitt Peak but also the National Astronomy and Ionosphere Center, best known for its radio telescope[39] in Puerto Rico about 12 miles south of Arecibo; the National Radio Astronomy Observatory with observing sites in West Virginia at Greenbank; and the National Center for Atmospheric Research in Boulder, Colorado.

There were lots of reasons to be excited, with the most dramatic being the stunning insight in the earth sciences that the earth is hardly static but geologically very much alive. It was an insight that could explain how mountains are built and continents move, the geography of earthquakes and volcanoes, the patterning of the Hawaiian Islands. That pieces of the earth—most obviously, the facing coasts of the Americas and Africa—could neatly fit together was an old observation and the genesis of the idea that continents drift. But it was an observation, with evidence for and against. Absent an explanation of the forces that moved continents, the notion made it into the Sunday supplements but not into mainstream science. The answer as it emerged in the early 1960s is that continents don't move; plates do. By the late 1960s, the fundamental ideas of plate tectonics were in place. The earth's lithosphere, which includes surface rock, is broken into plates that slide past, collide, or separate from each other. And as the plates move, they carry their passengers, the continents, with them. Plates are created at seams in the earth—such as the mid-Atlantic ridge—by upwelling of new rock and then recycled where they collide, in a continuous cycle of creation and renewal. The emergence of this idea—and the observations to back it up—transformed geology. It was revolutionary but also a continuum in that "from time to time in the history of science a fundamental concept appears that serves to unify a field by pulling together diverse theories and explaining a large body of observations. Such a concept in physics is the theory of relativity; in chemistry, the nature of the chemical bond; in biology, DNA; in astronomy, the Big Bang; and in geology, plate tectonics."[40]

Geology wasn't the only science in revolution. But as the notion of plate tectonics unified one science in the 1960s, so the discoveries in astronomy in the same decade bewildered another:

> The previously well-organized universe, which for ancients, was a planetary system centered on the earth, exploded into a bewildering universe of new types of objects, large and small, with exotic new names and marvelous new natures. . . . One central theme of the last seven years is the discovery of a high-energy, explosive universe. Two universes coexist—hot and cold. The "hot" involves phenomena of explosion; very high temperature; energetic cosmic rays; strange events in galaxies; new types of hot, dense, possibly young galaxies. [The "cold" universe] is the ordinary stellar and interstellar matter. . . .[41]

In my first year as NSF director, astronomers could say for the first time in the long history of the science that they could "observe the universe in virtually any part of the electromagnetic spectrum" as they moved their telescopes off the earth with "rockets, balloons, and stratospheric airplanes." They could observe X-ray stars, the diffuse X-ray background, very "hot" infrared galaxies, "cool infrared stars (that may include planetary systems in the process of formation)," and through rockets with ultraviolet detectors find hydrogen molecules in the regions between the stars.[42] And soon the chemical laboratory in space became quite well stocked, astronomers finding not only hydroxyl radicals in the dust clouds between stars but also other diatomic molecules, such as carbon monoxide, silicon oxide, and carbon sulfide. These weren't simply curiosities, but, as our radio telescope became more powerful in extracting details, they became tools for probing the nature of interstellar dust clouds, their role in the formation of stars, and the mechanisms for forming even more complex molecules.

The rest of physics hadn't been exactly idle either. The internal structures of the proton and neutron came into much greater focus. That queer particle, the neutrino, with neither mass nor charge, was found to come in two forms, electron and muon. And an audacious effort begun in 1968 to validate ideas about how the sun shines by detecting solar neutrinos was under way several thousand feet down in a South Dakota gold mine.[43] We recently had come to better understand the electronic structure of solids, work that set the table for the major advances in fashioning new materials in the decades to come. The advent of semiconductors and their extraordinary demands in materials properties led to techniques for creating materials of exquisite purity and as nearly perfect single crystals. That led to turbine blades for jet engines made of single metal crystals that were much more ductile and stronger.

The barriers that not only physics but also other sciences faced in being "optics-limited" became more porous. "That is, the speed or accuracy with which a measurement could be made, a device controlled, an object detected, or a chemical analysis completed often was limited by fundamental optical problems of intensity, resolving power, stability."[44] The signal event in breaking through the limits was of course the laser. Einstein (who else?) first suggested the possibility of the stimulated emission of light in 1917, but it wasn't until 1954 that the first microwave laser (the maser) was built, followed in 1960 by the first optical laser. The applications that emerged in the 1960s turned hollow the early skepticism that this was an "answer in search of a solution." Monday morning quarterbacking made it obvious that the special talents of the laser would make it an extraordinary tool. Properly stimulated, the laser could emit very high-energy beams of coherent light—that is, light composed purely of one color or frequency. Moreover, the light emitted was many times more powerful than the light used to stimulate the lasing action. Here was not only a source of pure light but also an optical amplifier. The effect was enormous, on pure science and on technology. The laser was used in the 1960s by elementary particle physicists to probe the structure of electrons; by chemists to probe the structure of materials and to do wholly new types of chemical reactions; by planetary scientists to measure the exact distances to the moon and other bodies; by geologists to measure to an exquisite precision the movement of tectonic plates; and by biologists to probe the structure and even do surgery on bodies within cells such as chromosomes. And the laser enabled holography, this seemingly magical (even now) tool to create images that appear three dimensional without lenses.

The list is endless, and, of course, the uses of laser are ubiquitous today, from compact disc players to eye surgery to holographic images on credit cards. The point is that this single tool transformed and pushed forward an enormous spectrum of sciences. In the same vein, in the early 1970s the National Science Foundation was putting in place new instruments and facilities that also changed forever the nature of fundamental sciences and their applications. A prime example were our facilities for oceanographic programs, such as the very substantial ocean sediment coring program, which had within it the Deep Sea Drilling Program. Started in August 1968, the aim was to explore the age, history, and development of the ocean basins and their seas by drilling through the ocean floor and taking out cores. On June 24, 1966, NSF and the University of California signed a contract for the Deep Sea Drilling Project, to be sited at the Scripps Institution of Oceanography and with Global Marine, Inc. doing the actual drilling and coring. The centerpiece was what became the first and then famous ocean drilling ship, the *Glomar Challenger*. Its keel was laid in Orange, Texas, in October 1967. In March the ship sailed down the Sabine River to the Gulf of Mexico. It operated in all the world's oceans for 15 years, recovered over 19,000 cores taken from over 600 ocean bottoms. It reached down to as deep as 23,000 feet of ocean water and almost 6,000 feet into ocean floor. It tied dock the last time in November 1983. Parts of the ship, including its positioning system and engine telegraph, are now in the Smithsonian Institution.[45]

The cores brought up from 17 holes drilled at 10 sites along an oceanic ridge between South America and Africa brought up powerful support for a fundamental tenet of plate tectonics: that new sea floor was made at the rift zone in the ridges. Indeed, the age of the ocean floor turned out to be 200 million years compared to the age of the earth of about 4.5 billion years.

As fine as the science was, just as astounding was the technology to get it. Doing the drilling without breaking the pipe meant that the ship had to be kept very stable in oceans home to ferocious and sudden changes in current, wave, and wind conditions. That was solved with a computer-controlled positioning system reliant on sound waves coming from acoustic sources on the sea floor. Acoustic positioning also dealt with what seemed to me the even harder "needle in the haystack" problem. Drill bits wear out, which meant pulling up perhaps several thousand feet out of the ocean bottom and then reinserting the pipe with a new bit in the same hole. It was done for the first time on June 14, 1970, off the New York coast in 10,000 feet of Atlantic Ocean water.[46]

Just as the voyages of the *Glomar Challenger* transformed forever the geological sciences, so did new instruments that looked up and not down, in particular the advent of radio telescopes. Radio waves being at the lowest energy end of the electromagnetic spectrum barely interact with atoms and molecules. The upshot is that the earth's atmosphere is transparent to radio waves day and night, making radio astronomy a 24/7 science. Karl Jansky, a radio engineer, located radio waves coming from the center of the galaxy in the 1930s, but it wasn't until after World War II, and the enormous investments and gains made in radar and radio technology, that the first radio astronomy telescopes were built. In 1958 we had the first radar echoes from another planet, Venus, sharply establishing its distance. The structure of the Milky Way was probed by measuring radio signals originating from hydrogen atoms distributed in the spiral arms of our home galaxy. By 1972 the United States was operating over 15 radio telescopes, with NSF supporting the major ones, including a large telescope in the northern mountains of Puerto Rico, about 12 miles south of Arecibo. It originated in a 1958 paper by Cornell scientists; construction began two years later, in the summer of 1960; and dedication was on November 1, 1963.[47] Within six months there was radar contact with Mercury, and a year later that planet's rotation rate had been accurately determined. In 1974, Arecibo found the first pulsar in a system of binary stars, a discovery that in turn enabled the confirmation of Einstein's theory of general relativity and a Nobel Prize.

Unique among radio telescopes, the 1,000-foot dish nestled in a natural bowl[48] doesn't move, and it is the transmitting and receiving detectors hung overhead that are manipulated to "point" the telescope. Although first conceived as an active radar telescope—that is, sending out radar signals and catching them on their return—it quickly also was adapted as a passive radio telescope, catching radio waves from space from regions beyond the eyes of optical telescopes. I gladly inherited the task of having the Congress approve funds for a major upgrade of the reflecting surface, to enable detection of shorter wavelengths, which meant ever finer geographic details of planets, and also a 10,000-fold improvement in "hearing" radio signals from space, enabling the telescope to detect signals that may have started toward earth some 10 billion or more years ago.

My portfolio as director of the National Science Foundation also included the ends of the earth, for NSF had programs in both the Arctic and the Antarctic. Those programs embraced national research efforts and also the very sizeable facility at the South Pole, at McMurdo Sound in the Antarctic. An example of research within a national program, in this instance the International Biological Program,[49] was the work on tundra ecology at Barrow, Alaska, 5 degrees above the Arctic Circle. The growing season in this tundra is short and cold; the rest is Arctic winter. Alaska, Finland, Norway, and Sweden are about one-third tundra; yet at the time we knew very little about it, most especially how tundra responds to different kinds of stresses. That issue was especially acute when I became NSF director because the North Slope oil fields at Prudhoe Bay, about 200 miles from Barrow, were close to operation.[50]

The foundation also had another role in biology, that of seeing to the health of fields neglected by other research agencies, typically because they were not within their portfolios. The plant sciences are a prime example, where after 1970 both the National Institutes of Health and the Atomic Energy Commission drastically cut if not eliminated their support in these sciences, sending areas such as photosynthesis, plant physiology, and plant pathology into a financial hole. The U.S. Department of Agriculture, which didn't have a competitive grants program at the time, didn't step in. But NSF did. And in the 1970s it built a strong program in this area, focusing on such fields as photosynthesis, nitrogen metabolism, plant cell culture, and plant genetics. Ultimately, plant science and ecology became the majority of its metabolic biology program.[51]

RUNNING A CONTINENT

And in a way I inherited a continent, for the NSF in 1972 took responsibility for the entire U.S. program in Antarctica, except for icebreakers.[52] Antarctica is the highest, coldest, driest, and windiest continent. With very low precipitation, most of the continent is technically a desert, with the icecap containing almost 70 percent of the world's fresh water and almost all of its ice. There were four research stations, and that austral summer 142 scientists were working on 51 research projects at the U.S. stations and in the field. That research intensity was even more impressive against the logistical[53] problems of a very short time span to do any work, move people and goods in and out, and the loss at the time of two of the five ski-equipped aircraft and the need to "retire" two more. The research work on the ice was complemented by work on the research ships *Eltanin* and *Hero* and on two icebreakers.[54]

None of my three predecessors as director of the foundation had been to Antarctica,[55] even though it was a substantial part of the NSF budget. Senior Navy and Coast Guard officials went down there, but not the NSF director. I changed that, going to Antarctica in early December 1972. Before going, I read journals and accounts of early expeditions, including the one by Roald Amundsen, who got to the South Pole on December 14, 1911. That was 162 years after Captain James Cook circumnavigated it in 1773 and just ahead of Robert Scott, who made it early in 1912 but died trying to get back.[56] I also read about the flight over the South Pole by Richard E. Byrd[57] and two colleagues on November 28, 1929, in a Ford Trimotor, the *Floyd Bennett*. Those achievements were especially close to me because in 1926 I saw and touched Byrd's polar airplane when it was displayed at Wanamaker's department store in Philadelphia. I was 10, and visiting my cousins, Guy and Nell Ford, to see the Sesquicentennial Exposition celebrating the signing of the Declaration of Independence in 1776.[58] As a young Boy Scout, I was very aware of the competition for an Eagle Scout to go with Byrd on his South Pole expedition, a competition won by Paul Siple.[59]

I was well briefed for the trip by foundation experts, including Philip M. Smith, program officer for the Antarctic program who had been to the continent 14 times. On December 7, 1972, I flew via Los Angeles to Auckland, New Zealand, and then on to Christchurch, where we were met by the commander of the naval station that was the jump-off point for Antarctica. We got our cold-weather clothing and safety briefing, and I went on to press interviews, including one by a New Zealand TV station.

The next day there was a jet-boat outing on the Waimakariri River with John Hamilton, son of the inventor of the water jet boat, hosting. The Waimakariri River arises in the mountains on the west coast and flows through beautiful gorges across the whole South Island and into the ocean on the east coast. The river is an ideal place with lots of rapids, some shallows, and some swift-running water—just exactly the kind of stuff that water jet boating was invented for. The water jet sucks up water in a scoop and then ejects it at high speed out of the rear for a water jet thrust. The small boat that we had, good for four or five people, could really fly up the river, skimming the tops of rocks that were only 2 or 3 inches deep.

As we unloaded for lunch in a gorge, we heard a "hello" from the top of a cliff. A fellow started winding down a serpentine-like path to get down to the river and join us. It turned out to be a friend of John Hamilton and one of the leading sheepherders of the South Island, with thousands of sheep pastured all over the lands bordering the river. He came down and we started a conversation and he said, though I cannot imitate his thick New Zealand accent, "Say, didn't I see you on the telly last night?" And I said, "Well, I was on the telly broadcasting about the National Science Foundation." He said, "Yes, that was a very interesting broadcast. Was I correct that you said that the annual budget was over $600 million?" I said, "Oh, but it's nothing; those are in U.S. dollars, not New Zealand dollars." He observed: "That's just about the entire national budget of New Zealand."

Another former colleague of mine at MIT, Bob Seamans, then Secretary of the Air Force, joined us in New Zealand.[60] We had a deal that he provide us an Air Force C141 transport plane to ferry our researchers to McMurdo Sound if we invited him to the South Pole with us. Also Grover Murray, a member of the National Science Board, and some members of the New Zealand Antarctica Authority accompanied my NSF party. We arrived in Antarctica, and after dinner at the McMurdo Station we went out in the continuous daylight to visit the Shackleton and Scott huts[61] at Cape Evans and Cape Royds, respectively. The bunks, sleeping bags, the place where they slaughtered seals were all remarkably preserved by the cold. We were told that the canned foods were so well preserved that we could eat some. We declined.[62]

We left for the South Pole the next morning, on a ski-equipped C-130 Hercules plane, landed midmorning at the Scott-Amundsen base, the first U.S. research station at the pole, and launched into a hectic schedule. First up was a lesson in what happens to buildings at the South Pole. A building sitting on top of the ice is buried by the ice and snow, slowly sinking deeper and deeper. The ensuing pressures slowly destroy the building; indeed, several of the buildings we saw were visibly buckled and had been declared off limits. We were allowed in and entered a room with three desks, one of which had been Paul Siple's. We saw efforts to deal with the pressure problem, by building a very large dome serving as protective cover against the drifting for the laboratories housed in it.[63] After that briefing, intended to show me why operating at the South Pole costs so much, we had briefings on the science being done at the pole, concentrating on glaciology and meteorology.

Then lunch, addressing cards to friends postmarked at the South Pole, and then finally a visit to the pole itself, for pictures and the obligatory circumnavigating of the world by running several times around a small stake representing that year's exact location of the pole. Bob Seamans and I chased each other around the world three times, quite an exhausting feat because of the rarified mile-high atmosphere and our heavy clothing and booting. (Each year that stake in the ice is reset several tens of yards because the mile-thick ice cap flows steadily over the land.) Bob and I had a chance to tighten an immense nut on an immense bolt used in the construction of the first geodesic dome at the South Pole Station. I could hardly lift the heavy wrench used by the Navy Seabees in the construction.

Then we had a bit of luck. The weather was good enough to fly to the Soviet Union's South Pole station at Vostok. The station was a few hundred miles from the pole, at a higher altitude, at the geomagnetic South Pole, and the coldest place on earth, −126° Fahrenheit having been recorded there. We landed after six tries in wind-driven snow and were warned that the engine and hence the propellers would keep going so that the hydraulic fluids didn't freeze. We were warmly met by the Soviets, and while we were there to hear science, formalities still had to be observed. That meant speeches and vodka toasts after every speech.

We finally got to the science, especially their work in boring through the ice. The Soviets had already bored through at least 10,000-year-old ice, taking out cores 5 inches in diameter. One we looked at showed clearly that about 13,000 years ago there was a major volcanic explosion somewhere in the world. We of course drank another vodka toast, this one cooled with 10,000-year-old ice, which fizzed as the ice melted, releasing tiny bubbles of gas trapped over 10,000 years ago. We saw more science, but then the pilot said it was time to go, else we risked the hydraulics freezing up and having to wait for the next plane, whenever. We were sympathetic to that argument. Still, the head of the laboratory insisted on us seeing his experiment, and that of course meant more vodka, this one a truly fine Georgian version. Finally, we had to go, not least because the vodka was winning, we being at the equivalent of some 11,000 feet and hence not too well supplied with oxygen. The local snowstorm into which we had landed was still going, and we used the sound of the motor to find the plane. The Soviet leader came out to say "goodbye" and headed directly for the plane right between the fuselage and the turning propellers! My heart stood still, but he made it, and we made it, all of us leaping aboard, helped along by some unceremonious shoves. The Soviets had done something very thoughtful. They gave us several 3-foot sections of what was at a minimum 10,000-year-old ice packaged in special containers to keep them from melting. The ice was then shipped back to the United States, for my colleagues and I to use at our cocktail parties.[64]

There were many more visits and tours and then the return trip to Christchurch, where I got taken out on the Selwyn River to fish for brown trout. It was fabulous dry fly-fishing, but while I had several on the hook, I didn't have any in the net. Then back to Washington, where I had one week before I went on my annual two-week jaunt to Randolph over Christmas and New Year's. Back at 1800 G Street, there were Christmas parties, meetings with new ambassadors to Poland and Spain, where NSF had cooperative research programs, and a meeting with the foundation's executive council to report on a brand new document called *Science Indicators*. The report has since 1973 been published biannually and has become indispensable in understanding the state of American science and technology. I also recommended that each year the board have at least one member visit Antarctica to ensure the quality of research and the exercise of its oversight.

My first year as NSF director was done. I had survived. I enjoyed it. No political blowups. But I had a nagging feeling that this was going to change.

8

Tumult

1973–1976. Three jobs in three years: director of the National Science Foundation, science advisor to Nixon, science advisor to Ford. All this in the midst of watching science tossed out of the White House, the first energy crisis, a fling with détente, and joining Jacob Javits and Hubert Humphrey as "one of the greatest menaces to the United States!"

New Year's Day 1973 was benign enough. Our entire family was at our vacation home in Randolph. New snow. I went skiing at Wildcat Mountain with three of our four children—Roy, Sarah, and Margo. Then in succession the Rose Bowl game, a party at nearby friends, and a game with all four children, Guy, Jr., included this time.

Guy, Jr., was in his third full year of teaching English at Berlin High School 15 miles away from Randolph. Sarah was in her third year of doctoral work at the University of Michigan on Italian renaissance history, relatively newly wed, the only one of our children then married. Margo was close to graduation at the University of Pennsylvania, majoring in painting. Roy was enrolled at Cornell University with interests in natural resources and their management.

That wonderful time of family, friends, snow, with nary a thought about the National Science Foundation (NSF) ended Tuesday afternoon, January 2, when I returned to Washington, driving over to Portland, Maine, and flying down from there. Bunny would follow in the car later, spending some more time in Randolph with the children before they had to return to their work and their colleges.

EVICTION

I knew I was flying into a gathering political storm. Nixon was over-whelmingly reelected to a second term in a rough campaign. Watergate was still low-level noise with few aftershocks at the time from the break-in in June 1972 at the offices of the Democratic National Committee. Yet changes were in the wind. I knew Ed David, the White House science advisor, was planning to resign and did so on January 3, the day after I returned to Washington. Bob Seamans, a good friend from the Massa-chusetts Institute of Technology (MIT) who had been with me on that wonderful trip to Antarctica, was leaving his post as Secretary of the Air Force. But George Shultz, also a former MIT colleague, was strengthen-ing his dual roles as secretary of the treasury and special assistant to the president. Roy Ash came from industry to run the Office of Management and Budget (OMB).

Science itself faced problems. I had asked for $675 million for fiscal year 1974 but wound up at $640 million. This was going to be unpleas-ant to explain to the scientific community, especially the academic re-search part of the community, deeply dependent on NSF to fund research, support graduate students, and, not too incidentally, help the rising stars get tenure. It would be even more unpleasant to explain the budget to congressional committees with a special responsibility for science. This budget continued the declines since 1967 in overall federal support for basic research. That decline continued in 1975—about 20 percent in real terms since 1967; in constant dollars it went from $1.7 billion in 1967 to $1.4 billion. Most fields—aside from some areas of biology, engineering, and oceanography—declined, with the sharpest cuts in the physical sci-ences, especially physics and chemistry.

The strongest imperative for NSF in hard times was then (and still is) to protect its core raison d'être: the support of individual researchers, principally at the universities. We favored that in our internal adjust-ments, reluctantly, by deferring maintenance for scientific facilities, cut-ting back support for instrumentation, and reducing grant sizes and stretching them out. It was painful to do but unavoidable if we were going to protect basic research.

Budgets were in some ways the least of the problems early in 1973. It became apparent that the Nixon White House was determined to get rid of its entire science apparatus—the president's Office of Science and Technology (OST), the President's Science Advisory Committee (PSAC), and a resident science advisor. And it looked like the foundation was going to inherit some of these functions.

Why? An immediate reason was certainly the one always mentioned: that former or current PSAC members and consultants, ostensibly speaking as private citizens, had publicly challenged several of Nixon's programs, most dramatically proposals for building an antimissile defense and earlier the proposal for a supersonic transport. It came to a head when in early 1969 Nixon requested funding for an antiballistic missile (ABM) program, *Safeguard*, a successor to the *Sentinel* program. Moreover, Nixon made it known that he relied for technical judgments not on the PSAC and his science advisor, Lee Dubridge, but rather on the Department of Defense.[1] In congressional hearings, a parade of former PSAC members and consultants and past science advisors criticized the administration's ABM plans. Indeed, one senator commented that he was "unable to find a former presidential Science Advisor who advocates the deployment of the ABM program."[2]

It's simplistic, however, to believe that the ABM was why the Nixon administration tossed science out of the White House. Animosity had been building up for a long time. Up to 1975, the most successful science advisors arguably had been the two who had served Dwight Eisenhower—James Killian and his successor, George Kistiakowsky. Their accomplishments were considerable, including reassuring a country terrified by the implications of *Sputnik* that it had a strong and robust missile and space program. Overall:

> . . . the scientific advisory apparatus of this period had a seminal force both on the organization of the government's scientific and technology effort. . . . The creation of NASA [National Aeronautics and Space Administration] in 1958 to pursue an independent civilian space program was one early accomplishment, but the White House scientists also made the initial recommendations to set up the Arms Control and Disarmament Agency in the State Department. They recommended major improvements in the ICBM [intercontinental ballistic missile] program, including important new emphasis on solid-propellant rocket engines, acceleration of ballistic missile early warning capabilities, and advances in anti-submarine warfare capabilities and photographic reconnaissance from espionage satellites.[3]

Further to the expulsion were shifts in the structure of the White House. Killian and Kistiakowsky worked dominantly on issues of defense and national security, areas in which they were giants. And they were backed by people who had worked with them at Los Alamos, the Radiation Laboratory, and other places where science established its centrality to U.S. security. The PSAC was composed of physical scientists who knew each other well through their shared experiences during and after World War II. And, not least, the science advisor and PSAC were virtually the only forces on national security in the White House. That changed sharply with the arrival of the Kennedy administration when McGeorge Bundy substantially enlarged the National Security Council's staff. That organizational change was mirrored by an enormous broadening of the agenda for a science and technology advisor, to encompass environmental, food, and in time energy concerns. The upshot was that national security issues on which the science advisors and PSAC had built their reputations moved to the National Security Council, while the agenda for a White House science apparatus became more diffuse, less susceptible to hard analyses and decisions (e.g., the Eisenhower-era decision on whether to emphasize solid rocket boosters) and more dominated by "softer" issues where analyses were increasingly perfused by the "on the other hand" syndrome and decisions had even wider economic and political implications (e.g., automotive fuel economy and emission standards). At the same time, the growth of the national security apparatus in the White House was soon joined by other agencies in the White House, each claimants on a piece of the nominal agenda for the science advisor: the National Aeronautics and Space Council, the Council on Environmental Quality, and the Federal Energy Office. These accretions collectively diluted the influence and resources of the White House science office. Science advisors responded to these changes. Jerome Wiesner[4] faced the need for more staff by creating the OST. The funding had to be provided by the Congress, which of course meant congressional oversight. That chimerical structure—one office, that of science advisor not subject to congressional inquiry, and the other, the OST, subject—sometimes caused confusion, but it worked and produced successes. However disenchantment grew as the Vietnam War continued, and the politicos in the White House—not least, Lyndon Johnson—

BOX 8-1. Postwar White House Science Advisors, 1957–1973

James R. Killian, Jr. (Eisenhower)	1957–1959
George B. Kistiakowsky (Eisenhower)	1959–1961
Jerome B. Wiesner (Kennedy)	1961–1964
Donald F. Hornig (Johnson)	1964–1969
Lee A. Dubridge (Nixon)	1969–1970
Edward E. David, Jr. (Nixon)	1970–January 1973

began to lose confidence in the White House science structure. Eviction of science from the White House came in the fall of 1972 when the PSAC learned it was history at what turned out to be its last meeting, where the topic was to be how PSAC could improve its relations with the president. I arrived late to the meeting but in time to see the PSAC members leaving. "We've been fired," Luis Alvarez blurted as he erupted out of their last meeting.[5]

The full dénouement came when George Shultz, now secretary of the treasury and also a special assistant to the president for economic affairs, asked me to come to his office in the Treasury Building on Friday, January 12, 1973, at 9:30 a.m. Shultz told me that the White House Office of Science and Technology would be abolished and many of its functions transferred to NSF. He asked me to meet with Roy Ash, director of OMB, and we made a lunch date for that Saturday, the 13th. Shultz and Ash questioned me at lunch about what was on the plate for the White House science office, and I responded in part that they ought to pick up Ed David's considerable success in setting up science and technology exchanges with the Soviet Union. They certainly agreed to that because détente with the Soviets was something the Nixon administration wanted very badly. But when I said that we still needed something like the PSAC that had just been disbanded, they said that was not in the cards.

They also told me not to discuss it with the National Science Board, but I did talk about it with my deputy, Ray Bisplinghoff, the next day, Sunday. On Monday I met with David Beckler of the OST.[6] The National Science Board was coincidentally meeting that week, and I invited

George Shultz to meet with the board and describe the proposed reorganization. The board was for the most part enthusiastic about the proposal, some reminiscing that this was supposed to have been the role of the board when it started. It was a return, at least on paper, to the role that Vannevar Bush had envisioned for the NSF, of guidance over research programs across the government.

I began to think of myself stuck in the middle of a large snowball rolling downhill out of control with arms and legs sticking out. Nixon was sworn in for his second term on Saturday, January 20. The next day I went to a White House worship service, feeling much in need of divine inspiration. Realism intruded that day as I was briefed by Hugh Loweth of the OMB not just on the NSF budget but also the full panoply of federal research and development budgets. Friday, January 26, 24 days after Shultz broke the news to me that I had a second job, I told the senior officials of the National Science Foundation. I briefed the press the next day, both on the NSF budget and the science and technology budgets across the government.

And that week, on January 26, Nixon publicly announced what was coming down, in his Reorganization Plan No. 1 of 1973. For good measure, it abolished not only the OST, transferring its functions to the NSF, but also the Office of Emergency Preparedness and the National Aeronautics and Space Council.[7] Reactions were mixed. Congress had 60 days to react to the plan but didn't; I was close to shock that Congress swallowed the reorganization with little comment. One congressman, after I had testified on the changes, remarked, "Well, I guess the president can organize his White House any way he wants." It was different with the outside science communities. Quite a few were upset that the science advisor no longer had direct access to the president—although in recent years that had become a rather weak reed—but that the science advisor now had to channel his thoughts through George Shultz. Some were upset that the reorganization plan had moved OST's responsibilities for national security and defense matters to the National Security Council and the Department of Defense. They had a point: the OST, especially under Killian and Kistiakowsky, as I've mentioned, had an enormous effect on military technology, not least in killing some truly bad ideas.

The administration reacted to these complaints. On July 10, 1973, it appointed as science advisor the director of the National Science Foundation (i.e., me), made the science advisor the head of the Federal Council for Science and Technology, and asserted that the science advisor would have a direct line to the president and would in fact take on national security issues in classified areas. But that didn't dampen the belief of many people that the NSF director suddenly had too many things to do, with the danger that neither job—running the foundation or advising the president—would be done well. How could I respond to that? After all, I couldn't say that I could handle the added work with a simple twist of the wrist or that I'd just spend less time running the foundation. But it wasn't all that complicated: I worked a lot harder, had many more appointments, and lots more problems on my daily agenda. One of the first things I worked hard at was organizing for the dual task, for which, reasonably enough, I set up parallel structures at the NSF, albeit without PSAC, the very name of which was anathema to the White House. I tried to entice several of the OST staff to join me at the NSF. Not much luck. Many were understandably bitter. Others thought going to the foundation from the White House was a step down. I did have one lucky break, when one of my senior people said: "Guy, you need a talented assistant to the director, with technical, organizational, and political savvy." I heartily agreed to that, and he suggested Phil Smith. I accepted that suggestion and never regretted it.[8] To deal explicitly with the tasks left behind by the dissolution of the White House science office, we set up within the NSF a Science and Technology Policy Office. And I persuaded Russell Drew, a naval officer then directing the London office for the Office of Naval Research, to lead it.[9] And I had a splendid speechwriter in Stan Schneider.

DÉTENTE

I plunged in and was quickly in the thick of what was, aside from the Vietnam War, the major international preoccupation for the Nixon administration: détente with the Soviet Union—finding ways to coexist despite obvious differences and strong distrust of motives. As usual, when the United States decides to improve relations with a country, especially one that has been hostile, the first—and mostly easy—agreement is on science and technology exchanges. Ed David had started that work in the summer of 1972, going to the Soviet Union to explore what kind of exchanges might be made. I was part of the planning group, and the next meeting with the Soviets was set for spring 1973. That was now very much on my plate. There were tensions: the Soviets were pushing for exchanges on computer science and technology. That was strongly opposed by the National Security Council and the Department of Defense for the coupled reasons that we were far ahead of the Soviets in these fields and that they were becoming very important to the military. The Soviets persisted and a small concession was made by setting up a group to discuss computer systems and their uses. This was attractive to the Soviets because with their centrally planned economy they thought computers were key to their future. Little did they know that democracy and free enterprise were better answers. We also agreed on exchanges in fields less sensitive to military and economic security: high-energy physics; intellectual property, where we had serious problems with the Soviets; and materials science. Various agencies had also moved ahead on their own; for example, the National Aeronautics and Space Administration (NASA) worked for many years with the Soviets on the Apollo-Soyuz hookup program.[10]

Tuna Sandwiches, Milk, and Cookies

We of course wanted to do a very good job on this, certainly not least because it was of great interest and importance to the White House. It was a part of the first real attempt at a "thaw" of the Cold War, and science and technology played a useful role in opening avenues of cooperation between the United States and the USSR. We were determined to be very good hosts, which meant, for example, that we would go to New York to meet the Soviet delegation at Kennedy Airport. Bunny in her diary wonderfully captured how it went:

> Sunday, March 18th. Windy, cold day—had an early roast beef dinner. At 1:30 Dick Neureiter and John Thomas (State) arrived for a briefing on the trip for us. They had had no luncheon so in all the confusion I gave them tuna sandwiches, milk and cookies.
> Left at 2:30 for Andrews A.F.B. picking up Yevgeny Belov at State on the way. (He's the Russian Embassy Science Attaché.) Takeoff from Andrews was

historic in the scariness as a heavy gust of wind hit the plane at liftoff and it fishtailed in the most horrifying way. Thought I was going to be catapulted back to the Capital!

Such milling about at Kennedy when we arrived but the Russian delegation arrived not too much delayed at 6 something and we duly met their Academician and Madam Trapeznikov and 14 other delegates. The ride back to D.C. was fortunately smoother—we drank OJ and ginger ale—the ladies were presented with orchids and we conversed lamely through John Thomas. We were met at Andrews Air Force Base by Ambassador and Madam Dobrynin and Madam Belov who gave Madam T. bunches of posies. The ladies were driven back in town to the Statler Hilton where we left them and returned home—bone tired!

Monday, March 19th. A.M. madly spent making crab casserole, salad and all the fixings while the Soviets toured the White House.

At 11:45 my guest, Mrs. Thomas, arrived. Mrs. Dobrynin arrived at 12:15, she is most attractive and of course speaks very good English. At 12:30 or a bit after, Madam T. arrived escorted by Pat Nicely[11] (NSF) and our interpreter, Ms. Irina Kieraeff, a very nice 24-year old girl. Pat had sent a camellia corsage for everyone so after a tour of the house we had camellias, Dubonnet or sherry and then luncheon which turned out to be good indeed! After the luncheon we, Irina, Mrs. T., and Madam Trapeze went to Mt. Vernon which obviously impressed her, especially the breathtaking view from the front portico.

Tuesday, March 20th. Met Madam T. and Irina at hotel at 10. Driven to the Children's Hospital where we had a real cook's tour. Very informative. At 12 down to the Rayburn Building, Madam T. having trouble with her hearing aid and so completely missed all the slum area! Pat Nicely met us— we got some scotch tape and the aid was fixed, thank goodness for Irina's sake!

We went over to Congressman Hannah's office (Cal.) where he and his wife and Congressman and Mrs. Mosher (Ohio) had a very nice luncheon in one of the hearing rooms.

An ex-Congressman Spangler took us around the Capitol and a marvelous tour into the private Senate and House reception rooms not open to the public—to the chapel—but on the whole he sure didn't plan the leg work very well. We went back and forth exhausting Madam T. Madam T. was delighted to catch a glimpse of Ted Kennedy on the Senate Floor: she knows as much about the Kennedys as I do! Back to hotel at 4.

7:30. National Academy of Sciences for dinner and what an evening—it was fantastic. By now I am Madam T.'s closest American friend and she cuddles right up to me: I think Madam Dobrynin and Madam Belov are as overwhelmed with her as I—but she is amusing and fun. I sat on Academician Trapeznikov's right at dinner—Harvey Brooks on my right and the Ambassador and Guy across the table. It was a fantastic fascinating dinner. Mr. T. is a very pleasant, friendly man and though his English is a bit difficult—we hit it off. Dobrynin is very pleasant (at the moment) and apparently was in high form that evening, more relaxed and friendly than ever witnessed.

Apropos of that—Guy remarked that they had called off their "watch-dogs"—they were there on Monday but not thereafter—things are going well with the meeting.

Golly, I almost forgot to put in that Ambassador Dobrynin, Academician Trapeznikov and Guy went to see the president at noon: were received in the Oval Room and had a very good 25 minutes!

Well, back to the dinner.

Phil Handler led off with a toast and then remarked that Dr. Stever had a special toast so Guy toasted the visitors and told them about his visit to Vostok, the Soviet station 300 miles from the South Pole.[12] He told them about the vodka and the ice and said we would have some with our champagne.

Sure enough with dessert, more toasts and Guy said that he had a new law for international cooperation—vodka (or champagne) and 40,000 year old ice. So we all drank our fizzy with the ice and of course the Russians were delighted! It made a tremendous hit, very clever of our HGS!

Home to bed.

Wednesday, March 21st. Met Madam and Irina with Pat, driven out to Montgomery Mall where she was overpowered by all the affluence and variety of shops. Looking for Lady Lee Rider dungarees for her 17 year old Toddy! She got a tie at Sears for Papa Trapeze and we finally found the Levis at a little shop on Wisconsin Avenue.

We exchanged gifts in the car. We gave her mahogany Williamsburg placemats which she wouldn't undo because of the pretty red, white and blue ribbon! Pat gave her some maple syrup and kitchen gadgets. She gave me an amber bracelet and necklace and a pretty shawl. All very gay and fun.

12:30. Met Madam Belov at the Cosmos Club and Gigi Neureiter—had a very pleasant luncheon. Then off to the National Gallery where we met Dr. Cooke who gave us a special tour of the impressionists and deep dark secret—a private advanced tour of the coming Russian visiting exhibit which had not even been hung but were just propped against the walls. What fun—though I hardly took a breath with all those canvasses at foot level. Matisse, Van Gogh, Gauguin, etc. At home Guy, Trapeznikov and the Commissioner signed the treaty agreements though the Russians unfortunately refused to open up everything to visitors as had been hoped. However, it's a milestone and a great success. How extraordinary to think that Guy had undoubtedly played a part in the bettering of Soviet/ American relations. Long may it last!

Home at 4:15. Off to the Russian Embassy at 7. Another gay party and we were received like royalty! Dave Langmuir[13] was there!! We bid our fond adieus and bon voyage to the Trapezes in a real football huddle! And staggered home triumphant, totally weary but euphoric. Also clutching a gift box of caviar and vodka from the embassy. Well, well, well, what an unforgettable three or four days.

Guy has been given a lovely group of rocks on a teak tray,[14] very handsome indeed.

March 22nd, Thursday. Not quite over—at 8 the phone rang—Mrs. T.—

all of atwitter about the flowers in her room—I must have them and they
must be picked up at 8:45. A special trip by NSF car and secretary!!
    March 23rd. Oh death and damnation. I ate some of the crab Thursday
p.m. but it had been left out too long on Monday. Got terrible ptomaine and
spent a day in bed, flat with only coke, slept almost all day.

Bunny was right: the meetings indeed went very well. You have to
remember that they came in a time of tumultuous events. A year before
our talks, Nixon visited China. Four months later, in May, he was the
first American president to visit the Kremlin. He and the Soviet Brezhnev
signed a set of agreements that included a start toward limiting nuclear
weapons and in other areas, including exchanges in science and technology. As *Time* observed:

> The meeting underscored the drive toward détente based on mutual self-
> interest—especially economic self-interest on the part of the Soviets, who
> want trade and technology from the West. None of the agreements are shat-
> terproof, and some will lead only to future bargaining. But the fact that
> they touched so many areas suggested Nixon's strategy: he wanted to in-
> volve all of the Soviet leadership across the board—trade, health, science—
> in ways that would make it difficult later to reverse the trends set at the
> summit.[15]

The Soviet delegation was led by Academician Trapeznikov,[16] the director of a major laboratory on information systems and computers and
number two[17] in the State Commission on Science and Technology. That
our talks were focused on science and technology didn't equate to nar-
rowness, and we dealt over several days with topics such as intellectual
property, magnetohydrodynamics, energy sources and distribution, computers for large systems analysis, high-energy particle physics, materials
of many kinds, and, not least, human rights, criticizing the treatment by
the Soviets of specific scientists. The latter soured things a bit. The Soviets started talking faster and more excitedly and, predictably, told us in
effect to butt out of "internal matters." They clearly had the human rights
script down pat, and we made no obvious progress, which was par for the
course in these exchanges.
    Academician Trapeznikov, Soviet Ambassador Dobrynin, and I met
with Nixon, who was very upbeat and enthusiastic about the proposed
exchanges. He told the Soviets that we were anxious to cooperate and all
of us to "make this work." Both the Soviet press and the U.S. press gave

our meetings a decent reception, although agreements on science and technology paled compared to some other events, such as the negotiations on nuclear weapons reductions. The next meeting was arranged for that fall, in Moscow. We arrived on Tuesday, November 27, 1973, several weeks after the momentous event of the October oil embargo, imposed by the Organization of Petroleum Exporting Countries (OPEC) in response to the Yom Kippur War. We continued the discussions started in Washington in the spring, often arranging for data or expert exchanges, but the most important happening was outside those discussions. When I was scheduled to meet with President Nikolai Podgorny and my counterpart in the science exchange, the deputy prime minister for science and technology, academician Vladimir Kirillin, Spike Dubbs, the ranking officer in our embassy in the absence of the ambassador, asked if he could accompany me to help in the translations. Since I was the first presidential appointee to arrive in Moscow after the OPEC oil embargo, this was Spike's first chance to meet with a high-ranking leader of the Soviet Union.

Trained as an engineer and president of the Soviet Union since 1965,[18] Podgorny first questioned us closely and sharply about our proposed program, being very insistent that the meetings have, in a phrase I heard often, a businesslike atmosphere. But things became much more interesting when he suddenly widened the scope of the questions he was asking me, and Spike began madly scribbling notes. Podgorny moved from science and technology, to my views on how the Vietnam War was going, to the energy situation, especially the impact of the Middle East turmoil. I quickly realized that the root message through me to Nixon was that détente was primary and that the Soviets would not seek to exploit the Middle East problems nor would they embarrass us in Vietnam. Spike Dubbs spent the night summarizing the meeting, checked it out with me the following morning, and posted it to Washington. He gave me an exaggerated reputation as a skilled international negotiator in fields far from my experience and expertise except for energy. For years afterwards, as I visited embassies, they would ask if I was "the Stever who had the interview with President Podgorny."

We moved off the heady plane of high-stakes international diplomacy back to negotiating science and technology exchanges, echoing many of the issues that had come up the past spring in Washington, still trying to make progress on human rights and intellectual property. We also got to visit some of the Soviet science institutes, some quite good and others not. The off hours were glorious, filled with a visit to the Hermitage Museum in Leningrad, the Kirov and Leningrad ballet companies, and even a visit to the rooms of Dimitri Mendeleev, who conceived the periodic table of elements, making sense of what then seemed chaotic.

And though hardly intentioned by our hosts, we learned something about why they had problems innovating. When we visited what was by their standards a quite modern electronics plant, they showed us a very high-resolution film. We asked where it had been developed, and they said, proudly, "We did it here because we needed it." In the United States, if a researcher needed a particular kind of film, he wouldn't create it himself but simply call someone, say, at Eastman Kodak. And Kodak would provide it, from stock or from its laboratories. Much more efficient, with the U.S. researcher bearing down on his particular task and expertise, getting the tools he needed from others who were much better at it. Not so in the Soviet laboratories. They simply didn't share our cross communication among our science and technology laboratories. We went on to two days in Poland, with a major item the new Nikolaus Copernicus Astronomical Center under the Polish Academy of Sciences.[19] The National Science Foundation was going to provide partial support, and the Poles were anxious to have us appreciate its importance.

On a later return to Moscow my interview was with Prime Minister Aleksei Kosygin. His message for me to carry back was that the science exchanges needed some highly visible symbol such as a world-class hospital, built and staffed jointly by the two countries. This never came to pass. Returning from that meeting we stopped in Romania, where I had a one-on-one talk with the Dictator Nicolae Ceaucescu,[20] who pleaded for help in building a commercial nuclear plant. That, too, never got started in his remaining time.

It wasn't all East-West. The U.S.-Israel Binational Science Foundation had started up in 1973, using funding available for cooperative programs. The funds had first been used to support the translation of Soviet science literature by Russian-speaking Israelis. Most of these were of course recent emigrants, and it kept them employed. The Israelis weren't all that interested and wanted us to use the funds to support joint research. With the guidance of Herman Pollack at the State Department, we set up with the Israelis an analog to the U.S. National Science Foundation, the U.S.-Israel Binational Science Foundation, a granting agency using peer reviews and the like to select among grant applications for joint projects from Israeli and American scientists. I chaired the first-year operation of the overseeing Board of Governors, with an Israeli taking the second year and then alternating. That program has done very well and in fact is still going.[21]

My work with the Soviets, the Poles, and the Israelis was only part of it. My agenda became very heavy with the gathering intensity and interest by other countries in exchanging with us, and not least getting some of our dollars. The dynamics also changed, as tools emerged for comparing the science and technology efforts in different countries. Notable was the publication by NSF early in 1973 of its first Science Indicators, which compared how different countries were doing in supporting science and technology by, for example, comparing the percentage of their gross national product invested in science and technology programs.[22]

In any case, as NSF director I was up to my ears in science exchange programs and the like. It made for special moments, such as when a visiting delegation from the Soviet Academy of Sciences—about a dozen—trooped into my office right after Nixon had tossed science out of the White House and I became the science advisor. I said a few pleasant words, and their leader answered with, "So, Dr. Stever now you are czar of all of science in the United States." I disputed that: "We don't have anyone in this country who has that role. And, besides, we all know what happens to czars." There was a brief pause, and then the delegation members who understood English started laughing really hard, echoed by others when I was translated.

GET BACK TO THE WHITE HOUSE ASAP

The Eighty Sixth Congress established the National Medal of Science in 1959 as a presidential award to be given to individuals "deserving of special recognition by reason of their outstanding contributions to knowledge in the physical, biological, mathematical, or engineering sciences."[23]

Up to 20 awards were to be given in a year. It took the usual fiddling around so that the first award wasn't given until 1961 in President Kennedy's first term, and then, fittingly, I thought, it was given to only one person, Theodore von Kármán, the great pioneer and leader in aeronautics and astronautics.[24] With the disappearance of the White House science office, the awards could also have disappeared. But they had become very important to the science community, and I was determined that they would go on. And they did. In October 1973 after the expulsion of science from the White House we had medalists and a ceremony to honor them. There was a fine lunch in the diplomatic reception rooms of the State Department, and George Shultz gave the speech. I introduced him and sat down just to his right. He gave a very good talk on science and technology and the importance of these medals. Before he was through, however, one of his Secret Service agents came in, walked behind the podium and reached up and put a message in front of him. Being in a better position than anyone else, I kind of leaned over to see what the message was. It read: "Please return to White House as soon as possible. Vice President Agnew has resigned." George didn't miss a beat in his speech, but I noticed he hurried it up, summarizing rapidly. He then said that he unfortunately had to leave for an important meeting at the White House.[25]

There was of course a darker shadow over the administration than Agnew's troubles and departure: Watergate. What started with the arrest on June 17, 1972, of five men at the Watergate offices of the Democratic National Committee was slowly but surely eviscerating Nixon's presidency. In January two former Nixon aides, G. Gordon Liddy and James W. McCord, Jr., were convicted of conspiracy, burglary, and wiretapping in the break-in. On April 30, Nixon's top aides in the White House, H. R. Haldeman and John Erlichman, and Attorney General Richard Kleindienst resigned, and another White House aide, John Dean, was fired. The televised Senate Watergate hearings began on May 18. On October 20, about a week after the National Medal of Science ceremony, was the "Saturday Night Massacre": Nixon fired Archibald Cox, the special prosecutor for Watergate, and both the attorney general and deputy attorney general, Elliott Richardson and William D. Ruckelshaus, resigned in protest.[26]

"If the American People Don't Shoot Me First"

Thankfully, the NSF, science, and I remained untouched by this mess. But we saw some of its effects close hand. I saw President Nixon in the Blue Room in the White House just before the awards ceremony for the winners of the National Medal of Science. I used the private moment to ask if he planned to attend the summer camp at the Bohemian Grove.[27] His response shocked and depressed me: "I'd like to come to the Bohemian Grove, if the American people don't shoot me first." His mood changed, he became pensive, and gazed out the Blue Room to the gardens below. But he came back to give a splendid talk at the ceremony, with smiles and gentle kidding when some of the citations I read got too deep into technical jargon.

Watergate was the immediate and eventually lethal problem for the Nixon presidency. For the country, there was still Vietnam, but also a serious energy crisis and a faltering economy.

In 1973 our natural gas production, little more than one-quarter of our total energy consumption, had leveled off. Coal production was about one-sixth. Nuclear energy, which had become a factor in supplying energy in the mid-1960s, had grown to still only a small percentage of our total energy use.

The energy crisis was an oil crisis, either not enough oil or enough oil but too costly. Through much of the postwar period leading up to the crisis, the demand for oil exploded, in part to drive the rapid industrial expansion of the developed countries, including the United States. In 1973 almost half of the total energy consumption by the United States was petroleum, and only about half of that came from domestic sources. Since in 1973 energy consumption was increasing by about 4 percent per year with no domestic energy sources of any substantial nature that could handle that increase, we were becoming steadily more dependent on foreign supplies. We were hooked on foreign petroleum, and our dependency was growing rapidly.

At the same time, the petroleum exporters faced a 40 percent decline in the purchasing power of a barrel of crude oil. Yet by 1971 the potential power to control oil prices had shifted to the OPEC countries. The raison d'être for breaking the pincer of more demand and less profit came with the onset of the Yom Kippur War in October 1973 and the support of

Israel by the West. OPEC cut production by 5 million barrels per day, resulting in a net loss of free world (i.e., non-Soviet bloc) production of 7 percent.

Clearly, the major alternative to fossil fuel energy for at least the remainder of the century was fissionable nuclear fuel. We had lots of coal, but it was a serious polluter. We had a reasonable supply of natural gas and domestic oil but not enough. We either had to stop increasing our energy consumption by using it more efficiently or using less of it or building alternative supplies such as nuclear power plants. In the FY 1973 $770 million budget for energy research and development, almost two-thirds was for nuclear.

Nuclear in principle seemed a logical way out. But efforts to build nuclear fission power plants were handicapped by a growing opposition alarmed by an uncertain safety record. The record was spotty, with some nuclear fission power plants very well run but others built, maintained, and operated poorly, handing opponents lots of ammunition. Another serious problem for nuclear power was our failure to come to terms with long-term nuclear waste management. The disposal of nuclear waste has haunted the nuclear energy program continuously and at no time has the country put enough effort into solving it.

Although coal was one of our most abundant sources of energy, the emerging environmental and mining health and safety constraints had greatly limited its use. Improving methods of developing and utilizing coal in harmony with environmental and social objectives became a national priority. So the program in coal was aimed at coal gasification to produce synthetic gas from coal, coal liquefaction to convert coal into liquid fuel for transportation uses, and technologies to remove sulfur compounds both before and after combustion. Research on oil shale conversion and demonstration programs carried out with companies were also part of the strategy. Then as now this potential resource awaits a breakthrough technological process that makes it cost competitive.

The National Science Foundation had been involved in energy issues for some time before I became director in 1972. In addition to sponsoring research and development on fossils fuel technologies, it pursued what have come to be called alternative energy technologies: solar, wind, geothermal, and hydropower. Taking parts of these existing programs in August 1973, I created the NSF Office of Energy Research and Development Policy, responsible for giving advice on energy research and development to the director in his role as science advisor to the president.[28] Using especially the work done through the RANN (Research Applied to National Needs), we constructed scenarios using a spectrum of assumptions on energy demand and supply. We also felt that it was important in our long-range energy research priorities to start dealing with U.S. energy supply and demand as a complete system, not least because changes in any one part of it quickly spread and ramified throughout the whole system. We needed to look at the total energy system to gauge the impacts of changes in energy production, use, and conservation. We needed to see how the system could be "gamed" to minimize the impact on health, safety, and environment. On the last item, we looked at the consequences of such key issues as the control of emissions from fossil fuel combustion—in stack gases, automobile exhaust, and many other sources; new coal mining techniques leading to improved health and safety conditions in the underground operations; intensified research on nuclear safety problems, including processing, transportation, storage, and disposal of radioactive wastes; and techniques for restoration of land destroyed by strip mining and the like.

We also wanted to begin quantifying the social cost of various energy strategies (e.g., the genetic and health costs of exposure to low-level radioactivity). And we wanted to develop techniques for assessing and defensibly comparing the risks in different energy strategies and options. That was a big menu of complicated matters. Yet it was very important for us to start looking at them in a systematic and holistic way. And that we did. That's not to say we solved them; we didn't. But we got a good start. These studies in time wove their way into U.S. energy policies and regulation. For example, the Federal Power Commission revised its usual ways of predicting energy demand by a more or less straight-line extrapolation to include conservation scenarios. More widely, the central idea of looking at energy in the United States as a complete system has endured and is now considered obvious. It wasn't obvious in 1973.

ALCOHOL AND NICOTINE

The year 1973 was also when environmental issues forced themselves into presidential policies. One of the first things I did as science advisor was to pressure OMB to release an important report on chemicals and health done by the late Office of Science and Technology. The White House worried that the data on the environmental costs of technologies described in the report might further wound an economy already stumbling. Still, we did get the report out. Reporters at the press conference were a bit startled, however, when John Tukey[29] of Princeton University, chair of the report panel, told them that "although there are many problems of health and environment due to the many, many chemicals that are being conceived at all times, the problem that society had with almost all of those were minor compared with the two great ones which they hadn't yet been able to harness, namely alcohol and nicotine." [30]

I was deep into energy issues in the summer of 1973, digging harder into being science advisor and running NSF, including getting its FY 1975 budget into shape. Arrangements for the U.S.-Israel Binational Science Foundation were progressing; we'd had our first meeting with the Soviets on science and technology exchanges; plans for the fall meeting were moving along; and the new energy office in NSF to help me in my role as science advisor was in place.

It was a hectic summer, but we still managed an extended stay at our summer home in Randolph, New Hampshire, although my travel continued to be very heavy. I gave many talks with many purposes, not least trying to assure the academic science community understandably distressed by a research budget that had declined significantly in real terms since 1967. In a talk at a place quite familiar to me from my previous life, the Lincoln Laboratory in Lexington, Massachusetts, I did acknowledge that "in general science faces tight times in the days ahead, but the picture is not as black as many of us make it. The basic reasons for the pressure are the administration's desire to work with a realistic 'full employment budget' without increasing the national debt or raising taxes; the need to allocate resources to meet first priority national programs based on urgent economic, social, and environmental demands; and a public attitude that is less sympathetic to the needs of science than to the need to solve our most pressing problems soon." I also tried to assure this audience of friendly skeptics that even though I had two full-time jobs, that the foundation was in good shape and that my role as science advisor was a real one, that it had both substance and impact in the White House.[31]

On Camera and Wet

NSF was indeed running well. The budget was in good shape, although we could always have used more. We innovated in, for example, creating a grants program for science and technology museums; reorganized the facilities programs to align them with their disciplinary "homes" within the foundation; and created a new directorate for biological, behavioral, and social sciences, appointing Eloise Clark as its first director. Our twenty-fifth anniversary arrived, and we celebrated in the usual ways with a convocation run by the National Academy of Sciences and in at least one exceptional way: by creating the Alan T. Waterman Award, after the foundation's first director, "to recognize an outstanding young researcher in any field of science or engineering supported by the National Science Foundation."[32] Not least, the NSF administration was reorganized, and in particular I reversed the previous introduction of professional administrators from other agencies such as NASA in favor of restoring control to true science administrators (i.e., who understood the science for which they were responsible). When Ray Bisplinghoff resigned as deputy director to take a university presidency, I chose Dick Atkinson, a "hard" social scientist from Stanford University with excellent results.

I think I have earned a reputation as a nice guy, always willing to help where I can. But that can get me into trouble. During the summer of 1973 I was in Aspen, Colorado, to give a speech, but this one had a special pleasure to it: I had been asked to go fly-fishing on the Roaring Fork of the Colorado River. However, just before I was to leave for a day of fishing, I was asked by a TV station in Denver for an interview on the energy crisis and on science and technology more generally. I said nothing doing. But being a nice guy I didn't stop there, and said that if they wanted to interview me on the banks of the trout stream, I'd do that. Sure enough, the next day after I had already caught six nice trout, a TV truck lumbered up a logging road toward me. There was nothing to do but do it. They put a radio pack on the back of my fishing vest, a microphone on the lapel, and told me to head into the middle of the stream to start fishing and answer questions. So I stood in the stream, caught fish, and answered questions about the new Prudhoe oil fields in Alaska, about solar energy, and especially about coal shale of which western Colorado had many deposits. They wound up with two reels of film that became a half-hour broadcast, titled "Stever—Man of Science." I never saw it but lots of friends did. It was one of my better payoffs for working away at energy problems.

NOT THERE

C. P. Snow in his *Strangers and Brothers* has a senior official offer several rules for survival in the upper stratum of the British civil service. One rule was "Be There."[33] In late 1973 the science advisor was of course not "there." Not in the White House, not in a fancy office in the Old Executive Office Building. But I felt in 1973, and certainly do in hindsight, that it was a good thing science was "not there" in the White House. I'm not sure that it would have made a difference; the budget was going to be squeezed hard wherever I sat, and I certainly had access to the budget people, including senior OMB officials. I was directly involved in key issues for the Nixon administration, notably détente through my negotiations with the Soviets on U.S.-Soviet exchanges on science and technology. But I was not involved in the worsening Watergate scandal. By the end of 1973, Nixon had told the country "I'm not a crook," and the White House blamed an 18-minute gap in an Oval Office tape on a "sinister force." I told friends that it was a good thing for the science advisor and science to be "out of the White House" and even once said so publicly but was not reprimanded, even though my remarks were published in a major newspaper. As Watergate moved from what seemed at time the dirty politics by a few overzealous junior operatives into the Oval Office, many of us came to realize that a tragedy was unfolding, one that the Nixon administration would not survive. Many people, including senior science people in and out of government, began to seek out Gerald Ford, whom Nixon selected as his vice president after Spiro Agnew resigned.[34]

I was also pressured to talk with the new vice president and did so, using as my entrée the publication of the new *Science Indicators,* which reported that our investments in science and technology were at best flat if not declining, in absolute terms and relative to rising investments in other countries such as Japan and Germany. He pointed out quite rightly that the governments of Japan and Germany were not saddled with large military budgets and could focus almost exclusively on civilian investments, including science and technology. He also asked me how things were going, an opening I used to argue quite strongly that the White House needed its own capacity for science advice and also something like the president's Science Advisory Committee. He agreed on the first point but not on restoration of PSAC. He was well aware of the political problems that led to its end.

On July 24, 1974, the Supreme Court ruled that Nixon had to turn over all the tapes of conversation in the Oval Office, denying his claims of executive privilege. Three days later, the House Judiciary Committee voted the first of three articles of impeachment, this on obstruction of justice. Nixon resigned August 8. Gerald Ford was sworn in as president the next day. Four days later the new president held a reception for all top presidential appointees. As I was going through the receiving line, he said to me: "Say, I'm delighted to talk with you, Dr. Stever." (He called me Dr. Stever not Guy until very late in his term.) He said he wanted to reestablish the White House science office but by a congressional act. That would mean in principle that the office couldn't be summarily dismissed as Nixon had done.[35] From that point on, our relationships with the White House thrived. Everyone in the government was now supportive, including the OMB, because they knew the president was interested in the well-being of science.

"A PRESIDENT WHO IS LISTENING"

Work on creating a new science office in the White House began almost immediately in the Congress. For me it was "sausage making" at its most graphic. (And it wasn't until almost two years later—May 11, 1976—that President Ford signed into law a new Office of Science and Technology Policy within the Executive Office of the President.) Sausage making started in the Senate when Senator Edward F. Kennedy introduced a bill to establish a Council on Science and Technology in the White House.

That got strong support from science society presidents who thought, as I wrote in a memo to senior Ford officials, "a White House science mechanism must be used by the president partly as a foil to excessive military domination." I went on that "the continued absence of comment on the subject by the Administration is creating an impression that it is not interested" and, showing my frustration, vented my strong feelings that "preemptive action by the Congress would create still another mandated organization which may or may not have functional effectiveness." And I added for good measure that, while I would continue to be the good and loyal soldier, "I am reluctant to be a public advocate, for it places me in the position of being a constant advocate for the status quo which can be viewed as completely self serving, not unlike the old Office of Science and Technology."[36]

I wasn't alone. James Reston of *The New York Times* wrote that, while the government had at its disposal a vast amount of scientific knowledge and expertise on matters then quite important, such as energy problems, increasing world food production, and the like, that information was quite dispersed. "It is not brought together, with all its potentialities for the future, and put before the president as a vision of the possible and the basis of his policies, which is too bad, because we have a President who is listening."[37] And journalists were perceptive in understanding that the role for a science advisor had changed radically from the glory days of Killian and Kistiakowsky. "The task of advising a President on science policy has shifted from a relative simple one of supporting or opposing proposals to develop and deploy defense and space hardware to a more complicated job of initiating and monitoring R&D programs in energy, transportation, and other domestic problem areas."[38] The role of science and technology had since Eisenhower and his science advisors moved from responding to the Soviet threat—whether in missiles or space satellites—to issues much more central to American life. Jim Killian himself in describing a report he chaired on restoration of science advice to the White House pointed out that "a new science advisory mechanism would have to deal with a range of problems quite different from those which had confronted Eisenhower's science advisor [i.e., Killian], and that its foundations for a new structure should not nostalgically seek to repeat the 'good old days' of the Eisenhower and Kennedy administrations when space and national security were so high on the presidential agenda."[39] And Vice President Rockefeller[40] in a "Memorandum for the president" quite correctly pointed out that "the dissolution of science advisory apparatus in the White House in 1973 was greeted with great dismay by the scientific community. Pressure is growing steadily from the scientific community leaders for action to restore some science presence to the White House." The same memo was blunt in its assessment of what had gone wrong. "The failure of the Office of Science and Technology staff to relate to the White House policy formulating procedure made it difficult to integrate that Office's recommendations with those of other advisory functions in the White House. . . . As the Office of Science and Technology allegiance to its constituency grew, its effectiveness in serving the president diminished."[41]

Yet public recognition of the importance to the nation of science and technology was mirrored by rising criticisms, that while science and technology were public goods, they could also be misused. However unfair, many no longer thought of science as an unalloyed good. The heady days of *Apollo* were over. "We are," I told a journalist in 1975, "going to be involved as never before in the economic success or failure of this country, and the rest of the world, and we are going to be taking the praise and the blame for far more than we ever bargained for, not only economically, but ethically and socially." The same journalist also captured the spirit of the scientific community: "Many U.S. scientists feel bewildered. They know American science is at the peak of its powers. Yet they see their research support tightening, their job market dwindling. They sense public hostility to what they are doing."[42]

President Ford on June 9, 1975, forwarded "proposed legislation to create in the Executive Office of the president an Office of Science and Technology Policy headed by a Director who will also serve as my Science and Technology Adviser." Many things intervened between the president's wishes to create a new OSTP in August 1974 and its formalization two years later. At the root were fundamental political and administrative disagreements over control and power. The political part was the reluctance of the congressional branch, principally in the Senate and even more particularly by well-placed staffers, to establish a strong science and technology office in the White House, favoring a weaker structure controllable by the legislature. The administrative barriers were principally within parts of the president's domestic council, many of whom were holdovers from the Nixon domestic council staff. They were reluctant to in effect cede some of the council's control to a new office replacing the office they had helped disestablish.

The domestic policy staff was the first hurdle. Valuable weeks passed between President Ford's signaling his intent to have the science office reestablished legislatively and sending the Hill a proposal for such an office. There was internal debate about the structure and functions of the office, the size of its staff, and its budget. Finally there appeared to be a consensus on a decision memo for the president, but it did not go forward over several additional weeks. Vice President Rockefeller, as frustrated as I was by the seemingly endless discussion, took the decision memo on the science office into the Oval Office. When it came out signed, staff could only fall in line.[43]

The relevant House committees were generally supportive. But, as I mentioned, the Senate side was another question. The Senate approved on October 11, 1974, Senator Kennedy's bill to establish a three-person Council for Science and Technology as part of the Executive Office. The council would, among other things, annually appraise science and technology in relation to national needs, do policy studies, and through the National Academy of Sciences examine the federal organization for civilian science and technology. In contrast, the House Committee on Science and Astronautics, chaired by Olin "Tiger" Teague, a Democrat from Texas, basically accepted the president's proposal for a full-fledged Office of Science and Technology Policy led by a single director but added several mandates: confirmation of the director by the Senate, a declaration of national policy for science, and a federal science and technology survey committee to examine within two years of enactment a survey of the totality of the federal science and technology effort—missions, goals, funding, etc. This latter provision was a compromise within the committee itself. A subset of the House Committee on Science and Astronautics led by Rep. Mike McCormack (D-Wash.) championed the establishment of a department of science and technology, which of course would have been a nonstarter with the administration as well as most in the scientific community. The survey was a way to get that group on board with the House bill.

The administration wasn't totally happy with the House version—

we didn't, for example, need a national declaration on science policy—
but it was even less happy with what the Senate wanted to do. The gap
widened in late November when, after the House had approved its version, Senator Kennedy surfaced a new bill empowering the OSTP director to determine funding levels and priorities for federal science and
technology programs, with the president obligated to give his reasons for
not accepting these recommendations. I and all the senior officials in the
administration objected to the bill, especially to OSTP's preempting the
president by setting funding levels. Vice President Rockefeller asked Senator Kennedy to accept the "House bill intact, without further alterations
in the Senate." No dice. Senator Kennedy and his colleagues shot back
with: "that there are a number of areas in which the House bill should be
strengthened, and that it is in the national interest that we attempt to
improve the legislation in the Senate."[44] Happily for the sake of compromise, the Senate Committee on Space, chaired by Senator Frank Moss, a
Utah Democrat, weighed in with views that paralleled the House's version of the legislation and the administration's position and prevailed on
Senator Kennedy to modify his views.

New legislation closer to the administration's liking passed the Senate on February 4, 1976, and, with some promoting from the president
to get it done, the House and Senate conferees reconciled their differences and agreed on a final bill. On May 11, 1976, the president signed
into law the bill to create the Office of Science and Technology Policy
and quoted Thomas Jefferson that "knowledge is power; knowledge is
safety; knowledge is happiness." Just so.

"EVERY DIRECTOR HAS TO HAVE ONE SCANDAL"

I was somewhere between spectator and player in the serpentine route to
restoring science to the White House. There was of course the National
Science Foundation to run, not to mention continuing being science
advisor, now for a new president. The NSF side was going well, and I
think we were handling ourselves quite creditably in various special tasks,
such as providing an analytical base for the nation's energy programs,
continuing to work with the Soviets on science and technology exchanges,
maintaining strong contact with the science community, and represent-

ing the administration's role in science to the Congress, to the scientific community, and to my counterparts in many countries. And I paid another visit to Antarctica to dedicate a new station at the Amundsen-Scott base at the South Pole. One downside was the departure of my deputy director, Ray Bisplinghoff. We had been colleagues at MIT and shared the NSF leadership throughout my years there. However, the pain of Ray's loss was salved quite a bit by the arrival of Richard Atkinson as deputy director. Dick was a social scientist at Stanford University and more specifically a "hard" social scientist who used mathematics and computation heavily in his research on memory and cognition. He also applied his fundamental work to develop, for example, one of the first computer-controlled methods for instruction, which subsequently went commercial. Dick came to the NSF in 1975, and after a stint as acting director, as of October 1976 when I moved full-time to the White House, was appointed NSF director by President Carter.[45]

Turning to a very able social scientist as deputy pirector turned out to have some unexpected pluses, as I was hit full-bore by the mixed views, if not downright hostility, that some people—especially in the Congress—held of the social sciences. That had a history, of course. Vannevar Bush was dismissive, and didn't leave room for them in his postwar plans for "The Endless Frontier." The original enabling legislation for NSF didn't mention the social sciences, only referring vaguely to "other sciences" besides chemistry, mathematics, etc.[46] That they grew within the foundation was due largely to my predecessor, William McElroy, who introduced the perspectives of the social sciences as the agency turned more of its resources toward the applied sciences, especially in its RANN program. It was also due to the considerable effort in the 1960s by the chair of the House Science and Aeronautics Committee, Emilio Q. Daddario (D-Conn.). I was part of an advisory group to the committee and strongly supported Representative Daddario, which earned me barbs from my fellow advisors, who said, in effect, that "the social sciences aren't really sciences at all, and, worse, they'll take money away from the physical and biological sciences which are really important."

Ironically the Vietnam War probably helped the social sciences. "The reaction within the universities against the Vietnam War has been accompanied by an increase of interest—especially among younger scientists—in dedicating their research to social policies that they consider liberal and humane. . . . The main result of this has been to give new emphasis and status to the social sciences."[47] But the gain was two-edged, for the implicit linkage of social action and concern with the social sciences also aroused political hostilities. It aroused suspicion that the social sciences were being used to drive a political agenda; that value judgments and beliefs were being promoted under the cover of science; that federal funds in the form of grants and contracts were paying for an attack on fundamental values. The social sciences were attacked as drivers for social change, using school "reform" as a stalking horse to turn children against the beliefs and values of their parents. And these weren't simply teacup disputes. There had been violence in some of the more bitter disputes, as in West Virginia where parents verbally attacked the "godless" and "dirty books" their children were studying in the schools and physically attacked local defenders of curricula using these books.[48] When those beliefs were held by the political powerful, the upshot was trouble. A very direct example was that the splendid budget the foundation got for FY 1975— $768 million, $100 million more than the previous budget—didn't apply to the social sciences, whose budget was held flat. In somewhat the same vein, the budget for science education was cut.[49]

Doris McCarn was my secretary at the foundation, having done the same job for every NSF director.[50] She was a very wise woman, never more than when she observed that "every Director has to have one scandal in his directorship." My scandal arrived abruptly in 1975. We had been hearing some criticism from the press and some in the science communities about a foundation science education curriculum study program for fifth and sixth graders called MACOS—Man: A Course of Study.

Some background. Science education at NSF was embedded in its organic act of 1950, which empowered the new agency "to initiate and support basic scientific research programs to strengthen scientific research potential and science education at all levels." True to that, through much of the foundation's history, research has accounted for about half of its budget and science education about a quarter.[51] By 1975 the science education programs at NSF had evolved considerably. Originally, the programs focused on undergraduate and graduate education. With *Sput-*

nik the program reached downward, into science education at precollege levels. And with the growth of the foundation's budget, Congress pushed science education for "institution building": helping smaller colleges and universities strengthen their undergraduate science programs and bootstrapping science capability in second-tier graduate schools. Tensions between the administration and the Congress built up toward the end of the 1960s. The administration, arguing that scientific and engineering manpower was sufficient for national needs, terminated institution building in 1972 and cut funding for direct support of graduate and undergraduate education, believing that the support could more effectively come if indirectly through research grants and contracts. Congress fought those changes, and the administration's 1972 impoundment of science education funds didn't help. Strong differences on strategies intensified the tensions. For example, should the science education programs at the foundation be experimental, fostering new ideas until they either wither or become self-sustaining? Or should they simply be a means for moving public funds into education programs, using standard grant practice? To what extent should the foundation—indeed, the federal government—market the educational materials developed with its support? Where is the line between a federal role in science education and an intrusion into values and social beliefs?[52]

This strong stew boiled over with MACOS. The foundation had since the late 1950s as part of its mandate supported some 53 precollege curricula projects spanning almost all the sciences within its purview.[53] And it had done so without much controversy, aside from some criticism of evolution theory taught as part of a foundation-supported program, the Biological Sciences Curriculum Study. MACOS was different.[54] Developed between 1963 and 1970, MACOS was a one-year course intended to help children "think about the nature of human beings and what is unique about being human."[55] To probe these questions, the course included the life cycle of the salmon, the habits of herring gulls, the group behavior of baboons, and the daily life of the Netsilik Eskimos. It was the last part that got us trouble. The Netsilik live in the Pelly Bay region of the Canadian Arctic, with the nearest town being Taloyoak, the most northernly settlement on the Canadian mainland. The program showed traditional Eskimo life before the European acculturation. The

Netsilik Eskimos had long lived apart from other people and had depended entirely on the land and their own ingenuity to sustain life
through the rigors of the Arctic year. Examining the Netsilik offered insight to the unique qualities shared by all human beings no matter how
harsh their lives, and the lives of the Netsilik are indeed harsh.

MACOS had been extremely successful, adopted since its introduction in 1970 by over 1,700 schools. The feedback had been by and large
very positive. Moreover, schools and their teachers had a choice of which
modules in the series to show. As I looked into the course, I found many
grade school teachers who thought it was a great course. And I also learned
that teachers had wide latitude to decide what they showed their kids and
what they didn't.

When the rumblings about MACOS started, I asked for a briefing,
which turned out to be poor because it made MACOS content seem
palatable. I then learned that one feature of family life it showed was that
anyone who was too old and infirm for the vigorous life of these people
was put on an ice floe and sent out to sea. That was not very pleasant,
especially teaching it to fifth and sixth graders. But there was worse. For
example, if the wife of a hunter was ill and couldn't go on a hunt with her
husband, he borrowed someone else's wife. And, finally, the films showed
vivid hunting scenes in which the hunters killed seals as they came up to
their breathing holes in the ice. Butchering the seal on white ice was a
pretty colorful scene for young people in grade school, never mind other
scenes such as popping the eyeballs of the seal out to eat as a delicacy,
eating the still warm liver on the ice, and drinking the blood. Then there
were scenes of children stoning a snared gull to death.

I was told that in some classes where this was shown boys and girls
fainted. I have no doubt that the film and all that went with it were
faithful to the reality of life of the Netsilik and as such a major contribution to showing with care the life of a culture quite different from ours.
At the same time, while the science was right, the politics was not.
MACOS was bound to cause trouble at a time of rising suspicion of the
social sciences. There were unbelievable misjudgments on what should
be given to grade school students.

"Civil Servants Lie All the Time"

Things got worse, much worse, when an internal investigation by NSF turned up serious irregularities in approving the grant for MACOS. And even more lethally the irregularities subverted the foundation's peer review system. It struck at the heart of how NSF gained and kept high standards in its funding decisions. Unlike the National Institutes of Health, NSF gave its programs officers significant latitude in selecting reviewers, vetting and summarizing their comments, and making a final judgment. The MACOS grant—in this case, an extension of an earlier grant—had gone through peer review and also, because of the size of the grant, required approval by the National Science Board. Eleven reviewers had vetted the grant proposal and, according to the program officer, all 11 supported it. Not quite: 8 reviewers supported it, 1 was neutral, and 2 were negative. I told the program officer that it seemed she had lied. And she said, in effect, "oh, yes, because my supervisors told me that we had to have a perfect case or the grant would not be approved." I was shocked but even more so when we put her up for disciplinary action and a Civil Service Commission officer in effect dismissed the charge of lying with "civil servants lie all the time" and that the real and only question was whether she made any money by lying. I thought there was a higher standard for civil service.

It got worse. All this was leaked to certain members of Congress. At first I was puzzled how the Congress learned that one of our own program officers had violated our peer review principles. But then I found out. One day as I was leaving the foundation building at 1800 G Street, a car belonging to one of our leading congressional critics, Representative John B. Conlan (R-Ariz.), stopped at the entrance and out popped one of our senior staff members. A few days later, as I was walking early in the morning down the corridor to my office, the same fellow jumped out from a nearby office. He was excited and a bit incoherent: "Dr. Stever, do you know what they're doing out there? All the things they're doing?" I could only respond that we had put together a good internal review committee[56] and hoped to learn more. This fellow, I found, was very conservative, and I imagine his friends and family had put great pressure on him to do something about the "awful things" the foundation was doing. I never took any action against him. To have done so would have only increased the problems in the NSF and the Congress.

Next up was testimony on NSF's FY 1976 budget, which meant private visits with key people in the Congress. This included Tiger Teague, chair of the House Science and Technology Committee; the ranking member, Charles Mosher (R-Ohio); and James W. Symington, Jr. (D-Mo.), who chaired the subcommittee dealing directly with NSF. Those private conversations made it clear to me that they were very concerned about what they were hearing on MACOS and the larger issue of the foundation's peer review system.

I was flabbergasted when I sat down to testify before Representative Teague's Committee on Science and Technology to present our new annual budget proposal. Every member's seat was filled. Most of the time the seats are empty, with maybe one or two members in attendance, often on a revolving basis, as one member makes a cameo appearance and another departs for yet another hearing. I knew of course why the members were there and why the audience was packed. I couldn't avoid what they wanted to talk about, nor did I want to.

So I began by simply saying that before I started my formal testimony on the foundation's budget for the next fiscal year "I would like to take a few minutes to discuss a very serious problem." I then briefly discussed our recent problems with MACOS and the questions on peer review and that I had put together an internal committee to look at what happened and what lessons and correctives were to be drawn from it.

I could almost hear an audible sigh of relief. I wasn't ducking, and the issue was on the table. I got lots of questions, some detailed—on peer review, on MACOS, on the role of the social sciences in the foundation. I did not comment on the egregious behavior of the program officer who handled the MACOS grant but did discuss pretty much everything else during the question period.

A rather strange thing happened a few days later when I testified before the Subcommittee on NSF Appropriations, chaired by Edward P. Boland (D-Mass.).[57] A congressman, not a member of the committee, after requesting permission to quiz me asked, "Any of your staff at the National Science Foundation told you [of] things going on in the Education Directorate that weren't right?" I instantly remembered the fellow who had caught me alone in the hall with charges of nasty doings in the Education Directorate. So my answer to the Congressman's trap was: "Of course. I recall meeting in the hall a staff member who was very seriously concerned, and I told him that I was going to tell the Committee I had appointed to look into this." I went on that one of our senior secretaries at the foundation had told me her daughter had seen MACOS films in school, and that she couldn't stand them. I also said that we have about 1,000 staff members at the foundation, that they are hardly lock step in their beliefs, and that they certainly don't universally agree with the decisions made by program officers, the National Science Board, or by me. But, I added, we do our best to listen to all points of view before a decision is made. Chairman Boland then interrupted: "I think we've had enough along this line, we've had frank and very open answers, and that we had better turn back to the other material"—that is, our appropriations for the next fiscal year.

That was welcome, but the Congress, or some of its members, weren't finished with me. Two congressmen, John Conlan (R-Ariz.) and Robert Bauman (R-Md.), led the attacks. It came to a head when the foundation's FY 1976 appropriation of $755 million came before the entire House. In the debate, Representative Conlan, a very articulate gentleman, described MACOS as "a course for 10-year olds mainly about the Netsilik Eskimo subculture of Canada Pelly Bay Region. Student materials have repeated references to stories about Netsilik cannibalism, adultery, bestiality, female infanticide, incest, wife-swapping, killing old people, and other shocking condoned practices. . . . It is absolutely unacceptable for NSF to continue using taxpayers' money for aggressive promotion and marketing activities for their own preferred social studies courses, undercutting competition from regular textbook publishing houses."[58] The House debate on MACOS was acrimonious, the upshot of which was that the House rejected, 215 to 196, a Conlan amendment to provide for congressional review of completed NSF curriculum projects but approved a Bauman amendment that the Congress would review all NSF research proposals! Thankfully, the amendment didn't survive House-Senate conference, but we were hardly out of the woods. Much of the spring and summer of 1975 dealt with the foundation's peer review system. There were multiple inquiries—by the congressional General Accounting Office, through hearings before the Symington Committee; by a "science curriculum implementation review group," appointed by Representative Teague; and, of course by our own internal review committee. The upshot of all this was to tighten our procedures—for selecting curriculum developers and peer reviewers, for monitoring reporting of peer review results, and the like. While at times quite painful, I believe the episode strengthened the foundation. Another silver lining was that we learned that when the chips were down the foundation, a small agency in the world of political Washington that had quite firmly avoided lobbying its own interests, had powerful friends in the Congress. Congressmen Teague, Mosher, and Symington, Senator Kennedy, and others gave us a vivid demonstration of "tough love"—arranging for unsparing examinations of the foundation's procedures but when mostly positive verdicts came in strongly supporting us against attacks by their colleagues.[59]

Those positive verdicts were reflected in substantial increases proposed for the foundation's FY 1977 budget, and I was especially pleased that a new directorate for biology and the behavioral sciences that I had created got strong budget support from the Congress. The NSF wound up with a $792 million budget, $180 million more than its FY 1973 budget, the first one I defended before the Congress.[60]

In the White House Rose Garden on a lovely May day, President Ford signed into law the the National Science and Technology Policy, Organization, and Priorities Act of 1976. Science was back in the White House. The new office needed a new director who would also be the president's science advisor; appointment of the OSTP director required senatorial approval while that of science advisor did not; the director was subject to congressional oversight; the adviser was not and could in principle claim executive privilege. The peculiarity in all this was that one person had both jobs.

Who would that person be? I was not anxious for the position, having enough to do and having already determined that, with the end of President's Ford term, I would leave government, no matter who won the election. That meant that my tenure if I were to become OSTP director and science advisor would be short indeed. And I had been preparing myself for a bit easier load until the end of this presidential term. No deal. A few days later after the signing, Vice President Rockefeller asked me to come see him, to discuss possible candidates. A little while later a White House personnel officer called to ask if I would be interested.

I was hardly a safe choice. President Ford was the obvious Republican candidate for the 1976 presidential election.[61] But the extreme right wing of the Republican Party not only didn't care for President Ford, it didn't care for the National Science Foundation or for me. It got pretty vicious and, if one had a sense of the macabre, sometimes funny. One day, Pat Nicely, of the foundation's congressional liaison office, came in wearing a big grin and carrying a newsletter. It turned out to be a right-wing publication and on the front page was my picture, along with that of two senators, Jacob Javits, a Republican liberal from New York, and Hubert Humphrey, Democrat from Minnesota. The headline explained that "these three men are one of the greatest threats to America." I did then (and still do) consider it a great honor to be included with these two fine people and distinguished public servants.

Still, the vicious attacks were hard. And I faced for much of 1975 and 1976 attacks on me and much testifying on our social programs, on our peer review systems, on MACOS, and our science education programs. It was beginning to get me down rather badly. It was easy for me to decide that I was really going to leave government at the 1976. I even looked at the family exchequer, realizing that we had in fact seriously dropped off in saving for the education of our children because the pay wasn't very good at that time in government; it was a lot worse than it is today. I wanted to get back and do other things, get away from government service.

An attack on me by four Republican senators[62] didn't help my mood, but the response did. In a letter hand delivered to President Ford on June 9, 1976, they wrote that:

> We are most concerned about reports that H. Guyford Stever, Director of the National Science Foundation, may be appointed to the newly-established position of science advisor to the president.
>
> The General Accounting Office recently reported to the Congress that NSF officials have seriously manipulated and abused the NSF grant award process in connection with a multi-million dollar curriculum project long supported by the foundation. Prior to the GAO report, Dr. Stever and other top NSF officials had repeatedly denied before Committees of Congress that these abuses had occurred. Now, with evidence that top NSF officials did

know about the wrongdoing when they denied it to Congress, the GAO is again down at the foundation investigating official cover-up at NSF.

It would be inadvisable, and in our judgment an affront to the Congress, for Dr. Stever to be appointed to another high position before this bad NSF situation has been completely investigated, and the full extent of official involvement is known. Such an appointment would bring great controversy and inevitable opposition to Dr. Stever's appointment by the Senate

Moreover, both Rep. James Symington and Sen. Edward Kennedy, NSF Subcommittee chairman respectively in the house and senate, failed to get to the bottom of this NSF matter, despite repeated insistence by Republican members that they do so, or to act firmly against wrongdoing in the awarding of Federal grants by this agency under their direct jurisdiction. Your appointment of Dr. Stever as the president's science advisor will make it most difficult for Republicans to call these Democrats politically to account for their error in judgment and lack of initiative in this important matter.

Whew. Quite a letter, and in its attacks on Congressman Symington and Senator Kennedy quite outside how members of Congress treat each other. The reaction was swift and blunt. Representative Teague in a letter to the senators called their comments "inaccurate," "an affront," and "untrue." Fellow Republican Charles Mosher, ranking member of the House Science and Technology Committee, regretted that the four senators had "accepted very inadequate, selective, and distorted information as the basis for the judgments you expressed." And he added that "I hate to see the president and Dr. Stever publicly harassed by allegations that I am convinced are blown far out of proportion to the realities of the situation." Senator Kennedy also weighed in with language as blunt as theirs, citing "unsubstantiated and unfounded allegations" in their letter, "which can only be viewed as an irresponsible attempt to undermine the bipartisan effort to restore this urgently needed function to the White House." And he twisted the knife a bit with: "Also welcome would be your interest, in a legislative forum, in the six-tenths of one percent of the NSF budget devoted to curriculum programs [i.e., MACOS and the like] and in the remaining 99.4% of the NSF budget." But maybe the strongest rebuttal was the letter from Elmer Staats, the comptroller general, of the General Accounting Office, stating that the letter was simply wrong in how it characterized the GAO's findings.[63]

The upshot of this senatorial melee, which of course found its way into the press, was that I was confirmed, not unanimously as I had been for the NSF job, but by a very lopsided margin: 78 to 6, with 16 abstentions. But it came after some nine weeks of delay. Given the controversy, I had suggested that my appointment be reconsidered, and the president in effect put it on hold a while until it was clear that he had enough delegates to win the Republican nomination against Ronald Reagan.

Vice President Rockefeller swore me in on August 12, 1976, after a generous speech by the president, affirming that science was back in the White House. And despite the fact that by my own choice I would have a very short tenure—leaving office no later than January 20, 1977—I set myself an ambitious agenda for this short time.

Even before the act of 1976, in late 1974 and early 1975, Vice President Rockefeller and I determined that we needed to make a study of the kinds of issues such a new office as was being bandied about in the discussions of the new legislation should do. So President Ford, in early 1975, issued a directive that we should set up something to look at how the office might conduct its affairs if it were established. This led immediately to the proposal for two committees, which were established, one led by Bill Baker of the Bell Telephone Laboratories and the other by Simon Ramo of TRW. One of these committees looked at the important future scientific and technological advances broadly and the other at relating scientific and technological advances to the health and strength of the economy. These committees were quite active in 1975 and into 1976. An early report was ready for me when I assumed full-time responsibility for the White House office.

Then the new President's Commission on Science and Technology was chosen, with Si Ramo as chairman and Bill Baker as vice chairman. We went to work on a broad range of issues, including scientific and technical information and the systems for handling it government-wide; improved technology assessment in the executive branch of the federal government; improved methods for effective technology innovation, transfer, and use; stimulating more effective federal-state and federal-industry liaison and cooperation in science and technology, including the formation of federal-state mechanisms for the mutual pursuit of this goal; reduction and simplification of federal regulations and administrative practices and procedures; strengthening the nation's academic institutions' capabilities for research and education in science and technology; maintenance of adequate scientific and technological manpower in quality and quantity; technology for local and individual needs to consider the technological needs of communities and individuals; and long-range planning for future national problems.[64] The committee had a short tenure because all of its recommendations were accepted in March 1977, less than a fourth of the time it was supposed to be in business. A short but effective life.

Additional items that I worked on included new directions for energy supply and demand, policy for exploiting the oceans, and means for safe disposal of radioactive wastes. Also in the energy field I became a member of an economic council to help the Saudi Arabian Kingdom in its economic development. This, together with a military council to help the Saudis develop their military strengths, was devised to get the most powerful member of OPEC on our side. I met several top Saudi Arabian civilian and military leaders when Prince Faud led their teams to Washington. I elected to stay away from national security matters, in part because I thought they were in capable hands, in part because I thought that the science advisor should deal with issues not clearly "owned" by one agency, and in part because I didn't want to spend my few months in office on bureaucratic struggles with 800-pound gorillas such as the Department of Defense.

And so it came to pass that I left the National Science Foundation in October 1976 with a tear in my eye. I had enjoyed my time there very much. We had many battles, losing some, winning many more, substantially raising our budget, and becoming a stronger agency. I was fortunate in having two very good people from the foundation join me, Philip Smith and Russell Drew. And I was also able to persuade two of the nation's leading scientists to join me: Donald Kennedy[65] of Stanford University to deal with biological issues and William Nierenberg, director of the Scripps Institution of Oceanography, for the physical sciences, especially geology, which was very prominent in those days.

Election Day finally came, and Jimmy Carter narrowly defeated President Ford.[66] The day after the election at the senior staff meeting we were instructed to cooperate fully with the incoming administration, including summarizing all we were doing so the new people could pick up what they wanted without too much trouble.

My office dutifully did as it was told. The first visitor from the Carter transition team was a young woman, who breezed in and announced that she was in charge of transition for, among other things, the Office of Science and Technology Policy. No talk about anything substantive. Next was a young election worker from Alabama or Georgia, who announced he was in charge of OSTP transition and could I tell him what OSTP was and did. I soon realized that he was illiterate in science and technology. He did ask me for the name of a Democrat who would know all about this, and, after catching myself, I suggested that he get in touch with Jerome Wiesner at the Massachusetts Institute of Technology, who of course had been President Kennedy's science advisor. He then asked me who Wiesner was and where. I knew then something had to be done and said so to the Carter transition team. Fortunately, a few weeks later a perfectly capable fellow arrived, and things went much more smoothly.

We had a very nice Christmas party, a white tie ball for the White House staff. Then at the end, early on January 20, 1977, we went to the White House for a breakfast reception with President Ford, Vice President Rockefeller, and other leaders of the administration. It was a moving occasion. President Ford had to leave to dress for the swearing-in of his successor. I walked back to my office with Alan Greenspan, the president's economic advisor, who observed, "I've lost some in my life and won some. It's more fun winning." I sat in my empty office until shortly before noon, walked out of the office, handed in my badge, and left the building. It was over.

9

End and Start

1977–1989. A time of national introspection on U.S. competitive prowess, on why academic research mattered to the country, and for me on how to support my family on approaching 65.

My time in government was indeed over. But I was only 60, Bunny and I were healthy, and I easily imagined at least another 10 active years. It was again time to redirect our lives. As an undergraduate at Colgate University, I was determined to be a university professor. Yet given the choice after I earned my physics doctorate at the California Institute of Technology of joining the Stanford University physics department or the Radiation Laboratory at the Massachusetts Institute of Technology, the locus of American work on microwave radar, I chose the Rad Lab. And I didn't hesitate for a minute when offered a chance to go to wartime London to strengthen the links with British radar work. Coming back to the United States in 1945, Vannevar Bush wanted me to stay in Washington as a special assistant. I turned that down and returned to MIT determined to become a professor. I did, establishing a research career in aeronautics and astronautics. I hesitated a bit when called to the presidency of what was then the Carnegie Institute of Technology. That turned out to be a terrific challenge if not without its difficulties, but creating a new institution, Carnegie Mellon University, was tremendously rewarding. Then four years of government service, a rather dizzying experience, leading the National Science Foundation and serving as science advisor to two presidents.

By 1977, when I left the White House, Bunny and I were pleased with our four most important investments—our children. Educating our children had almost been completed and was beginning to pay off. Guy, Jr., had gotten his B.A. from Colgate and an M.A. from Middlebury College in Vermont, and was now teaching English in Berlin, New Hampshire, and living in Randolph. Sarah had a B.A. from Sarah Lawrence and a Ph.D. from the University of Michigan and was on the faculty of the University of Detroit. Margo received her B.A. from the University of Pennsylvania and her M.A. from Indiana University in Bloomington, had just started teaching at the Massachusetts College of Art, and was en route to becoming an independent artist in Oakton, Virginia. Roy had his B.A. from Cornell University and was preparing for graduate work in natural resources management, which he did later with M.A.'s from the University of New Hampshire and McGill University. He joined Alcan, eventually being assigned to its headquarters in Chagrin Falls, Ohio.

DEPLETED COFFERS

Nationally and internationally, it wasn't an auspicious time. The U.S. economy was not in good shape. President Ford inherited a very high inflation rate that he tried to deal with by slowing down the economy. The result was a severe recession in 1974–1975, which indeed trimmed inflation some but at the cost of a double-whammy labeled a stagflation: the economy slowing, unemployment rising, prices going up. Energy supply and demand, together an issue that first bit the nation with the 1973 energy crisis, were a continuing problem. The terror of the two superpowers bristling with nuclear weapons sitting on long-range missiles was very much with us—in effect, a "Mexican standoff"—two guys pointing guns at each other, except the "guns" could annihilate most of humankind and change the planet's climate in a way that no one could predict. Federal support of research was an issue, not only funding but also purpose. It was a time of rampant self-analyses, of a torrent of studies of what was wrong with our economic engine, of predictions of America's economic decline, of "hollowed-out corporations," and of the seemingly relentless and unstoppable power of "Japan, Inc." The MIT Commission on Industrial Productivity asserted that "American Industry is not producing as well as it ought to produce, or as well as it used to produce, or as well as the industries of some other nations have learned to produce. . . . If the trend cannot be reversed, then sooner or later the American standard of living must pay the price."[1] Academic research was hit by these doomsday predictions: "While acknowledging U.S. strengths in academic research, the education of scientists and engineers, technological development, and venture capital financing of technology-based start-up firms, studies observed serious weaknesses in the capacity of American corporations, compared with their Japanese competitors, to turn these first-class assets into advanced processes and commercially successful products."[2] In other words, what was the country getting for all this wonderful science? Was it just, as Henry James put it, "private fun at public expense"? The Japanese, of course, had few of these assets, but they were superb at production. Nonetheless, there was "growing dissension about whether the large-scale U.S. support for basic research, primarily at universities, was worth the cost. Increasingly, suggestions were made that university research support ought to be more closely targeted on areas and activities that are deemed likely to feed directly into technological innovation."[3] What was it getting for the some $80 billion dollars of taxpayer money going into research and development?[4]

These questions—energy problems, widespread doubt about the nation's competitiveness, a crisis of confidence in the utility of research— did much to shape my life after I walked out of the Old Executive Office Building at noon on the day of Jimmy Carter's inauguration. While I shared these national worries about the status of American industry and the effectiveness of research universities as a source of innovation, I also had more prosaic matters to think about, like earning money. At MIT and then at Carnegie Mellon, I had added to my fairly healthy salaries with outside consulting that increased my financial intake by 50 percent and at times more than a 100 percent. These funds were needed, for the education of our four children and for our second home in Randolph, New Hampshire, indispensable to renew our spirit and as a retreat when the world sometimes got too clamorous.

The money situation changed hard in 1972 when I went from Carnegie Mellon to the directorship of the National Science Foundation and then to the White House as science advisor. There I was on a government salary, lower than what I'd earned as a professor and university

president and cripplingly lower than my total earnings, since con-
sultancies and directorships were ethically and legally not possible. I spent
lots of time while in Washington trying to figure out how to make ends
meet, and we had to cut significantly into our savings there. So it was
imperative that I had to produce money when I left the government. I
knew I wasn't going to go back into academia. I had offers, but it was
time for a new beginning.

Over and above earning money, I also wanted to do pro bono work
to begin to pay back a bit for what I had been given. That was the fun
part. But pro bono work would not replenish our depleted coffers. We
wanted to continue to help our children through college and for Bunny
and I to continue to have a joyous life together. And we certainly didn't
want to become wards of our children; we loved them very much but
also knew it would be a lot better if we could go it alone on the financial
side. I decided that the best course would be to work with some corporations, either as a consultant or board member. It would be a way to make
contributions based on my accumulating years of experience with various research and technology-based institutions, and at the same time it
would renew my technical contacts. It also was a way to save and invest
for the future. In the late 1970s and throughout much of the 1980s, I
worked with five major corporations, either as a consultant or a member
of their boards. I believed that in at least a small way I could help these
companies with their problems and, of course, gain needed compensa-
tion. I also took on pro bono assignments described in the next chapter.

My industrial board and consulting work brought me into direct
contact with the economic and industrial realities of the 1980s. With
TRW, I saw at close hand a company that prided itself on always being
at the forefront of new technologies and was constantly remaking itself to
stay competitive. TRW had to thread its way through the gauntlet of
energy crises, tectonic changes in its historic missile and space businesses,
and the competitive problems of the American automobile industry. At
Bethlehem Steel there were the problems of a classic, old-line industry as
it faced fierce competition from overseas steel makers whose minimills
were producing specialty steel "on demand." TRW and Bethlehem were
windows into the radical differences in dealing with competition in electronics versus steel. At Schering-Plough, I had a still different window

into the global nature of competition in the pharmaceutical industry but with broadly applicable lessons. Participating in the infusion of fundamental biology as the seed of a new industry, I helped Schering-Plough lay the groundwork for its growing pharmaceutical business. Caterpillar, for whom I consulted, and Goodyear, where I was a member of the board, faced different issues. Technologically, Caterpillar was a world leader, but lower labor costs meant that Japanese heavy equipment manufacturers could undercut Caterpillar's market. Goodyear was ready to adapt a new technology, radial tires, but it was captive of a conservative American automobile industry that did not want to adapt to the new tire technology, even though it had swept the European automotive industry and markets in Europe and Latin America.

TRW

I joined TRW in February 1977 as a consultant and board member and stayed with the company for almost 12 years until I reached the mandatory retirement age of 72 in 1988. I had consulted with TRW earlier when in 1960 I chaired a committee looking at the future of the Space Technology Laboratories, a TRW component. Our role was to throw some cold water on the Space Technology Laboratories, to remind STL that its glory days of overseeing the development of the first generations of intercontinental ballistic missiles (ICBMs) for the Air Force were over, that it no longer had a unique position, and that it needed to refashion itself if it was to compete in a drastically changed environment. As I pointed out at the time, "the uniqueness of the STL capability in systems engineering and space and ballistic missile fields is gone. . . . The field has lost some of its glamour, the very best of the STL people have gone on to other (usually higher) jobs in STL and TRW; some contractors really have some talent of their own in this field now; the whole technical fraternity of the country has learned a fantastic amount about ballistic missiles and space in the last few years; the Air Force generally has more capability now." We reminded STL that it still had enormous capabilities in the missile and space business, along with enormous admiration for its accomplishments but that at the same time it had to turn to new directions. The new directions we pointed to included communication satellites, especially if TRW linked up with companies such as AT&T, and the soft lunar landing program, which we thought could lead to "twenty years of business." Those recommendations were indeed taken to heart and in fact proved to be a substantial part of TRW's future business.[5]

Ruben Mettler[6] became chairman and chief executive officer of TRW[7] in December 1977, shortly after I joined the board, and my first meeting with him to talk about what I should do as a consultant was quite revealing. He pointed out that a major business reason for merging Ramo-Wooldridge, a high-tech operation on the West Coast, with Thompson Products with its automotive and aeronautical operations based in Cleveland, was to diversify so that if one business sector had problems another could keep up the revenue. In fact that's what happened in the early 1980s, when the United States and other industrial economies were hit by the worst recession since the 1930s:

> When its automotive and industrial businesses turned down, in part because of the second big surge in oil prices in six years, its energy businesses picked up. The different timing of the recession around the world enabled the company's strong position in Europe to mitigate troubles in North America, and subsequently vice-versa. When commercial businesses faltered, defense operations boomed under the huge American defense buildup.[8]

Rube wanted me to help strengthen science and technology in all major business units, and in particular to get more synergy out of the marriage between unlikely mates. He was interested in trends and predictions in the new sciences and technologies in the fields of interest to TRW and TRW strengths and weaknesses in them. His interests included the technologies for new products such as electronics and safety restraints for automobiles, energy products, spacecraft and aircraft components and systems, and the technologies to aid their work and to gain productivity such as computer-aided design and new manufacturing technologies. Rube also wanted a reading on TRW's professional climate as seen both by its employees and outsiders, including academics who could be of value to TRW. Transfer of science and technology information from universities and governments here and abroad to TRW especially was an important subject. With a keen understanding of technology, Rube was first and foremost a brilliant corporate business strategist. He developed a plan of attack for TRW—and for my reviews—that evaluated productivity in terms of *return on assets employed* by the TRW unit. This business model had much to do with our success in merging cultures in TRW and in cross-fertilizing technologies between all the business units.

That charge led me through a dizzying tour of virtually all of TRW's facilities and operations, and senior TRW officials—Si Ramo, Dick Delauer, George Solomon, and Johnny Foster[9]—told me about the company's technological strengths and weaknesses. Soon we developed the idea of putting together in the Cleveland part of the company small information exchange groups of people working on different products but in fact coping with the same kinds of technologies; for example, while they may have been working with the same materials on the way to different products, they all still had to cope with their heat treatment and machining. Or from another perspective, machining and forming materials, toxic wastes handling, and better material properties for many automotive and aerospace products advance when they are informed by cutting-edge understanding of material properties. We developed a system of grouping engineers with common technological issues no matter the particular product. That produced interactions. That took a lot of my time, not so much from the standpoint of the technologies but from the standpoint of the people.[10] As I had learned from my MIT, Carnegie Mellon, and government years, technology flows and cross-pollinations are often not the issue; the people who have to embrace institutional change are. Our "experiments" in Cleveland were a test bed for our later and larger efforts. We broadened, for example, to bring in more of the automotive divisions, such as Ross Gear at Lafayette, Indiana, and the TRW group at Pontiac, Michigan. Soon those interproduct technology groups were expanded to include the California operations at Space Park.[11] At the same time these technology reviews were widening, Mettler asked for a list of candidates that could be moved up the corporate ladder and also moved between the West Coast units and Cleveland. Johnny Foster led my list and many others agreed, and Rube convinced him that he should move from the West Coast to the Cleveland headquarters to be corporate vice president for science and technology. Johnny had been a leader in TRW's energy businesses from the time he joined TRW. He was a good choice to help evaluate the core technologies and foster their spread across all divisions. Once in Cleveland, Foster started recruiting some new talent. Arden Bement, a former colleague of mine at MIT, who also had helped jump-start the Department of Defense's research program came aboard.[12]

The Stever River

As we started bringing the TRW European units into the cross-TRW technology review, I made a number of trips to the continent, as a board member and on some trips as the consultant to Mettler on the cross-TRW technology review. At dinner in Cleveland with European managers during one of our TRW technology meetings, I mentioned spotting the Stever River in Germany in an atlas. Several German managers attending the Cleveland meeting knew it, promised to send me details, and did. So Bunny and I decided to see it during a visit of the TRW board to Düsseldorf in North Rhine, Westphalia. Bunny and I with a German TRW colleague, his wife, and their children spent a day touring the Stever River from source to end. The river's source is a scummy pool on a farm about 10 miles west of Münster. It drains into a large and very wet field where cattle graze. Photographing this made for wet feet, but we didn't mind. It was a momentous occasion! A small stream emerges from that field, crosses underneath the Dortmund-Ems Canal at Senden, and meanders south to cross the canal again at Lüdinghausen and Olfen. It then turns west to Haltern, where it forms the Haltern Staussee, a small recreational lake. The Stever River flows into the Lippe River and then into the Rhine. Once the Stever River grows, it becomes quite beautiful, with great "water castles" built on islands in the river. In the Stever Valley hard by the river we stopped for lunch at the Gast Haus Stever Tal, the Stever Valley Guest Inn, where the proprietor made a great deal over the visiting Stevers. The marquee of the Gast Haus had a large painting of a leaping blue trout, indigenous to the river. Bunny chose that for lunch on the basis that she had already feasted on one Stever trout!

My TRW board membership continued until 1982. Over the 12 years I got a great deal of pleasure from my work as a board member and consultant. TRW repositioned itself well to global competition and the dramatically changed nature of the defense and aerospace industry.

BETHLEHEM STEEL AND CATERPILLAR

As I was learning about TRW, I also took on two industrial consulting projects, one of which—at Bethlehem Steel—was short term. The second was a three-year engagement for Caterpillar. Bob Seamans, head of the Energy Research and Development Administration in the Ford administration, asked if I would like to join him and a small number of others to advise the leaders of Bethlehem Steel on new research and technologies. Don Blickwede, Bethlehem's vice president for research and a member of the National Academy of Engineering, had convinced top management they needed experienced outside advisers to evaluate Bethlehem's research effort. This was a tough time for an old-line industry: it faced serious environmental problems with very high price tags for fixing them, intense competition from overseas, rapidly changing markets, severe labor problems, new technologies costly to put in place given the high sunken costs of its existing facilities, and the emergence of minimills that could outpace the larger integrated producers in the markets for specialized steel. Companies in South Korea, Turkey, and Spain, among others, in addition to Japan, were competitors. The U.S. steel industry by the early 1980s went from world dominance to about 10 percent of the market, with Japan now the leader. It was not a good time to be optimistic about research. In fact, throughout the 1980s, budgets for internal research and development in the steel industry declined up to 75 percent.[13]

In June 1977, Bob and I began our two-week review of Bethlehem Steel's research. We got a very good tour of the Bethlehem manufacturing and research facilities, in Pennsylvania and at the Sparrow Point plant near Baltimore, where there were immense coke ovens as well as some ship-building operations. All of Bethlehem's facilities, and the Baltimore coking oven plant especially, had enormous environmental issues to deal with, which by the 1980s could no longer be avoided. Most of the basic automotive, steel, and other heavy manufacturing industries were in the environmental spotlight throughout the United States. Bethlehem had to change its production if it wanted to survive. We made recommendations across the board, including strengthening research and development work that we felt Bethlehem Steel needed to do a first-class job and regain its economic footing. We also commented on personnel strengths and weaknesses and the quality of the laboratory facilities. Our report was well received but limited in its impact. The funds Bethlehem Steel could generate for the research and development operations it needed were getting low. And anything the company did that changed its operations involved such a large capital investment that progress was unavoidably slow. Moreover, the environmental regulatory climate for a company like Bethlehem was uncertain as government policies seesawed in the 1970s and 1980s.[14] It was a striking lesson for me on the differences between steel and electronics, the latter not requiring very expensive plants to absorb new technology.

I did not know a great deal about Caterpillar when I was approached about being a consultant. As a red-blooded boy of the twentieth century, I knew full well the logo of Caterpillar and the patented yellow of the Caterpillar products, and I knew something about the performance of the stock as one element of the Dow Jones stock portfolio because of my trusteeship of a trust for the benefit of my wife from her father, in which the Caterpillar stock was one of the strong performers. I am not quite sure from whom Caterpillar learned of me as a possible consultant to a program the company was going to conduct. I got a call from and arranged a lunch in Washington with Don Fites, later to be chairman and CEO of Caterpillar. He wanted to arrange a three-year consulting deal with me to participate in developing a major new business plan for the corporation.

I met with Caterpillar officials—Lee Morgan, chairman, and Bob Gilmore, president—in Peoria and learned their problems. The Japanese were competing heavily with Caterpillar in all the world's markets. Technologically, Caterpillar and Japanese manufacturers had similar products, but Mitsubishi and other manufacturers not only had lower labor costs but also favorable currency exchange rates. But it was not just the Japanese competition that prompted Caterpillar to make a study of every aspect of its business. Lots of things were changing. The Information Age was coming in. The capability of close control of machinery, manufacturing machinery, operating machinery was coming in. The use of information technology in machine control, through computer chips, was emerging.

Caterpillar decided that it would make a thorough study of every aspect of its business. It was a pioneering management decision on Caterpillar's part, a 360 degree examination of the business. I was one of several consultants, and my role was to use my experience to advise on meshing new technologies, especially information technologies, with the company's business. I signed on for three years. My visits to Peoria in spring and summer of 1982 were primarily aimed at getting me fully aware of the current status of Caterpillar's technology and research. I visited the company's most modern tractor plant and was overwhelmed by the size of the tractors and the handling of such heavy assembly production. I visited the research laboratories, where Caterpillar was interested in more accurate control of the performance of a tractor or an earthmover, how you could automatically drive an earthmover, tractor, or other piece of heavy equipment over a field and then repeat it.

The Dog That Didn't Bark

There was a thought in the company that it should become involved in the new high-tech electronics business, chip design, laser design, and so forth. In my TRW experiences with the West Coast operations in the electronics and high technologies of space versus the Cleveland-centered automotive technologies, I recognized, as did the top people in TRW, that they were dealing with two different cultures.[15] I believed it would be too much for Caterpillar to try to become a first-class competitor in the culture of producing chips and other high-technology electronics. I recommended that the company concentrate on learning to be a good adapter of information and control technology and use it throughout its design, production, and products. That is what the company did. It was a variant of the dog that didn't bark. Caterpillar properly and profitably resisted the siren call of new technologies, sticking rather to its core business and infusing it with the new technologies albeit without changing its face to the outside world. Indeed, Caterpillar became an industrial partner at the National Center for Supercomputer Applications at the University of Illinois and has remained so. Many advanced computer and visualization technologies developed at the center are now integral to Caterpillar's assembly line and equipment control technologies.

SCHERING-PLOUGH AND GOODYEAR

TRW dominated much of my time in the early 1980s, but I also joined two other corporations as a board member: the Schering-Plough Corporation, a medium-large pharmaceutical company, and the Goodyear Tire and Rubber Company. Like TRW, Caterpillar and Bethlehem Steel both faced momentous changes to which they either adapted or risked being history.

Schering-Plough[16] faced the early days of what is now called the biotechnology industry: the use of highly sophisticated biological techniques to ease the manufacture of pharmaceuticals and to create entirely new drugs and even a new sort of medicine dependent on deep understanding of genetics at the molecular level. Moreover, the innovation process was new, a new style that the drug companies had to learn, and fast. If a company were to compete, it would have to form close alliances with the very best in American molecular biology, which was in the universities. And it would have to find effective ways, compatible with both academic and corporate cultures, to move frontier research into the marketplace. This meant having people who could understand the research and were themselves good enough at it that their work had the respect of academic peers. It meant recognizing not only the possibilities of a new discovery but also the talent that could transform it into a commercial product. And it meant investing the money—large amounts largely directed to research with speculative, longer-term returns—to make it happen while keeping investors happy.

When I joined the board, Schering-Plough was under enormous pressure to come up with new products. Its patent on a major product—gentamicin, the first antibiotic useful against gram-negative organisms—was running out. Led by Chairman Willibald Conzen, a giant in the pharmaceutical field, and backed by Robert Luciano and Richard Kogan, who were being groomed for future leadership, Schering-Plough started a much heavier research and development program. Soon it had contact with a group of Stanford University professors—Charles Yanofsky, Arthur Kornberg, and Paul Berg, the latter two earning Nobel Prizes—who had started a company, called DNAX (DNA and "experiment"). Stanford was an entrepreneurial hot bed, not only in the silicon technologies that drove the computer revolution and launched Silicon Valley but also in much of the work that launched biotechnology. Schering-Plough, in a real coup, created a working relationship with DNAX, and indeed today DNAX is a major part of the Schering-Plough research capacity. My role was to help mesh academic and industrial culture in DNAX with Schering-Plough's strategic investment in advanced research in pharmaceuticals. I became part of a three-man team to look over DNAX's performance and how well it had met its performance goals. The company was doing very well.

The success of the DNAX venture emboldened Schering-Plough to invest steadily and heavily in the then-new recombinant DNA technology. Like all pharmaceutical manufacturers, Schering-Plough is today a corporation transformed by its business decisions to embrace the "new biology" made in the 1980s. It is equally important to remember that Schering-Plough's older core businesses in health care products, cosmetics, and other of the less "high-tech" products in its overall portfolio remain profitable. Many chemical and pharmaceutical companies have learned that a diversified product mix has served them well as the "old-line" products have remained profitable and supported the new biotech-based product development or were sold off to provide capital when they had had some scientific and financial downs.

I had not anticipated an appointment to the board of the Goodyear Tire and Rubber Company, but I learned later that Charles Piliod, the Goodyear chairman, found out about me through Rube Mettler. Piliod had run international activities for Goodyear and knew of the changes in tire making that were sweeping European tire manufacturing and their huge acceptance in Europe and America. Piliod invited me to meet him in Akron, where he gave me a pretty impressive sales pitch, strongly flavored by his international perspectives. He saw me as an ally who could help "engineer" the needed technological and cultural changes in the company and offered an appointment to the board. I thought about this for a few days, a little concerned with my accumulating corporate responsibilities and also my pro bono work, which was accelerating, but decided to accept. Two weeks later I was voted onto the board, so I now had three boards to juggle. But it worked because many meetings were within one day's reach of Washington, in Cleveland, Akron, and New York.

Goodyear faced three problems: the probable loss of global business unless it shifted to production of radial tires, quality control on the production lines to reduce imperfectly molded tires or "blems" as they were called, and integration of best technological practices and new technologies throughout its worldwide operations. I was not going to be of much help on the first or second of these issues, but I could and did contribute to the infusion of new technology and its adoption throughout the company. Tire making had begun to change in the 1970s, with the manufacture of tires with reinforcing fibers wound radially into what became known as the radial tire. U.S. automakers, unlike their counterparts in Europe, clung to the reinforcing fibers wound at an angle for a bias tire. Fortunately, U.S. automakers, then at the height of external criticism and internal corporate reevaluation and reeling from the growing numbers of imports, realized that they should sell vehicles with radial tires.[17] Piliod campaigned relentlessly with the recalcitrant companies and had a good deal to do with their changed stance on radial tires.

Goodyear solved the radial tire problem by vastly expanding production. It built a radial tire plant in Lawton, Oklahoma, that was fully automated. Whenever a problem was spotted anywhere along the production line, the line was shut down, and a team from all over the plant quickly assembled to make sure the problem was fixed. Soon Goodyear was producing 25,000 tires a day in that single plant, with practically no slightly damaged seconds, or "blems."

My approach to the third Goodyear issue—integrating best corporate technological practices throughout the worldwide operations and introducing new technologies—was similar to what I was doing at TRW. We set up teams to evaluate technologies being used in rubber making and tire production in the United States and at plants across the world, and we asked what science and engineering could be used to make production more effective. These were important contributions, but Goodyear's premier position in the world tire and rubber business today owes most to Charles Piliod's successful campaign to get the radial tire accepted in the United States.

TRAVELING TOWARD 65

I have mentioned TRW's international operations and the trip Bunny and I made to Germany when we saw the Stever River. TRW believed strongly that, if it was to be a successful global company, the board needed to see all its operations firsthand and to hear about the business environment in nations where its operations were located. Schering-Plough and to a lesser extent Goodyear also arranged for their boards of directors to inspect plants and operations around the world and to learn about the foreign business climate. These international trips were a rare opportunity for Bunny and me to share exciting new experiences abroad, and we valued the travel. Much of my travel for MIT, the Air Force, and the government had been solo. Bunny would not have been able to be with me in the earlier decades in any case because of the children.

Schering-Plough had an even more active foreign visiting program for its board of directors than TRW. It had many worldwide centers under the name of the Essex Corporation, with activities that ran from manufacturing to sales and marketing to joint activities with other companies. In 10 years from 1980 to 1989, there were nine trips: England, Puerto Rico, Japan, Italy, Ireland, Portugal, Argentina, France, and Spain. In late fall of 1981 all three corporations had an absurd concatenation of meetings that resulted in our making a month-long trip around the world. The trip started with an early morning shuttle flight to New York City and then for me an all-day meeting of the Schering-Plough board. In the evening we left New York to fly, eventually, to the Far East, where the Goodyear Tire and Rubber board was to meet. Bunny and I broke up the trip by stopping in Greece for a lovely stay in Athens with side trips to Delphi, the Peloponnesus, Corinth, and Mycenae. We left Athens late at night, stopped in Dubai, breakfasted over the Indian Ocean flying to Singapore, and then on to Jakarta.

Goodyear arranged briefings of its board by government officials, visits to major Goodyear manufacturing plants near Jakarta, and discussions of the overall status of the Southeast Asian economy.[18] In the evenings there were social affairs, including fine programs of native Indonesian dancing. After three days and evenings, the Goodyear board and senior executives left by charter plane for Malaysia and Kuala

Lumpur, where we essentially repeated the routine of the Indonesian visit. Our Malaysian visits included the Rubber Research Institute, which was a major institution in that country. The institute's principal aim was to protect and improve the use of natural rubber. During World War II, when the major combatants could not get natural rubber, there were many successful programs to develop substitutes that continued to cut into Malaysia, Indonesia, and many other natural rubber sites. Again, like Indonesia, Malaysia wanted to change from exporting its resources to using them locally to manufacture value-added products, and then exporting these. The contributions that corporations like Goodyear made to the developing infrastructure and human resources base in Malaysia predated that country's more recent rise as a center of computer assembly and software facilities for many major global companies. Goodyear's overseas trip ended in Singapore, where there was a meeting with all of the Goodyear senior executives from their facilities in the Southeast Asian region. There we were treated to an impressive Chinese paper dragon dance.

Bunny and I had now been traveling for about two weeks and were about halfway through the month-long trip. We arrived in Japan at Narita Airport, already then in the 1980s overcrowded and with a slow two-hour trip into Tokyo. We joined TRW friends at the Okura Hotel. We had two days free for sightseeing, getting to the Imperial Palace, gardens, and shrines—something I had been unable to do at the National Science Foundation, or the White House, although I made several trips to Japan in the 1970s. The TRW portion of our trip in Japan and China was focused on specific objectives. Japanese companies, especially electronic manufacturers, were establishing world markets with their exports, some of which competed with those of TRW. By the early 1980s, however, there were also Japanese companies interested in joint ventures. In China the objective was different. China was just opening up more broadly to university exchanges and to at least a few companies that could potentially create large markets in China. If TRW was to be successful in China, it needed to understand the country's military and consumer markets and its suppliers. Over the succeeding 10 days, the TRW board covered a lot of ground intellectually, culturally, and geographically.

U.S. embassy staff, including Ambassador to Japan Mike Mansfield, the former senator, who was a keen and perceptive student of Japan, started our briefings. In addition to meeting with us at his office, he invited the board members and their wives and a number of Japanese business leaders to a reception at his residence. Japanese government and business executives and officers of trade associations such as Automobile Worldwide and the Chamber of Commerce shared their points of view. Three days in Tokyo were followed by a trip to Kyoto on the bullet train, to see first the Fujitsu Namazo computer complex, a center for production and research and for what we in America called an incubator park, although it was a new idea in Japan at the time.[19] But they had fantastically clean and efficient computer equipment manufacturing plants. The next day, a Saturday, gave us a chance to visit the historic shrines and temples in Kyoto.

The following day we flew to Beijing. We met with energy officials including the minister of oil and gas, the vice minister of foreign trade, and a number of business leaders who were starting companies (or transforming state-owned companies into private ventures). As in Japan, U.S. embassy officials, including Ambassador Arthur Hummel, gave us their perspectives on the fledgling but rapidly developing business environment. We were taken to the Ming tombs and the Great Wall, de rigueur for all high-level visitors in those days, and we saw the famous Beijing traffic jam of bicycles of thousands of commuting workers.

We returned to Japan, then straight across the Pacific to Los Angeles and Washington, to arrive very late on Saturday, the 24th, my sixty-fifth birthday, which we celebrated on Sunday. Then we went back to New York to the Schering-Plough board meeting, completing a month's cycle of board meetings all over the world!

In 1988, when I was 72, my board memberships and industrial consulting ended because I had reached the mandatory retirement age; indeed, I had exceeded it in a couple instances. From the time I first worked in the Corning Glass Works in the summer of 1935, I had worked for many more than a dozen industrial companies with a strong science, engineering, or technology component and visited many times that number for the government.

The United States emerged out of the 1980s still facing economic difficulties; witness the difficulties in the early 1990s of companies such as IBM. Yet, in retrospect, the work done in the 1980s to regain our competitive strength—for example, policies to encourage venture capital, closer relationship between universities and industries, and the like—began to change things. The country slowly regained its competitive competence after years of self-doubt, of questioning its ability to even survive against the rise of very able national competitors across a host of industries and technologies, and of incessant recriminations about its innovative skills. Only much after the fact did I realize that the end of my industrial work coincided with the start of an American economic renaissance. New industries were created, old ones radically transformed, and the United States, rather than bowed by national competitors, took the economic challenge directly to them. No small part of that was U.S. companies strengthening—if not learning for the first time—their meshing of the frontiers of fundamental science and technology with their businesses. I was proud to have had a small part in helping a suite of companies—from the most technological, such as TRW, to the classical, such as Bethlehem Steel and Goodyear; from heavy industry in Caterpillar to those engaged in human health, as in Schering-Plough—go through change and prosper.

10

Ending the Century

1976–2000. Out of government but hardly removed from it. In the 25 years after I left the Old Executive Office Building, I returned to familiar territory in international development, "Star Wars," space flight, and science and technology policy.

The triumph of science and technology in the Second World War reverberated in peacetime. At home, economic and other forces demanded a reexamination of how to maintain the vibrancy and productivity of our science and technology. Other countries, especially the emergent economies, wanted to follow our path and learn from us how to do it. There were uncertainties about how to defend the country against attacks in space and on the ground. And there were disasters, as slip-shod management cost seven people their lives when the *Challenger* shuttle blew up shortly after launch.

STAR WARS

"On March 23, 1983, Ronald Reagan announced in a televised address his Strategic Defense Initiative, the ultimate purpose of which, the president said, was to intercept enemy missiles in flight and thereby 'to give us the means of rendering those nuclear weapons impotent and obsolete.'"[1] He posed the question: "What if free people could live secure in the knowledge that their security did not rest upon the threat of instant U.S. retaliation to deter a Soviet attack, that we could intercept and destroy strategic ballistic missiles before they reached our own soil or that of our allies?" He called on the "scientific community in our country, those who gave us nuclear weapons, to turn their great talents now to the cause of mankind and world peace, to give us the means of rendering these nuclear weapons impotent and obsolete. . . . Tonight, consistent with our obligations of the ABM [antiballistic missiles] treaty and recognizing the need for closer consultation with our allies, I'm taking an important first step. I am directing a comprehensive and intensive effort to define a long-term research and development program to begin to achieve our ultimate goal of eliminating the threat posed by strategic nuclear missiles. This could pave the way for arms control measures to eliminate the weapons themselves."[2]

Like all things, this one had histories, including Ronald Reagan's and mine. In 1945 an Air Force officer—at the time, an *Army* Air Force officer—warned that the United States would need a missile defense. In 1954[3] I chaired a government panel to look at the problems of an anti-missile defense system. We found plenty, even though issues were in a way simpler, not least that intercontinental ballistic missiles (ICBMs) were still in their infancy. But even then we saw many problems in attacking offensive missiles on launch, much less detecting them, killing them before they struck, and dealing with decoys. The world in the 1980s was more complicated, with the two superpowers bristling with nuclear-tipped long- and medium-range missiles, extraordinarily capable guidance systems, and taut warning systems in the event the other guy tried a sneak attack. That made for a standoff—mutually assured destruction—that uneasily caged the nuclear beast for decades. "Star Wars" in principle took out the "mutually assured" since with it we could stop the other guy from hitting us. That was new and for a lot of people scary.

Ronald Reagan may have gotten his original ideas about a strategic missile defense in a visit in 1967 to the Lawrence Livermore National Laboratory, one of the country's major nuclear weapons laboratories. More than a decade later, at a briefing during his presidential campaign at the headquarters of the North American Air Defense Command, Reagan had it brought home to him that, while the United States could detect missiles about to destroy its territory, it had no way to stop them. The 1980 Republican platform urged "more modern ABM technologies." With Reagan's election the pressures for an ABM program mounted, with well-known scientists such as Edward Teller pushing hard. Not least, the forces for a moratorium on testing and production of nuclear weapons were getting stronger and more insistent. In 1982 a resolution for a nuclear "freeze" failed to pass in the House of Representatives by two votes. Clearly, the nation's notion of a strategic defense dependent on who had more weapons was being tested hard.

Ronald Reagan, according to George Keyworth, the president's science advisor, had studied the stability assumed in mutual deterrence, and concluded that stability was eroding and came to "believe that any incentive for a preemptive nuclear attack could be removed by even a rudimentary missile defense system. . . . The SDI [Strategic Defense Initiative] was President Reagan's idea; it was his initiative, and it was his faith in technology that drove it."[4]

Troubling to me was that the decision to go ahead with the SDI flew in the face of deep technical skepticism, most especially the heart of the purported system: using lasers—specifically, X-ray lasers—to intercept and destroy incoming missiles in space. One technically astute administration official told George Keyworth that relying on "lasers to shoot down ballistic missiles was like expecting laetrile to cure cancer."[5] The White House Science Council, composed of some very distinguished scientists and technologists, was highly dubious of a space-based defense relying on lasers. Finally, Dr. Keyworth was also skeptical of the technical merits of the various proposals for missile defenses and was "inclined to protest the conflicting and exaggerated claims being made on behalf of strategic defense."[6]

What was the reaction? There was a sharp loss of confidence in the White House science advisor and his office by many in the scientific and technical communities. George Keyworth saw it quite differently:

> Much has been written about the fact that the President's SDI speech was conceived in less than a week. In fact, the President began to articulate the rationale behind the speech in late 1981, eighteen months earlier. The President wrote virtually the entire speech himself. I advised him, helped edit the speech and offered him choices for restating key points. . . . The President had, from the beginning, a clear vision of where he was leading the country, and SDI was part of how to get there. I believe he was right, that he was courageous in undertaking SDI, and it was my privilege to serve him in that endeavor. He and he alone made the estimates of the risks and benefit. And he never wavered.[7]

What the Russians would do was uncertain, but the Central Intelligence Agency (CIA) in a then-classified assessment issued three months after the speech, on September 12, 1983, asserted that the Soviets would rely on political and diplomatic means to oppose the missile defense plans, "or, failing that, to negotiate them away." And, the CIA added, they could turn to technical countermeasures. Missiles can be upgraded with new boosters, decoys, penetration aids, and multiple warheads. "The signatures of these systems can be reduced and new launch techniques and basing schemes can be devised which make them less vulnerable to U.S. missile warning and defensive weapon systems. These systems can also be hardened or modified to reduce their vulnerability to directed energy weapons. The Soviets can employ other offensive systems, particularly manned bombers and long-range cruise missiles with improved penetration aids and stealth technologies, to assume a greater burden of the strategic offensive strike role and to exploit the weaknesses in U.S. air defense capabilities."[8]

The Congress indirectly put SDI on my plate. When the Department of Defense requested funds for fiscal year 1985 to establish a Strategic Defense Initiative office, the House Armed Services committee and the Senate Foreign Relations committee requested that the congressional Office of Technology Assessment (OTA) examine the opportunities and risks in accelerating research on missile defense technologies. In spring 1985, John H. Gibbons, the OTA director, asked me to chair the OTA committee. "Now, why would you want me to do that? I've never really done anything for OTA." He said, "Well, there are three reasons. First of all, you know a lot about rockets and missiles and space and military uses of them, even studies of ballistic missile defense." "Yes, that's correct. What's the second?" He replied, "You've chaired lots of committees, and so you have good experience in that." I replied, "Yes, that's true, too. What's the third?" He said, "As far as I can see, you are the only person in the country who hasn't come out on the front page with your views on the subject." I laughed and said, "Yes, that's correct."[9]

The first meeting of the committee[10] was in a classified facility near the Capitol. Once we got through the throat clearing, with introductions and the like, we entered into substantive talk and trouble. The first to speak was quickly interrupted. Committee members had strong positions on the issue and weren't bashful about pushing them hard. For example, when Robert McNamara at the outset argued for the current weapons balance and its mutually assured destruction policy, Lieutenant General Daniel Graham sneered "Yes, we call that MAD, M-A-D, MAD!" It became absurd. I gaveled the committee to a full stop and said, "No one can speak without my permission, and the way to do that is either raise your hand or put your nameplate up, and the secretary will note your names in order, and I will then call on you in that order. Everyone who wants to will be heard. Also, you have to limit your comments on any one subject to three minutes—one minute would be better." It worked. They felt a bit sheepish, knowing that they were all very distinguished people and that this was a stupid but necessary rule to impose on them.

We soon got down to substantive matters on subjects such as ballistic missiles and defense, then and now; deterrence, U.S. nuclear strategy, and ballistic missile defense; arms race stability and arms control issues; technologies; feasibility; and alternative future scenarios and alternative research and development programs.

Our report[11] was issued in September 1985. The committee without much controversy agreed on continuing research in ballistic missile defense, but without any deployment or setting up of a station that would abrogate the ABM treaty. It noted that national security strategy was composed of strong offensive capabilities and real weaknesses in defending against them and that, "unless this imbalance between the offensive and defensive disappears, strategic defenses might be plausible for limited purposes, such as defense of ICBM silos or complication of enemy attack plans, but not for the more ambitious goal of assuring the survival of U.S. society." And the report added that "assured survival of the U.S. population appears impossible to achieve if the Soviets are determined to deny it to us," echoing the CIA's report on possible Soviet countermeasures to SDI. We also added that it was impossible to price the system, since there was no system—no design and none of the key technologies even close to development. And we added that a decision to either push ahead or scale back an SDI program means, like many things in life, "balancing opportunities against risks, in the face of considerable uncertainty."[12]

That report dealt with missiles attacking us. What about weapons against our space satellites? As Jack Gibbons put it, "How can the United States respond to the potential threat to its military capabilities posed now and in the future both by Soviet military satellites and by Soviet anti-satellite weapons?"[13] Those issues were dealt with in a companion report to the one dealing with SDI, a logical move, since the issues, the technologies, and the possible implications were in many instances close or the same.

There was more. Left very uncertain was whether the daunting computer software needs for an SDI system could be met. Congress was also interested and asked for an additional report drilling down harder on computer issues. Our 1988 report acknowledged the impressive technical work done in some 30 years of work on a ballistic missile defense system, that using optimistic assumptions a "first-phase" system might be online in 1995 to 2000 (i.e., 10 years after our report) and that the software problem was indeed huge.[14] Indeed, the report commented that "no adequate models for the development, production, test, and maintenance of software for full-scale ballistic missile defense systems exist."[15]

Ballistic missile defense had been on my mind from 1944 to the end of the century and it was rearing up again. The committee on missile defense I chaired in 1954 for the Air Force to look at defenses against ICBMs carrying nuclear warheads concluded that any defense was possible only in the terminal phase of the missile's flight. The Air Force lost interest because it considered that an Army mission. Some 30 years later I chaired another look with much the same conclusion overall, that the overwhelming advantage lay with offense. So back to mutual assured destruction. The changes in the USSR and its economic defeat have made us feel comfortable with the problems in building a missile defense system, though the United States has continued to do research and has had some success at terminal defense of a few weapons from known direction. Now in the twenty-first century a defense against rogue nations is thought desirable. In my view this will prove to be more of a political issue very much as it turned out for the Star Wars proposal of the 1980s.

BISON IN ILLINOIS

It wasn't all Star Wars. I also got heavily if indirectly involved in the purest of science: high-energy physics. When I went in 1965 to Carnegie Tech to be its president, my predecessor, Jake Warner, mentioned that he was involved in the start-up of a new organization that had as its goal building a new particle accelerator much more powerful than what the country had at the time. That piqued my interest, not least because my doctoral work at the California Institute of Technology was on the decay rate of one of the fundamental particles of nature, the meson. The other fundamental particles at the time were the proton, neutron, electron, and positron. A lot had gone on in the three decades since I did that work in the late 1930s, but in truth high-energy physics in 1965 was a bit of a mess: lots of particles had been discovered, leading to many research papers and headlines but little clarity. How did those particles relate to one another? How did they account for nuclear and atomic behavior? Probing that meant having a more powerful machine with beams that could more completely shatter atomic nuclei and giant detectors to probe and measure the debris.

The new machine would be very large and very expensive. The tactic for getting the money out of the Congress was twofold: form a consortium of universities backing the accelerator, have them pony up $10,000 each, and not select the accelerator site until after the money was appropriated. That was successful. Universities Research Association, Inc.[16] was created to coordinate the universities; Batavia, Illinois, just outside Chicago, was selected for the site; the facility was dedicated in November 21, 1967;[17] and Robert R. Wilson of Cornell University was named the first director in 1967. Wilson put down a set of firm principles for the new facility: scientific excellence, aesthetic beauty, stewardship of the land, fiscal responsibility, and equality of opportunity. Wilson turned the words into reality by building the new laboratory on time and on budget, demanding architecture and design for its 6,500 acres that were highly functional and dramatic and establishing a bison herd to symbolize the place as another frontier, that of high-energy physics.[18] The herd is still there, with new calves each spring. Bob Wilson died on January 16, 2000, and is buried in the small Pioneer Cemetery on the northeast corner of Fermilab where some headstones date back to 1839. Perhaps his finest moment came when he had to defend the cost of Fermilab before a highly skeptical congressional committee. Asked what the new accelerator had to do with national security, he said: "Nothing at all. It has only to do with the respect with which we regard one another, the dignity of men, our love of culture. It has to do with: Are we good painters, good sculptors, great poets? I mean all the things we really venerate in our country and are patriotic about. It has nothing to do directly with defending our country except to make it worth defending."[19]

As president of Carnegie Tech cum Carnegie Mellon University, I became a member of the Universities Research Association, Inc. (URA) consortium and from that post watched as Fermilab opened for business and became a major player in dramatic advances in high-energy physics. In its first 10 years, theoretical high-energy physicists working at Fermilab and many other sites in the United States and abroad began to untangle the messy zoo of particles, symmetries, forces, and the like that made high-energy physics such a wilderness. The key was a "Standard Model" that offered a framework for describing elementary particles and their fundamental interactions. It also showed some gaps of as yet undetected particles. Fermilab contributed by finding some of the major missing particles predicted by the Standard Model, including the discovery of the bottom quark in 1977, the top quark in 1994, and in July 2000 the tau neutrino, the last fundamental particle (at the time) to be observed.[20] In fact, confirmation of the tau neutrino completed the experimental sightings of the fundamental building blocks of matter prescribed by the Standard Model.

I had to leave URA when I became director of the National Science Foundation in 1972 but did go back for a dedication ceremony for Fermilab, representing President Nixon. Almost immediately after I left government I was asked to become a URA trustee. And almost immediately after that, in 1978, we lost Bob Wilson. Operating budgets for facilities were cut in the late 1970s. Bob wrote a blunt letter saying something had to be done or he'd quit and asked the URA officers if he should send it. I think he might have been willing to try a more positive approach, though I am not sure what. His resignation was accepted by the Department of Energy (DOE). That was a blow, but we got a wonderful successor in Leon Lederman, who served two very successful five-year terms. In 1988, the same year he stepped down as Fermilab director, Leon shared[21] the Nobel Prize in physics for his earlier discovery of the muon neutrino.

I became part-time president of URA in 1982 and served out my three-year term during which serious planning began for the next particle accelerator. That was supposed to be the Superconducting Supercollider, the SSC. It would have enough energy to produce a "Higgs" particle, the notion of which sent the theorists designing the Standard Model of the nucleus into ecstasy. DOE agreed to start, and proceeded, as had the Atomic Energy Commission when Fermilab was built with Glenn Seaborg at the helm; that is, prepare a design, ask for proposals, and then choose a site. But DOE did not have a Glenn Seaborg nor did the Congress have a Hubert Humphrey to, as in the case of Fermilab, rally the midwestern states to get such a big plum. URA had to have a broadly based Board of Trustees with two Boards of Oversight, one for Fermilab and one for SSC. This resulted when DOE asked Robert Hughes, president of Associated Universities, Inc., and me to referee the competitions between and among the six main laboratories for high-energy particle physics as well as one from Texas with smaller programs in its universities, bringing to seven the organizations we dealt with. All the while jealousies between and among the six main laboratories for high-energy physics kept popping up, and Texas, with a small nuclear particle physics program, insisted it be part of the discussions. But we were able to select Maury Tigner, a Cornell physicist to lead the design effort at Berkeley. The National Academy of Sciences appointed a site selection committee to winnow down the number of suitable sites. The politics began and Texas was selected. I thought we would pick Maury Tigner to lead the construction and program development, but a secret ballot produced Roy Schwitters from Harvard, who at the time had charge of a major experiment at Fermilab.

The SSC didn't make it, aborted early in its building by the Congress for a farrago of reasons, including costs; loss of political influence by Texas where the SSC was to be sited; notable and very public failure to obtain foreign backing for the project; a succession of perceived overruns; and rough questioning of how funds were used, including by URA, which managed the project for the government. Not least, many in the scientific community outside high-energy physics bitterly opposed the project believing it would drain their own support. SSC began construction in 1990. Four years and $1.7 billion later, it was terminated by the Congress on October 21, 1993. I was out of the line of fire, having left my presidency in 1985 after serving out my full three-year term. I did

stay on the URA board through the debacle.[22] Of the many reasons for
the demise of the SSC, I believe there was one sine qua non: the loss of
political support when Speaker of the House Jim Wright from Texas was
forced to resign. We lost the equivalent of Hubert Humphrey, who earlier had rallied the Midwest Congressional Delegations to get Fermilab
for the Midwest.

When I chose to do pro bono work from a Washington base and
independent employment instead of returning to academe, I knew I
would miss one feature of the research university, that of the intellectual
ferment of comingling with faculty and with other talented scientists,
engineers, and many other professionals. To help replace that, I chose the
National Academies. My very close friends, President Court Perkins of
the National Academy of Engineering, and President Phil Handler of the
National Academy of Sciences, immediately set me to work on a great
variety of projects, many in my professional fields, some in science and
technology policy, some in international affairs, and some in the organizational infrastructure and operations of the NAS, NAE, and National
Research Council (NRC). Court's successors, Bob White and Bill Wulf,
and Phil's successors, Frank Press and Bruce Alberts kept it up. In the
archives there is a record of about 50 assignments to various commis-
sions, boards, committees, panels, and working groups. Although I can
tell tales about each, I will be mercifully selective.

TECHNOLOGY AND TEA

While the domestic issues—SDI, the death of the SSC, and the response
to the *Challenger* accident, about which more later—were tough enough,
they often got even harder on an international scale. It's one thing to
counsel on the health and use of science and technology in a culture one
fully understands and quite another when you're asked to do that in a
different culture, where motivations, goals, and management styles are
quite different. That certainly was impressed on me as I undertook in the
late 1970s and early 1980s a series of international science and technology missions.

Mao-Tse Tung died in 1976, the "Gang of Four"[23] was defeated by
the fall of 1978, and there had been determined efforts by the Carter

administration to "open" up China, mirrored by determined efforts by China to establish links with our science, engineering, and technology. A high-level government-to-government visit was arranged by my successor as presidential science advisor, the geophysicist Frank Press.[24] Before that a private exchange was arranged when the new Chinese minister of education asked the National Academy of Engineering[25] to send a group to look at two items: (1) the quality and effectiveness of engineering schools in China and (2) how effectively different industries used engineers. Joseph Pettit, president of the Georgia Institute of Technology, was the chairman, and I the vice chair. We were among the first from the "outside world" to visit China after the fall of the Gang of Four, this being September and October 1978. We started off in Beijing with briefings from the minister of education and his staff, with their thoughts on engineering education and industry. These were two of the five central priorities of the Chinese government—agriculture and food, industry, national defense, science and technology, and education. We soon learned that the Chinese were absolutely standardized in receiving visitors. Wherever we went, we would be taken to the top man of that institution. In his office would be a low table. We would sit around it and immediately hot tea would be served in a fairly large cup, with a neat top on it. I like tea, but having this happen to me more than once per day, I soon developed a tea jag. And we heard the five central priorities of the Chinese so often we repeated them in our sleep.

We did get out of the offices to see some of their engineering work. At Xingua University, albeit not until after another round of tea, we visited a solid-state laboratory, a computer laboratory, and a mechanical engineering student factory. The Chinese had prepared very well. They knew our curriculum vitae and whenever they could they had teachers and researchers who had studied at our universities. I met up with a group of teachers and researchers who had been to the California Institute of Technology, where I got my doctorate 35 years earlier. We also visited the Institute of Mechanics at the Chinese Academy of Sciences, where I asked about a very distinguished member of that Academy, H. S. Tsien, a close associate of Theodore von Kármán. I'd known Tsien at Cal Tech and at MIT and on the Air Force Scientific Advisory Board. In

1950 he was accused of being an alien communist and a danger to the United States. Two weeks in custody were followed by five years of house arrest under threat of deportation. He left in 1955. I asked about Tsien and was told that they did not know where he was. I knew exactly where he was: working on the most advanced technologies for the Chinese military and space programs.[26]

It wasn't all business. We visited the Imperial Palace Museum and were taken to the emperor's theater house (more tea). This was a relatively small, comfortable room that looked out over a moat, across to a theater stage for opera and plays. The emperor and his family could come and watch from this room. The emperor was no longer coming. The curator of the museum, a woman, very attractive and very well spoken, showed us some beautiful Ming pottery. But there was not a lot of it. To be pleasant, I said, "Presumably, if the Nationalists had not taken those freight car loads of things to Taiwan, you would have a much more complete display." I never saw a face turn to so much wrath and fury—not at me but at that event. She held forth for some time about what she thought of the Nationalists raiding the treasures of China.

We duly issued a report[27] that included detailed descriptions and our impressions of the dozen or so engineering schools and laboratories we saw, as well as an equal number of factories where engineers worked. If there is a leitmotif to the report, it is that judging Chinese engineering education in the abstract made no sense. As Joe Pettit observed in his introduction, "One cannot understand the current forms of engineering education without comprehending major trends and events since 1949, such as the impact of Soviet technical advisors in China in the 1950s, the onset of the Great Proletarian Cultural Revolution in 1966, the political conflicts of the early 1970s, and finally the current era which follows the death of Chairman Mao in September 1976." I backed that up in my assessment of Chinese research and development, commenting that "one is struck by the enigma of a country with so many assets in people and natural resources not having developed more fully . . . and that to account for this enigma one must infer that the nature of Chinese culture, history, politics, and social structure, together with the burden of too many people to support, have been obstacles to modernization."[28]

Historical Start

I gained a second perspective on science and technology, culture and industrialization, when I chaired beginning in April 1978 the Joint Consultative Committee on Egypt and made two trips to Egypt with the Egyptians meeting in the United States in the interval.[29] It got off to an historical start, for just as we got to our hotel in Alexandria the television monitor came on with the announcement and pictures of the prime minister of Israel and the President of Egypt signing a peace treaty. Somewhat like China, Egypt didn't fully exploit its scientific and technical resources: some 50,000 scientists and technologists and over 150 research and development institutes. There were many reasons, with a standout one being the weak bonding between these intellectual resources and industry. The Egyptian government had strict rules on foreign investments in factories, including marketing, which meant that they missed out on a great U.S. strength: the ability of research organizations to get the products they needed from industry, rather than having to create their own. They lacked that connection, and it hurt them.

At the same time, the Egyptians were receptive to new ventures and we were able in my three-year tenure as chair to launch new activities, including several research and development programs, which in common emphasized close cooperation between research teams and the beneficiaries in industry and agriculture. These included evaluation of Egyptian phosphate ores for wet-process phosphoric acid and phosphate fertilizer production, the fertilizer industry, and Egyptian farmers; corrosion causes and control for the oil refinery at Suez; improving the processing of wool scouring and wool wax recovery for an Egyptian textile processor and the export market; and development of Red Sea fisheries for Egyptian consumers of fish and seafood products.

Like China, and like the United States for that matter, Egypt didn't escape the downside of industrial development, in this instance that development would seriously pollute the Nile. And it didn't escape a second downside, of charging ahead with technology without looking at the social consequences, pro and especially con. A woman who was dean of the College of Social Studies at Cairo University was quite perturbed about government officials treating all new development as a pure good with-

Plate 17

Science is officially back in the White House: Legislation establishing the Office of Science and Technology Policy is signed by President Ford in 1976. *(The White House)*

Kissing Bunny, the best part of being sworn in in 1976 as President Ford's science advisor. To the right is Vice President Rockefeller. *(The White House)*

Plate 18

President Ford and I meeting with senior Soviet officials, including, third from left, V.A. Kirillin, of the Soviet Council of Ministers. *(The White House)*

Richard Atkinson (left) and Phil Smith (right), 1975, who helped the Foundation and me get through some rocky moments, MACOS not the least of them. Dick Atkinson succeeded me as NSF Director and Phil went with me to the Ford White House and then stayed on to serve my successor as science adviser, Frank Press. *(National Science Foundation)*

Plate 19

The late Roberto
Clemente and I, each
getting "Man of the Year
in Pittsburgh" awards.
*(Pittsburgh Junior
Chamber of Commerce)*

Receiving the National Medal of Science in 1991 from President Bush. *(The White House)*

Plate 20

Losing at chess to V.A. Kirillin, the Deputy Soviet Premier.

Ruben Mettler, TRW Board Chairman, and I on the Rhine, surely doing serious business.

Plate 21

Bob Seamans, then Secretary of the Air Force, and I circumnavigating the world in style at the geographic South Pole, 1972.

At the Soviet South Pole station at Vostok. Not sure how many vodka toasts had gone down by then, but clearly there was no shortage of vodka, 1972.

Plate 22

Just prior to the "longest two minutes" of my life: The full-scale, full-duration firing of the Space Shuttle's re-designed solid-rocket booster, which had intentional flaws, to ensure that the hot gases would in fact be contained, even if one or two barriers failed. *(NASA)*

A "few years later": Victor Neher, my Cal Tech doctoral adviser who also helped lure me to the Rad Lab, and I in 1995.

Plate 23

Randolph, New Hampshire, 1995: Seven grandchildren, two proud grandparents, and one very fine car.

Bunny doing one of her favorite things, gardening at Randolph.

Roy and Guy Jr. laid out a splendid trip for us in the Wind River Range right after I left government in 1976. The whiskers came off when I got home.

Plate 24

Stever Ridge in Antarctica. *(U.S. Navy)*

out considering social consequences. Unemployment was high and living standards were low, and while development could in time make for a better life, the transition was likely to be especially tough on the most vulnerable, the poor and the jobless.

I gained a wider perspective on the efforts of Egypt and China to modernize in the late 1970s through another NRC committee I chaired in support of a United Nations Conference on Science and Technology for Development, in Vienna in August 1979. The previous conference was in 1963, and in the interval, as I wrote in my introduction to the report, "the world has come to recognize new problems—finite energy supplies, environmental degradation, runaway population growth—when it was hardly conscious of them" in 1963.[30] Our committee offered a plethora of recommendations, with one typical example to find ways to reduce postharvest losses. The message was that the world didn't have a shortage of food but a shortage of means to get it to the people. "Conservative estimates of the amount of food lost between harvest and consumption as a result of pests, microorganisms, inadequate storage, poor processing techniques, and the like range upward from 10 percent for grains and legumes and 20 percent for perishables."[31]

The meeting was a mixed experience: a chance to savor Vienna, its music, and its food, offsetting a somewhat disappointing meeting. The general sessions were quite repetitive, with nothing new in them; it didn't even pay to take notes. There was a lot of required business, with luncheons and dinners held by groups all seemingly with different agendas. It was like a cotillion with everyone hurrying to get their dance cards filled.

Still, there were many pluses. Bunny and I were delighted by the people we met, especially so when just as we got to our hotel we met Fletcher and Peg Byron, old friends from Pittsburgh who we met when I was president of Carnegie Mellon University and Fletcher the chief executive officer of the Koppers Corporation. The U.S. delegation was led by Father Theodore Hesburgh, president of Notre Dame; the Department of Commerce had a strong group led by Jordan Baruch and Lewis Branscomb and including many leading industrialists of whom Fletch Byrom was typical. Roger Revelle, a leading scientist in the appropriate use of science and technology for development, was a strong participant.

He and I were asked by our leadership to draft a summary statement for the final report.

There were durable results from the conference. An organization of Third World countries, the Group of 77, created in 1964 at the end of the first United Nations Conference on Trade and Development came away from the 1979 UN conference with a clearer view of the value of market economies. There was indirectly an affirmation of support for the International Institute for Applied Systems Analysis. IIASA was born in a remarkable speech by President Johnson in 1966 in which he said that "it was time that the scientists of the United States and the Soviet Union worked together on problems other than military and space matters, on problems that plagued all advanced societies, like energy, our oceans, the environment, health."[32] Chartered in 1972, now located in a former imperial residence of the Hapsburgs in Laxenburg a few miles south of Vienna, IIASA has served as a nongovernmental institution that brought the sharp tools of sciences of statistics and decision analysis to bear on studies of environmental, economic, technological and social issues in the context of human dimensions of global change.[33] IIASA had its problems, financial and political. The Johnson administration's support for it died. The American Academy of Arts and Sciences offered to support it and did. Things weren't made any easier by the discovery that one of the USSR representatives to IIASA was using it to operate a spy ring in Europe.

CHALLENGER

On January 28, 1986, came the terrible explosion of the *Challenger*, 73 seconds into its flight. Seven people[34] died in the fiery mess. I was in New Jersey at a Schering-Plough board meeting. My first thought was for Christa McAuliffe, the New Hampshire school teacher. The National Aeronautics and Space Administration (NASA) had decided to put a schoolteacher on board, held a national contest to choose one, and Christa won it. One of the contestants was my son Guy, Jr., also an English teacher in a New Hampshire school, who had spent a lot of time putting together a pretty good case for what he would do as a teacher having orbited in space. I also thought of the second woman on that flight, Judy Resnik. I remembered that Judy and her fiancé, when they were students at Carnegie Mellon University and I was president, invited Bunny and me to a lovely fraternity dinner in a restaurant atop Mount Washington. It was a wonderful conversation with wonderful people. Beyond my grief for Christa, Judith, and their fellow crew members, I thought about the longer-term effects. The *Challenger* payload included TDRS-2, one of three satellites for the Tracking and Data Relay System, the telephone central for spacecraft communications.

A presidential commission to investigate what happened issued its report on June 9, 1986, less than six months after the explosion. Its hearings were public, and I and others soon were disgusted by the weaknesses in the design of the solid-fuel rocket booster for the shuttle, mistakes that likely caused the explosion. What was worse, we learned of NASA's misuse of its own rules on operating limits for the shuttle.

Slow-motion pictures of the shuttle just before it exploded showed a flame shooting out of one of the two solid-fuel rocket boosters and burning straight toward the main liquid fuel tank. The problem was in the two O-ring seals in the booster intended to keep hot exhaust gases from escaping. The O-rings were vulnerable to cold temperatures. And it was unseasonably cold at the Kennedy Space Center the night before the launch. Cold stiffened the O-rings, and they couldn't seal effectively. Hot gases got past the O-rings, burned a hole in the metal case, and gas jetted out the side of the booster. This jet weakened or penetrated the adjacent liquid fuel tank. It broke apart. Hydrogen and oxygen mixed uncontrollably, exploded, and destroyed the shuttle and seven lives. The late physicist and Nobelist Richard Feynman of Cal Tech memorably showed what happened as he dunked O-ring material into ice water in front of television cameras and his fellow commission members. It became hard and inflexible. Feynman was justifiably harsh in his "personal observations" included as an appendix to the Rogers Commission[35] report in urging the commission to:

> . . . make recommendations to ensure that NASA officials deal in a world of reality in understanding technological weaknesses and imperfections well enough to be actively trying to eliminate them. They must live in reality in comparing the costs and utility of the Shuttle to other methods of entering space. And they must be realistic in making contracts, in estimating costs, and the difficulty of the projects. Only realistic flight schedules should be

proposed, schedules that have a reasonable chance of being met. If in this way
the government would not support them, then so be it. NASA owes it to the
citizens from whom it asks support to be frank, honest, and informative, so
that these citizens can make the wisest decisions for the use of their limited
resources. For a successful technology, reality must take precedence over public relations, for nature cannot be fooled.[36]

It shouldn't have happened. The Rogers Commission called the safety
reporting system at NASA "silent." The agency had in effect ignored
prelaunch warnings that could have prevented the tragedy. Engineers at
Morton Thiokol, where the solid-fuel booster was made, found the design flaw in 1977 and reported it, only to have it ignored. That continued even after erosion of the O-ring turned up after a 1981 shuttle flight.
And finally, and courageously, two Thiokol engineers refused on the day
of the accident to sign a release for the firing of the boosters. They knew
full well that the cold temperature at the time was beyond design limits.
NASA's ignoring them disgusted me. They became heroes in the engineering profession.

The Rogers Commission coupled its harsh criticism with a recommendation that the National Research Council, of the National Academies of Sciences and Engineering, oversee the redesign of the solid rocket
booster. While I was in Europe, on tasks as foreign secretary of the National Academy of Engineering, Robert White, president of the NAE,
called and asked quite bluntly that I chair the panel overseeing the booster
redesign. I asked for time to think about it, had a sleepless night knowing
full well that this would be a very difficult and possibly thankless task.
Bunny, who was traveling with me, told me I should do it. I accepted.

I had a hand in selecting the panel[37] and also its senior staff officer.[38]
I asked at the outset how much time was needed and was told two
months. Two months turned into almost 30. The panel held 21 meetings
between June 1986 and December 1988; individual members went on
77 site and inspection visits and wrote seven interim reports and one
final report.[39] All this from unpaid volunteers! We got incredible amounts
of data—from NASA, from Morton Thiokol, and from United Space
Boosters, Inc. United Space Boosters was responsible for various parts of
the assembled rocket, including the skirt, but Morton Thiokol was responsible for the cases and for mixing and pouring the solid propellant.
We also produced our own data, including a bar chart tracking the thick-

ness of the agenda books. The average thickness of the agenda books was 1.783 inches, the worst 2.569 inches, and the thinnest .75 inches. We had lots of data.

The panel's first meeting was rough, with tensions thick between NASA and Morton Thiokol officials and between them and us. When NASA gave its version of what happened in the morning and Thiokol in the afternoon, one panel member bluntly interrupted the second presentation to state, rightly, "What you're now telling us doesn't agree with the morning presentation." Awkward silence, then rancor, and then agreement by the panel that I go see James Fletcher, NASA administrator. I did that and told him it wasn't our job to decide who got the blame but to make sure the booster redesign was done right. That ended finger pointing before the panel by NASA and Thiokol.[40]

We worked the redesign over hard. We looked not only at the redesign itself but also the means for testing and verifying the safety of new designs, material specifications, and quality assurance and control. Four different joints in the booster rocket had to be redesigned. Crunch time came when the panel insisted on a test not just of the individual components but of the total assembly. And it insisted that a realistic test be done by putting an intentional flaw in the new design so that hot gases could in fact reach the back-up secondary O-ring. Would it contain the hot gas? In fact with the redesign there were now three barriers for containing the hot gases. We believed that final approval of redesign and flight first needed a full two-minute firing of a rocket on the ground—two minutes because that was how long the booster burned in flight. We believed that a full-scale, full-duration test was "essential to the certification of the design.[41] The results of the test, including assessments of the performance of the joints that are intentionally modified to assure pressurization of the primary O-rings, must be available for review before the return to flight."[42] Some NASA officials felt that would cause too much of a delay in return to flight. We insisted.

The test came August 18, 1988. Most of the panel was there. By then I had developed back problems and used a cane to walk out to the firing. It was the longest two minutes of my life. But the test was successful and afterwards some NASA and Thiokol officials talking to the press who had watched the test, took full credit for it. One of the reporters said to me on the side: "Yes, they're delighted—after you people dragged them kicking and screaming to do this test." We concluded our seventh interim report to the NASA administrator and the American public by noting that "the redesigned solid rocket boosters have incorporated a large number of improvements that should result in considerably enhanced safety and reliability, hence reduced risk. Risks remain, however. . . . Whether the level of risk is acceptable is a matter that NASA must judge. Based on the panel's assessments and observations regarding the redesigned solid rocket boosters, we have no basis for objection to the current launch schedule for STS-26."[43]

Indeed, on September 29, 1988, over two years after the panel began its work, I was at the Kennedy Space Center for the launch of STS-26. I again felt like I had stopped breathing for two minutes. But I was confident. And the launch went beautifully. The panel's work did not end with the launch. We inspected the burned-out boosters that were retrieved from the Atlantic Ocean and used that information in preparing our final report, delivered to NASA almost exactly two and a half years after the first panel meeting and a month less than three years after the *Challenger* accident. In December 1988, NASA recognized the panel members and staff for our work with its public service awards. About a year later NASA brought the panel together for a "reunion" to review data obtained from six shuttle flights and four ground tests conducted in that time. During this meeting, the panel members and staff were all honored by the Corps of Astronauts, which gave us each its Personal Achievement Award for improving the safety of manned space flight.

Fine Bones

That ceremony, putting in effect a full stop to an exhausting if very rewarding part of my life, led to some serious introspection. It was 1988 and I was 72. While I was done with the booster redesign, many other things were still in my life and one very large task. Various activities included assisting with advice by the Academies, especially on space issues, for the next president; work on affairs of the National Academy of Engineering; involvement in foreign exchange programs with Japan and China; and the like. The large task, to occupy me for five years, was the new Carnegie Commission on Science, Technology, and Government. But there was also my worsening back problem. For two or three years I had traveled all over the world with recurring pain, but somehow managed throughout to keep a sunny face for the outside world. I knew surgery was ahead. Dr. Donlin Long, professor of neurosurgery at the Johns Hopkins Hospital, did that. The operation[44] went well, albeit not without complaints from Dr. Long and his associate who did the surgery about the hardness of the bones they had to break to get at the spinal cord. I thought that a plus: my bones were in really fine shape.

In the midst of these medical problems, I returned in a sense to my old stomping grounds: advising the U.S. president, albeit this time from the outside. In 1988 I chaired a committee of the National Academies of Sciences and Engineering that advised the incoming president, who turned out to be George H. W. Bush, on U.S. space policy.[45] And in that year I also began five years of work for the Carnegie Commission.

It's a Washington cliché that advice not asked for is dead on arrival. Against that, we had little reason to be optimistic that our efforts to formulate directions for future space policy would provide fertile. But in fact they did. Part of the reason may have been desperation. While the flight of the *Discovery* shuttle in September 1988 put the U.S. "back in space," it was in fact a program looking for a place to go. There were many questions that a new administration couldn't avoid: what to do with the shuttle, a very expensive and aging transport system; the role of manned space flight; the purposes, not to mention the stunning costs, of the proposed space station; and certainly not least why we were in space in the first place. Why should the United States pay for a very expensive enterprise that seemed to have lost its way?

We responded by setting out U.S. goals in space and then the elements of an effective policy to achieve them.[46] On goals we argued for continuance of the established preeminence of the United States in space, noting that unmanned missions would be more effective and that the number of shuttle flights could be reduced. We also argued strongly for a program of earth observations done from space under the rubric of "Mission to Planet Earth" and defined the components in some detail. Finally, we dealt with the touchiest part: human exploration of the solar system. We didn't offer a yes/no on sending manned missions into space, including manned outposts on the moon or landing on Mars but pointed out that we didn't have the ability to do it in 1988, that the cost of *each* mission was likely to be in the $100 billion range, and that if we wanted to do it, a manned space station was the essential first step—to develop the technologies for manned exploration and to understand what happens to humans on extended trips into space.

We joined these goals with some practical advice to the new president, advice much of which in one form or another entered space policy. We recommended two parts to the space program, a core program and special initiatives. The core program would include a robust fleet of manned and unmanned launch vehicles, a balanced space science and earth remote-sensing program, and an undergirding of advanced technology. That program would then provide realism and flexibility for special initiatives such as Mission to Planet Earth. We also recommended strong international partnerships, both as a simple recognition of the increasing space capabilities of other nations and to learn from the then much greater experience of the Soviets in manned space. Some programs such as Mission to Planet Earth virtually compelled international collaboration. Finally, and certainly not least, we addressed NASA management, some of whose deficiencies were sadly put into relief by the 1986 *Challenger* disaster:

> NASA has become oriented toward short-term, large-scale missions rather than the patient, far-reaching campaigns that must be sustained over decades. It is also an aging institution that urgently needs revitalization of its physical plant and human capital. . . . Much of the difficulty is systemic in character. We need to redefine the now-diffuse roles of the NASA field centers. . . . Space operations should be separated from space science. . . .[47]

Many of these things came to pass over time, some within the Bush administration and others afterwards. I learned a long time back that, if nothing happens to good advice, try waiting a while. True in this case, and many others. Being effective in Washington is neither for the impatient nor the cynical.

PRESCIENT ACT

The belief that good advice eventually takes root was tested even harder in my service on the Carnegie Commission on Science, Technology, and

Government. Why the commission? Why then? After all, this was a very large and costly effort—five years and several million dollars. The immediate "why" I suppose was that the president[48] of the Carnegie Corporation of New York thought it time and through this foundation could pay for it. But the "why" was more nuanced than that. The 1980s were an unsettling time for science. Again, the crisis of the 1980s in American economic competitivess begged the question of whether the huge federal investment in research and development was worth the money. After all, other nations, investing much less than we, were in some fields beating us like a drum. What were we getting out of all that money? Why support basic research, especially basic research that seemed to have at best a tenuous connection to the American economy, to national security, to our social problems?

Those kinds of questions entered political discourse, as the Congress especially, including the overseers of the National Science Foundation, demanded transparency in the reasons for federal support of fundamental research. How relevant were the sciences to addressing some very serious problems, such as chemical toxins? The launch of the commission was a remarkably prescient act, in effect anticipating the end of the Cold War: the Berlin Wall came down in November 1989, the Soviet Empire in Eastern Europe dissolved, and the Soviet Union itself ruptured in 1991. The Cold War for 40 years served as a major rationale for American investments in science and technology, not only for military security but also for the physical sciences taken very broadly. With the Soviet bogeyman seemingly gone, why spend money on science we didn't seem to need anymore? What now was the reason for the federal government not only to support science but also to expand it? Not least, it had become painfully obvious that the structure of the federal effort in research and development—how it was dispersed among many agencies and its arrangements within those agencies; the often substantial, confusing, and resource-draining overlap of responsibilities; the often baffling arrangements in the Congress for dealing with science and technology programs and their budgets—was out of sync with contemporary imperatives for the management and application of science and technology.

And these problems were forcing themselves onto the dance floor absent perspectives from the balcony. The last full-scale examination of the American scientific enterprise was done 40 years earlier, the five-volume report on *Science and Public Policy*, prepared under the direction of presidential assistant John R. Steelman.[49] The Steelman Report[50] established in effect the bona fides for the establishment of the National Science Foundation. Issued in 1947, it characterized the "extension of scientific knowledge . . . as a major factor in national survival," offered a data-rich argument for support of fundamental research at universities, and recommended a minimum spending level for research of 1 percent of the gross national product.[51] It was then and still is "the most complete and detailed description and analysis of the U.S. research system that had ever been produced."[52] It also sought to bring science back into the White House after the end of the wartime Office of Scientific Research and Development. It recommended the establishment of an Interdepartmental Committee on Scientific Research and Development and appointment of a presidential science advisor. President Truman took the first half of that advice but not the second.[53] It took the cold water of *Sputnik* to get a science advisor in place in the White House.

The Carnegie Commission[54] took five years of my life, beginning in April 1988. It was led by two cochairs, Joshua Lederberg, a Nobelist and then president of Rockefeller University, and William Golden,[55] a businessman who had devoted much of his life beginning in World War II to the continued health of American science. Of the 22 members of the full commission, including former President Jimmy Carter, half were like me, scientists and engineers with substantial governmental experience, and half were nonscientists with great interests and achievements in public affairs and science policy. And there was an advisory group with a similar mix, including former President Gerald Ford. It produced in its life 19 formal reports plus memoranda, special analyses, and the like. It was by far the most thorough and exhaustive examination of federal science and technology policy in all its dimensions since Vannevar Bush did it virtually by himself in 1945 with *Science, the Endless Frontier*[56] and John Steelman with a lot of help produced the 1947 report *Science and Public Policy*. I was somewhat involved in much of what the commission did, most deeply in two efforts, chairing the groups that produced commission reports on *Enabling the Future: Linking Science and Technology to Societal Goals* and *E3: Organizing for Environment, Energy, and the*

Economy in the Executive Branch of the U.S. Government (E³ Report).[57] I especially liked the "Enabling the Future" phrase, which came from Antoine de Saint-Exupéry's *The Wisdom of the Sands*: "As for the future, your task is not to foresee, but to enable it."

The influence of the E³ Report antedated its publication in April 1990 especially its then-novel focus on the close coupling of environment, energy, and the economy. That message, which we had briefed around the government before formal publication, was embedded in President Bush's speech to the Intergovernmental Panel on Climate Change on February 5, 1990, where he cited the need to review and revise our national energy strategy to include economic and environmental effects. This E³ triplet is of course still at the heart of national concern. And the commission faced the even larger question posed by the steady questioning in the 1980s of the wider relevance of scientific research. That was in my portfolio. Some people questioned whether it was possible to state goals. I thought that nonsense and told them so, reminding them that there are five goals[58] in the preamble to the Constitution and that at least three of them relate to science—insure *domestic tranquility, provide for the common defense, promote the general welfare*. Our task force indeed produced four major goals in supporting science:

1. quality of life, health, human development, and knowledge;
2. a resilient, sustainable, and competitive economy;
3. environmental quality and sustainable use of natural resources; and
4. personal, national, and international security.

We then set out particulars for each of these very broad hopes, 25 in all. And we were realistic enough to know that goals do evolve, and for that reason we suggested a standing group—a nongovernmental National Forum on Science and Technology Goals—to dynamically seek to refine and seek the implementation of goals in the context of evolving national and international circumstances and policies. That never happened, but two fora were held focusing on goals for science and technology as they related to the environment and the nation's economic future. For the environmental goals, and backed by a very considerable amount of discussion, commissioned papers, and the like, we examined in detail six topics: economics and risk assessment, environmental monitoring and ecology, chemicals in the environment, the energy system, industrial ecology, and population.[59] The second effort, on science and technology and the country's economic future, set out three major goals:

1. Over the next decade, achieve a sustained level of productivity growth that will allow rising living standards and noninflationary economic growth. Do that by increasing investments in science and technology; developing new mechanisms for international research collaboration to advance fundamental knowledge, drawing on the experience of recent years; and developing better metrics and understanding of science and technology trends and their connections to economic growth.

2. Increase the number and proportion of Americans prepared for science and engineering careers with a focus on underrepresented groups. Do that by having more scientists and engineers work with local communities to improve K-12 education; creating institutions and a supportive culture that facilitates lifelong learning; and, encouraging U.S. industry and wealthy individuals, particularly those who have gained great economic benefits from high-technology booms, to focus efforts and resources on improving education for a science and technology savvy work force.

3. Improve the domestic and global market environments for U.S.-generated innovations. Do that by adopting national standards for securities litigation and product liability, and examine trade, antitrust, and intellectual property policies with a view to improving global market access for U.S.-generated innovations.

Those were deliberately ambitious goals, but at the same time they are goals that in one way or another have been addressed since we wrote our report.[60] Whether that was cause and effect or we were simply leading where others were going anyway, I can't say. But these things are getting done.

A particular example—one that certainly hit on our call for developing better metrics and understanding of science and technology trends and their connections to economic growth—was the work of a

committee[61] on which I served in the late 1990s to develop a more realistic understanding and allocation of what the federal government invests
in science and technology. The scale in dollars of the federal effort in
research and development has historically included items—for instance,
testing of new weapons—that are not really part of the effort to push
new knowledge. As the report,[62] which became known as the Press Report, after its chairman, Frank Press, who succeeded me as presidential
science adviser and then went on to the presidency of the National Academy of Sciences, stated:

> Almost half of Federal research and development is spent on such activities as
> testing and evaluation of new aircraft and weapons systems in the Depart-
> ment of Defense, nuclear weapons work in the Department of Energy, and
> missions operations and evaluation in the National Aeronautics and Space
> Administration. Those activities are very important, but they involve the dem-
> onstration, testing, and evaluation of current knowledge and existing tech-
> nologies. Even when they are technologically advanced, these functions do
> not involve the creation of new knowledge and the development of new tech-
> nologies. The federal research and development budget as currently reported
> is thus misleading, because it includes large items that do not conform to the
> usual meaning of research and development.[63]

The committee confronted this distortion of what we really invest in
new knowledge with a new concept: a federal science and technology
budget that gave a much truer measure of federal investments in new
knowledge and enabling technologies. It was different, and predictably it
set off a storm of debate. And predictably while lip service was paid to the
idea, the administration and the Congress ignored it. But like many good
ideas, it's beginning to go mainstream: the George W. Bush administration in its first budget message highlighted and used the concept for its
fiscal year 2002 budget—six years after the report was published! Again,
policy work in Washington is not for the impatient.

THE OCTOGENARIAN TRANSITION

This intense look at national science and technology policy through the
Carnegie Commission and the Press Report finished in 1995. But Bunny
and I had a new beginning: we moved. I would be 80 in 1996. Since I
came to Washington in 1972 to be director of the National Science Foun-

dation we had lived in a wonderful house in Georgetown. Best of all, our Georgetown base meant I could walk or cab to many working places—to the National Academy of Sciences building near the Lincoln Memorial, to the headquarters of Science Service where I served on the board, to the offices of Universities Research Association, and to the Washington offices of the Carnegie Commission.

But we knew our time in Georgetown was over. The 200-year-old house was creaking. There was a lot of work to maintain it and always stairs to climb. And Bunny and I were also creaking: we both had medical problems. In 1995 I surrendered my gall bladder to surgeons, and Bunny had pneumonia. And in 1996 I was told I had prostate cancer. So a summer idyll at Randolph got traded for nine weeks of X-ray treatment at Georgetown University Hospital. Even so, we were at Randolph in October for a "geriatric jubilee" for my eightieth birthday. My capacity for outdoor sports was down since I had my back operation and we'd given up downhill and cross-country skiing, and in the cruelest loss I gave up fly-fishing on some of my favorite streams where the slippery rocks made wading too dangerous.

After a lot of hard work, including consulting with dear friends who had faced these decisions, we found a wonderful place in Asbury Methodist Village in a Maryland suburb. It was a house with all the living quarters—from kitchen to bedrooms—on one floor, with a very large basement and attic. It was commodious enough to accommodate our steady stream of visitors, most especially our children and grand children—in 1995 when we moved we had seven grandchildren—although unlike our summer place in Randolph, New Hampshire, not all at the same time!

Those hard changes lightened a lot when I was given a chance to think about what had been done by twentieth century technology. Toward the end of 1999, I was asked by William A. Wulf, our effective president of the National Academy of Engineering, to chair a committee to pick the 20 greatest engineering achievements of the twentieth century. There was one condition: it was supposed to make a substantial favorable contribution to the quality of life. That meant that we would not select one great achievement, excluding to my regret the moon landing in 1969 or the construction of the Panama Canal.

BOX 10-1 Greatest Engineering Achievements of the Twentieth Century

1.	Electrification	
2.	Automobiles	
3.	Airplanes	
4.	Water supply and distribution	
5.	Electronics	
6.	Radio and Television	
7.	Agricultural mechanization	
8.	Computers	
9.	Telephones	
10.	Air conditioning and refrigeration	
11.	Highways	
12.	Spacecraft	
13.	Internet	
14.	Imaging	
15.	Household appliances	
16.	Healthcare technologies	
17.	Petroleum and petrochemical technologies	
18.	Laser and fiber optics	
19.	Nuclear technologies	
20.	High-performance materials	

The list we produced[64] (see Box 10-1) was stunning to me at the time and still is. These achievements enabled by science and technology changed everything. No, they were hardly unmixed blessings—several in the list from highways to nuclear technologies give one pause—but as with me they gave Americans many more choices in how to lead their lives than were obvious at the end of the nineteenth century. What choices will this new century bring?

11

Afterword

T he train that carried me in 1938 across the country to Pasadena for doctoral studies at the California Institute of Technology took me into a terrain of science familiar yet remote. Growing up in Corning, New York, I had my first glimmerings of science in seeing and hearing the great men of Cal Tech as they visited to check on the casting of the mirror for the Mount Palomar telescope. Those beginnings amplified by my years of Colgate drew for me a picture of the world of science that I thought I knew, a familiar world.

But it was a limited knowledge, and I knew little of the wider world of science I was to enter. Looking back on that time from a new millennium, I'm stunned by the transformation of what was then for me a remote enterprise to one that is now central to our lives. The American scientific enterprise in the late 1930s was not only smaller but also different in character. Government research was dominantly intramural, the work done by agencies and bureaus in their own laboratories—those of the Food and Drug Administration, the Bureau of Standards, the Bureau of Ships, and many others. Most important to my own future was the National Advisory Committee for Astronautics, which in the face of low funding managed to keep the country current in aeronautics and to launch critical facilities, including the world's first full-scale wind tunnel at the start of the 1930s and the Ames Aeronautical Laboratory at the end of the decade.[1] Nevertheless, physics, the field I was entering, depended heavily until 1940 on money from state legislatures, industrial corporations, individual donors, and, not least, philanthropic foundations, such as the Rockefeller Foundation and the Carnegie Corporation.

Cal Tech had its hands in two of the most fundamental—and expensive—fields of physics: (1) astronomy, from the solar system to the galaxies, and (2) the fundamental structure of matter beset by a growing array of fundamental particles and the forces among them. Throughout the twentieth century, ever more costly, complex, and clever telescopes and atom smashers were paid for by private foundations and increasingly by the federal government, notably the National Science Foundation, the Department of Energy or its eponymous predecessors, and the National Aeronautics and Space Administration.

Equally remote to me as I settled into my quarters at Cal Tech— including bunking on a cot as did fellow graduate students in the open loggia of the Athenaeum, the faculty club—was the remarkable history of the place. It had transformed itself within a generation. In 1920 it was the one-building Throop College of Technology graduating about 10 engineers each year. When I arrived, it was one of the premier universities in the country and the "only institution able to break into the ranks of the major research universities in the interwar years."[2] It did that with the time-tested recipe of brilliant leadership, wise and timely philanthropy, and a bedrock principle. The leadership was first that of George Ellery Hale, an astronomer, who wanted a center of scientific research to support his observatory on Mount Wilson and to help him do it succeeded in persuading two national leaders in science to join him: Alfred Noyes, a chemist, and, of course, the physicist Robert Millikan (whose visit to Corning to look at the Mount Palomar glass started my journey to Cal Tech). The philanthropy was mainly local, as leaders in Southern California in banking, oil, electric power, and the like signed on to Hale's vision that a strong center of science was essential to the growth of the Southern California economy. Most remarkable, I think, was the third part of Cal Tech's emergence in the 1930s, the principle that Cal Tech was not to be, as the Massachusetts Institute of Technology was in the 1930s, primarily a school of applied science. Rather, its strength was to be in fundamental science that would in time seed new technology, innovation, and economic growth. It took as a guiding principle that "basic research and scientific training would yield discoveries in practical matters as well as pure science; and these discoveries would prove to have tangible benefits."[3]

What was becoming less remote to me than the history of Cal Tech and how it got to be was that the world was getting very dark. Germany invaded Poland a year after I started my graduate studies. The coming of war and then the war itself radically reshaped the landscape of American science and technology. Until the onset of war, the federal government funded university research at a low level and often restrictively. President Roosevelt authorized the Works Progress Administration to give grants for research but stipulated that 90 percent of those getting money had to be on relief, not as laughable as it now seems, because a lot of scientists and engineers were out of work and dependent on the government for shelter and food. However tentative these efforts, they became the pilot tests for how to move money to universities when the sums exploded in size. The lessons learned were soon to be applied.

The military was also struggling with how to mesh academic science with its needs. It wasn't easy. Military bureaus were notoriously ingrown, each service doing its own research no matter whether it duplicated or could learn from work in other services. Often they had no idea what the other services were up to for they kept their research secret not only from the taxpayers and possible foreign enemies but also from each other. "Thick walls of secrecy separated the technical bureaus of one service from their counterparts in the other. And while the army and navy technical bureaus employed civilian scientists and engineers, the armed services did not have the money to pay for much outside civilian technical help, either industrial or academic, and as prerogative conscious as ever, they did not like to rely upon it."[4]

There was no joint research among the services, no overarching organization, no ceding of research direction to outsiders, such as university scientists. That was so even when military leaders came to understand in the 1930s that the country was lagging badly in military research and technology. As Robert Millikan observed in 1934, defense research ought to be "a peacetime . . . and not a wartime thing. . . . It moves too slowly after you get into trouble."[5]

Millikan was right in 1934 but was proven wrong when the war came. That a predictable disaster became a triumph was due to the rise to power of military leaders who realized that their research system was disastrously insular; the emergence of acute technological needs, especially for locating military aircraft and directing fire at them; and the work of one very astute and politically adept engineer named Vannevar Bush. That mix established that first-rate research rapidly done and deployed to the military was possible in wartime. The means for doing it were straightforward though revolutionary: the military set the goals and the scientists figured out how to get there with new contracting mechanisms to move money to the universities without compromising fiscal control and oversight and a willingness to have the problem shape the institution that would attack it, not the other way around. Thus, the Radiation Laboratory at the Massachsetts Institute of Technology was set up in response to and within a few months after the British delivered the cavity magnetron that enabled microwave radar.

Wartime research became a cascade of creation, and the Rad Lab and many other places became "universities in a pressure cooker."[6] And I fell right into it. I was 24 and fresh out of graduate school, but there was no hesitancy putting me to work within the inner sanctum of a desperate technological race to turn the breakthrough of the cavity magnetron into devices to fight the enemy. While they seemed graybeards to me, the Rad Lab "faculty" was in fact young, their average age in the mid-thirties and I. I. Rabi, later a Nobelist, the elder statesman at 43. Contrast that with the Germans, who didn't trust their university scientists, never mind the rigid hierarchy that kept promising students from rising quickly to serious tasks and real responsibilities. German research was done within strict military controls, in military establishments, and a strong military culture. A senior German scientist complained just before the war in Europe ended that "Germany lost the war because of incomplete mobilization and utilization of scientific brains."[7] Except of course for rockets, airplanes, and torpedoes. Thank God they didn't figure out how to build nuclear bombs!

The most powerful lesson of wartime research in the United States was that it worked, spectacularly so. And while radar, the proximity fuze, penicillin, the atomic bomb, and antimalarials were all terribly important products, the most powerful one in transforming American science and technology was the generation of people in the universities, industry, and the military that set up the institutions, ran them, and did the work. Their battle-tested belief in the productivity of university science when managed right shaped the structure and style of postwar research in the United States. It seeded new agencies for supporting research in the universities, from the Office of Naval Research to the National Science Foundation. It established new research laboratories with particular if broad missions managed by the universities, such as MIT's Lincoln Laboratory, Cal Tech's Jet Propulsion Laboratory, Johns Hopkins Applied Physics Laboratory, and many others. It solidified new mechanisms for investing money from the public treasury in the very best science. And, certainly not least, it set the course for my own career as I determined to continue to serve the national security of the country but from the vantage point of a university post.

It is now commonplace to comment on the seemingly chaotic federal structure that settled into place after World War II. And it is undeniably a jumble, a dizzying mix of agencies with differing goals—from simply supporting the best basic research as does the National Science Foundation to furthering the health of Americans as does the National Institutes of Health to assuring technological currency in national security as do the research agencies of the Department of Defense and the military programs of the Department of Energy. That the aims, structures, and cultures of these and other agencies are not unitary is not a weakness but a strength. The history of postwar research is well seeded with stories of young researchers denied support by one agency finding support at another and going on to win national and international awards for their work. Governmental scientific officers have often catalyzed openings in new fields because they had the insight to recognize an emergent research frontier and the funds to support the best in exploiting it. Finally, and most fundamental, the postwar partnership of government and academia in some agencies, particularly NSF and NIH, had as its goal that of supporting graduate education built on blending teaching and research. That was the kind of education I got at Cal Tech before the war and what I returned to when I had my own graduate students and when at the National Science Foundation I had a national opportunity to strengthen that system.

But the best counter to the messiness of the multiple agencies composing the federal research system is the same as for the methods of wartime research: it works. By any measure you care to look at, American science and technology equal and often lead competing work in other countries. And they have survived and thrived in the face of severe stress—budgetary declines in the late 1940s and early 1970s. The potpourri of programs in the 1970s and 1980s intended, with mixed success, to link science more directly with applications, "relevancy," or "strategic goals;" the intense questioning if not disbelief in the value of fundamental research occasioned by a loss in the 1980s of the nation's faith in its ability to compete economically. Yet closing onto 60 years after the end of the war, the American research system in Faulkner's phrase has not "only endured, but prevailed."

How it happened is foretold in many ways—by Cal Tech and other research universities betting on fundamental science, by the military entrusting the nation's security to people in the universities hardly trained to military ways, by the virtual destruction of the prewar system of insular and often stifling government-run scientific agencies that largely did their research internally without the freshening currents of external work and criticism. The war shattered that and what emerged afterwards was a fertile partnership of government support for university and industry research performance. No longer heavily if not exclusively dependent on state and private support and with a growing federal purse now open to them, universities could expand research faculties and facilities. Research that was broader, deeper, more ambitious, and costly was now within reach. The American public and its political leaders carried over into peacetime their war-proven faith in the value of research to national aspirations, and that faith shaped the centrality of the universities to American science and technology.

I can't pretend I thought about all that in August 1945. I simply needed to get on with my life. But how to do that? Still, the *Zeitgeist* was about, and I and many of my equally young colleagues wanted to be at a university. I determined to return to MIT and as the first chills of the Cold War were felt to contribute what I could to the nation's security. Strong and immensely capable university scientists enriched if not dominated postwar military research—Theodore von Kármán of Cal Tech and his seminal work on the technical future of the Air Force; George Valley and his conception of the SAGE (semiautomated ground environment) defense system that saved Project Whirlwind, which in turn became a seedbed for the U.S. computer industry; Louis Ridenour and his work on missile development and defenses; John von Neumann and his seminal work with Oskar Morgenstern on game theory that was fundamental to planning U.S. strategy in the Cold War. They were giants and set the tone for me: stay rooted in the university, take on the toughest military issues, ask hard questions, get the job done quickly, tell the truth even when—especially when—it's hard. What made it work was that the military listened—and acted. It developed officers who were technically educated and trained at advanced levels and who could understand and apply within the military framework the advice we gave. Witness General Jimmy Doolittle and his unsurpassed standing as someone with enormous and daring military achievements and a very strong technical education. Witness General Bernard Schriever, who led the development of intercontinental ballistic missiles. That dialogue with the military— whether working on new missile systems, leading the first hard look at the feasibility of defending against incoming missiles, thinking about the implications of wedding missiles with nuclear warheads, taking an honest look at the notion of a nuclear-powered airplane—was a refrain in my life as I moved up the academic ladder at MIT. That dialogue played out on a much larger stage as a set of university and industry scientists and leaders—James Killian, I. I. Rabi, George Kistiakowsky, Jerome Wiesner, William Baker, Edwin Land, Simon Ramo, and others—provided the nation with the very best technical guidance for its national security. And it was not only what these enormously capable people guided into use but also the bad ideas they stopped that rendered the country extraordinary service. And that set the tone for the role of science at the presidential level: provide the very best technical advice you can and trust the political system.

From the time I returned to the United States in 1945 until 1965 when I became president of the Carnegie Institute of Technology,[8] I led a double life as a member of the MIT faculty and as an advisor to the military and industry. It was a symbiotic link: we applied our best technical and scientific thinking to formidable military and industrial problems and in turn were tutored in the hard practicalities of such matters as missile guidance, design, and production. This infusion of actual practice into pedagogy gave our teaching the currency to enable us to prepare a new generation of scientists and engineers who could enter the defense and industrial worlds as new but hardly naive recruits.

But I got much more from this double life than the symbiosis of teaching and science and technology advice to the military. I was and remain proud that I could contribute to the country's national security during the Cold War as I did during World War II. Bunny and I were just starting our family when the Soviet Union set off in succession an atomic bomb and then a hydrogen bomb and when both superpowers were furiously building intercontinental airplanes and ballistic missiles to deliver them. With other Americans, we thought about surviving a nuclear war. We were scared. Early on, I had witnessed two nuclear tests set off above ground. Feeling the heat course through my body as the high-energy radiation hit, seeing the fireball through colored glasses and then the mushroom cloud, being buffeted by the blast wave, and all the while knowing how many would die if a hydrogen bomb hit a city hardened my determination to do what I could during the Cold War.

At Cal Tech I had learned to do fundamental science at the highest level. At MIT I infused that understanding of what first-class science meant with lessons learned in the war of how in the midst of terrible pressures to turn knowledge into applications. And those lessons gained an even more cosmopolitan flavor when I moved to ever-larger stages, first to what became the Carnegie Mellon University and then to Washington as director of the National Science Foundation and science advisor to Presidents Nixon and Ford. My time at Carnegie Mellon reinforced for me that the world was a messy place and that it certainly wasn't going to change for me. I learned that getting things done when people were in intense intellectual, political, and emotional conflict was a special art form, that I had to work hard to understand why people acted the way they did even if, on rare occasions, I thought their ideas were nuts. I learned to listen hard.

The legacy Carnegie Mellon gave me of a world view attuned and even sympathetic to clashing ideologies served me extraordinarily well when I went to Washington in 1972. I was hardly a political innocent with service as chief scientist of the Air Force and as member, vice chairman, and chairman of the Air Force Scientific Advisory Board. But in those positions I was within essentially a closed military environment. This time it would be quite different and very public. In going from consulting for the government to running an agency, I was moving from life in the rear to front-line combat. But while the stage got bigger, the stakes higher, and the players more powerful and sometimes nastier, I held fast to two principles. The first was the love of my family, my wife and children, and the wonderful life we had living in Georgetown and enjoying the city and the frequent visits of our children, grandchildren, and many friends. At the end of the day, no matter how brutal the politics, there was my family and its devoted constancy to put things right, to help me regain balance and get ready for the next day.

The second belief that helped me survive Washington politics was never to get on the other guy's territory. I couldn't be a better politician than those who made a profession of it, and, thankfully, I wasn't stupid enough to try. Rather, I knew that for me the path to survival was to stake my positions and arguments on the best science and technology, expressed as forcefully as possible and fitted honestly to the audience. That served me well whether I was arguing for new facilities at our research station in Antarctica, proposing new initiatives such as the biome research program, arguing each year for more money for the National Science Foundation, defending the foundation (and myself) against attacks on its curriculum development programs, or building the analytical base for knowledgeable responses to the energy crisis of the 1970s.

My time in Washington closed the circle for me. I had as a very junior player seen the forcing function of World War II carve a new arrangement between government and science. And now in the 1970s I was in the thick of that legacy envisioned by Vannevar Bush but given political durability by President Truman. From the exigent beginnings of wartime, the U.S. scientific endeavor became an enormous public enterprise; rather than a relatively minor and internal enterprise of government, it was now large and external, with fundamental research dominantly, though not exclusively, in the universities, now the "home of science."

In those phenomenal decades my activities and contributions were part of the larger issues of war and peace and the opportunities that opened up. But they were also influenced by some of my personal character and strong predilections. First, all my life I have been almost an incurable optimist, with a confidence and determination about my work. Second, I like people of all kinds and am reasonably tolerant, seeking the positive in people. Third, I like to organize things and run things, many simultaneously. This predilection showed early when I was in high school. Throughout most of my professional life there were many other opportunities to satisfy that urge. They were all quite different but all depended on selecting and getting the best available people for the job, supporting them, making sure they knew they had the responsibility and freedom to act. Few failed me. It always helped if I remained well informed, was enthusiastic about their work, and was in effect the chief cheerleader. And as the organizations I led got more complex (and bureaucratic), I made sure that all the leaders of the different parts had a say in major decisions. Chairing those overview meetings taxed my abilities, especially when we were dividing the budget pie, but that is what I was paid for.

And how productive that "home" and the work of industry and governmental laboratories have been in my time. A major textbook in biochemistry in the 1960s did not even have a separate chapter on the nucleic acids, the first glimmerings of a profound revolution in our understanding of genetics at the level of molecules still uncertain and tentative. There was more excitement and seemingly more importance in decoding slowly and with enormous effort the three-dimensional structure of proteins. When I returned to MIT after the war, the arrangements of the components of atoms—protons, electrons, neutrons, various antimatter forms— seemed sensible. The universe seemed a simple place, a mosaic of understandable pieces—stars, planets, debris between the stars, lots of space. We were just on the piedmont of exploring the ramifications of what we had learned in solid-state physics in the 1930s; and during and immediately after the war, work had begun on using solid materials to control the flow of electrons. And our own planet, its land and oceans, seemed a stable place posing mostly uninteresting questions. Chemistry in many ways was still chemistry in the large: statistical examinations of the interactions of many molecules, the tools simply not available to "watch" their behavior as individuals.

That all this changed, sometimes at a dizzying rate, is obvious. But what happened? I suppose at one level the war happened. It was not so much the particular science and technology created in the hothouse atmosphere of wartime but the means for doing science and technology that carried over with extraordinary vigor into peacetime. What we learned about microwave technology through work on radar in turn when combined with other work led to tools that created new fields, from more robust radio telescopes to masers and then lasers. The intense effort in wartime to create new antibiotics such as penicillin left as its legacy an enormously fertile organic chemistry, becoming a cornucopia of new materials, techniques, and ideas. A nexus of work during and immediately after the war—from Project Whirlwind to investigation of the properties of materials such as silicon supported by the armed forces to new mathematical tools—formed the seedbed for the stunning arrival of universally available and very powerful computing power. The organizational skills learned by scientists and engineers in wartime for driving large cosmopolitan endeavors involving a range of research and technologies were in peacetime mapped onto global programs such as the International Geophysical Year that put nations, first the Soviets and then us, into space and into other extremes, such as the Antarctic continent. Scientists who served in these programs, often as military officers, came home determined to pursue suggestive clues that our home planet was not quite that uninteresting.

Everything changed since I first had a balky mule carry my instruments up the High Sierras and then almost drowned in Lake Tulainyo. Our earth is now seen as a quilt of plates that move. Individual atoms and molecules are probed and their behavior monitored and controlled to a degree by an immense array of tools, yielding new materials and new technological possibilities. The science of the ever smaller has gotten even smaller, and the notion of nanotechnology, devices on the scale of 10^{-9} cm is becoming familiar, with scientists creating, for example, a "camera in a pill" that can be swallowed for examining the gastrointestinal tract. We have explored in detail the chemistry and topography of our neighboring planets in the solar system. Astronomy and particle physics, the two extreme ends of matter, have cycled through chaos to hints of order. Astronomy found an incredible array of strange objects in the universe—

black holes, quasars, pulsars—and evidence of extrasolar planets. Particle physics enormously expanded its bestiary of particles but gained some control through models that gave them some sense and pointed the way not only to explaining them but also to the source and energetics of what astronomers found.

Another lesson for science from the World War II—that different sciences working together can be remarkably fruitful—has endured. As one of many examples, the sharp and historic line between the physical and biological sciences is ever more porous. Researchers are trying to apply the ways that biological molecules "recognize" each other to creating useful nanomechanical devices. Geoscientists are exploring the contemporary science of genetics for useful clues to solving problems in the evolution of our planet. Cooperative efforts by biologists and engineers seek to create new robots. It is possible that how cells "talk" to each other will be applied in computer science. Not least, some wonder whether the emerging discoveries of brain science will revolutionize information technology.

Although the substance of contemporary science and technology has changed, how it is done largely has not. The universities, industry, government laboratories, and more recently specialized nonprofits are still the performers of research. There remains great cooperation among these institutions leavened by independent counsel on problems, promise, and performance, with the National Academies[9] a leading example.

That compact of government, industry and universities created a world that transformed the science, technology, and economy of the last half of the twentieth century, and set the stage for even more remarkable things to come. The new world shaped after World War II changed everything—in how we communicate with the remarkable transforming power of the Internet the most spectacular example, how we move with the arrival of the jet age, and in how we assure our health through the amelioration of many dread diseases such as smallpox and polio and enormous progress against others, such as cancer and heart disease. There is much more to come. We are, I suspect, only at the beginning of exploiting our understanding of genes. New materials in structure and what they can do are in the offing, as we exploit tools that can shape them at the atomic level.

Of course, the postwar legacy of science and technology has its dark side. We now have terrifying and plentiful weapons of mass destruction in many countries. The imperative for self-preservation imposed by the concept of mutually assured destruction was demonstrably effective when the United States, France, Great Britain, and the Soviet Union were the only nations with nuclear weapons capability. But now other nations have them and have a somewhat limited ability to deliver them. And nuclear threats are now joined by the potential for chemical and biological weapons of mass destruction. The latter is for me the saddest part of our legacy, to see, for example, our achievements in biology potentially channeled into weapons of death and disease. That is a hard reality, but like the Nazi enemy that faced us across the Atlantic in 1941, it is one we will confront and defeat.

Central to the world's long-term future is what scientists and engineers did in World War II and its immediate aftermath—namely, help Americans understand the potential power for good and bad of the very rapid advances in contemporary science and technology. We must work hard to assess and control the bad.

There is of course a severe mismatch between political timescales—invariably the interval between elections—and the time needed to understand and solve complex problems. Selecting a longer time scale, Wm. A. Wulf, President of the National Academy of Engineering, suggested that the twenty-first century should be an "engineered century." I see no reason why science, engineering, and technology cannot reach new heights in the new century as they did in the twentieth century, surprising us continually with what can be done, first with new scientific and technological discoveries and developments, followed by wise applications to societal problems.

Immediate candidates for an engineered century are global warming and in a larger dimension that of sustained global development, balancing energy, environment, and the economy. Over the past three decades, ever since we got the shock of the oil embargo in 1973, we have already made great progress discovering the nature of the problems and starting toward solutions, but there is a long way to go. It will require massive cooperation by our science, engineering, and technology communities, our political leaders, and society generally. We must continue to develop international organizations, both scientific and diplomatic, and bring scientific and technical deliberations and advice to international diplomatic discussions.

In June 1994, Bunny and I went to D-Day Remembered, the fiftieth anniversary of the Normandy landings on June 6, 1944. Memories flooded me of wartime London, where I had worked with the British on their desperate need for radar to protect themselves and to attack their enemy; the technical intelligence missions following closely on the invasion; my "capture" by the Germans; and coming on the horrific slave-labor missile factory at Nordhausen. I went to my sixtieth reunion at Colgate University, where I was set firmly on the path toward science and toward Cal Tech and physics. Even more poignantly for me, I was elected to the Steuben County, New York, Hall of Fame. My childhood in Corning played out before me, the love my grandparents gave their orphaned grandchildren, my sister and me, and the belief and real help teachers in high school gave me to get to Colgate at all. My time as President of Carnegie Mellon University was replayed in an invitation to a meeting of the Andrew Carnegie Society, the alumni of Carnegie Mellon. And I received the Vannevar Bush Medal of the National Science Foundation, triggering memories of leading the foundation through turbulent times and of Vannevar Bush, who had sent me to London and who understood when I declined his offer to join him in Washington after the war and returned instead to MIT to teach and do research. And I was sharply pulled back to my doctoral days at Cal Tech, climbing mountains, wrestling ornery mules, and publishing my first scientific paper when a surprise guest showed up—Victor Neher, my senior coauthor and Cal Tech mentor who went to the Rad Lab just before I did and helped persuade me to follow him.

A reporter for the *Pittsburgh Press* in a long story about me when I was president of Carnegie Mellon called me "The Happy Warrior." That was right, I think, even though there were occasions then and since when I was decidedly unhappy. But time has washed much of that away, and I can look back on my life being "in the game," taking part and a few times even making a bit of the history of American science and technology in the twentieth century. I'm satisfied.

Notes

1 STARTING OUT

1. The company had its start when Amory Houghton bought a glass company in Cambridge, Massachusetts, added the Union Glass Company in Somerset, Massachusetts, and having moved the operation to New York City, then bought the Brooklyn Flint Glass Works. The company moved to Corning in 1868, first as the Corning Flint Glass Works, becoming the Corning Glass Works in 1875. *http://www.corning.com/inside_corning/150th_Anniversary/Our_story/index2.asp.*

2. My formal name is Horton Guyford Stever, but throughout my life I've always been simply "Guy Stever."

3. I didn't appreciate it at the time, but my aunt's shout had a political bite, since Smith was a "wet" on Prohibition, long favoring the repeal of the 18th Amendment, while Hoover was a "dry."

4. Some of the stuff got pretty vicious. One leaflet published in New York state had this: "When the Catholics rule the United States, And the Jew grows a Christian nose on his face, When the Pope is the head of the Ku Klux Klan In the land of Uncle Sam, Then Al Smith will be our President And the country not worth a damn." *http://www.suite101.com/article.cfm/presidents_and_first_ladies/39019.*

5. The astronomer George Ellery Hale approached Corning after unsuccessfully spending $1 million to have the mirror cast in California from fused glass. He proposed that Corning build the mirror out of Pyrex, then a new type of glass. Pyrex expands and contracts much less with changes in temperature than ordinary glass, so a Pyrex mirror would be less subject to the focus and distortion problems suffered by a smaller, 100-inch mirror. It took two tries, but in 1936 Corning successfully cast the 200-inch mirror, which was then shipped from

Corning to Pasadena by a train never going faster than 25 miles per hour, using a route that avoided low overpasses and tunnels, taking 14 days to make the trip, and with the mirror protected by guards on overnight stops. It took 11 years, from 1936 to 1947, to shape the mirror, with almost 10,000 pounds ground away. First light was in 1948. Full-time scientific observing began a year later. *http://astro.caltech.edu/palomarpublic/history/.*

Palomar was only one of Hale's (1868–1938) spectacular accomplishments. He founded and directed three great observatories—Yerkes, Mt. Wilson, and Palomar; founded the *Astrophysical Journal,* then and now the leading journal in the field; and was an amazingly effective money raiser. A major beneficiary of his intellect, leadership, and rain-making skills was the National Academy of Sciences. He was the animating force in building a home for the academy in Washington and through his leadership in establishing the National Research Council helped create the modern academy. He invented the spectroheliograph for studying the sun, and in addition to his many scholarly publications also wrote popular books, such as *Depths of the Universe, Beyond the Milky Way,* and *Signals from the Stars.*

6. After a distinguished career at the University of Chicago, he succumbed to lobbying by George Ellery Hale to become president of Cal Tech and, with Hale and chemist Albert Noyes, transformed it into the Cal Tech that I entered as a graduate student in the late 1930s. He made enormous scientific contributions, including accurate measurement of the charge carried by an electron and verification of Einstein's photoelectric equation. Millikan was awarded the 1923 Nobel Prize in physics "for his work on the elementary charge of electricity and on the photoelectric effect." He wrote many books, with a recurring theme the reconciliation of science and religion in such books as *Time, Matter, and Values* and *Science and Life.*

7. These included Houghton Stevens, Stuart Rice, Paul Cook, Helen Lovegrove, Winifred Stanton, Barbara Hungerford, and many others.

8. Charles H. Townes went on to share the 1964 Nobel Prize in physics for "for fundamental work in the field of quantum electronics, which has led to the construction of oscillators and amplifiers based on the maser-laser principle." *http://almaz.com/nobel/physics/1964a.html.*

9. W. K. H. (everyone calls him Pief) Panofsky served as professor and director of the Stanford Linear Accelerator Center from 1961 until his "retirement" in 1984, but continues to be extraordinarily active in many issues, not least arms control.

10. Arthur A. Noyes, professor of chemistry at Cal Tech from 1916 to 1936. *http://books.nap.edu/books/0309060311/html/26.html#pagetop.*

11. Carl David Anderson, by William H. Pickering, *Biographical Memoirs.* Vol. 73, National Academy of Sciences: Washington, D.C., 1998, p. 29. *http://books.nap.edu/books/0309060311/html/26.html#pagetop.*

12. Anderson received the 1936 Nobel Prize in physics at the age of 31 for his discovery of the positive electron, the positron.

13. Just to underline that high-energy physics is a complicated business, the mesotrons cum muons of my time were also called mu-mesons. However, it turned out that they aren't mesons but rather members of a different class of particles called leptons. *http://www.nobel.se/physics/laureates/1968/alvarez-lecture.html.*

14.   Yukawa received the 1949 Nobel Prize in physics for his prediction of the exist-
       ence of mesons on the basis of theoretical work on nuclear forces.

15.   You can see a picture of a faltboot at *http://home.t-online.de/home/fossil/
       faltboot.htm.*

16.   William H. Pickering became professor of electrical engineering at Cal Tech in
       1946, and then a great leader of the U.S. space program. He organized the elec-
       tronics work for Cal Tech's Jet Propulsion Laboratory, and served as JPL Director
       for over two decades, from 1954 to 1976, a time when JPL ran off a string of
       spectacular achievements: The first U.S. satellite (Explorer I), the first successful
       U.S. circumlunar space probe (Pioneer IV), the Mariner flights to Venus and
       Mars in the early to mid-1960s, the Ranger photographic missions to the moon
       in 1964–1965, and the Surveyor lunar landings of 1966-67. *http://www.the-
       cape.com/ccas/pickering.htm*

17.   Bill commented later that we were lucky we didn't drown in Lake Tulainyo. I
       wasn't worrying at the time about drowning but about an experiment in great
       peril.

18.   William "Willie" Fowler, professor of physics at Cal Tech, received the 1983
       Nobel Prize in physics, with Subrahmanya Chandrasekhar, for his theoretical
       and experimental studies of nuclear reactions in the formation of the chemical
       elements in the universe. Fowler had a very special style to him. For example, in
       a 1941 story *Life* magazine did on Cal Tech, he described his social life as a
       graduate student in the 1930s: "The main thing I did was play poker on Satur-
       day nights with the other graduate students. . . . There was one mathematician,
       Max Wyman, who cleaned us out all the time. . . . Father [Henry] Bolger sup-
       plied us with all the wine we needed. He was a graduate student who later
       founded the physics department at Notre Dame. . . ." I too played games with
       Max Wyman, but badminton and pool, not poker and not for money. He later
       became president of the University of Alberta. And I ice skated with fellow phys-
       ics graduate student Charley Townes, future Nobelist, and climbed in the San
       Gabriel and San Jacinto Mountains with Bob Wells, later engineering vice presi-
       dent of Westinghouse, and Tim Smith, later a professional meteorologist.

19.   J. Robert Oppenheimer was a theoretical physicist and scientific director of the
       Los Alamos laboratory—the Manhattan Project, which built the atomic bomb
       (1943–1945)—and was director of the Institute for Advanced Study in
       Princeton, N.J. (1947–1966).

20.   H. V. Neher and H. G. Stever, "The mean lifetime of the mesotron from electro-
       scope data," *The Physical Review*, 1940, vol. 58, 9, pp. 766–770. See also Eldred
       Nelson, "Notes on the Neher-Stever Experiment," *The Physical Review*, 1940,
       vol. 58, 9, pp. 771–773.

21.   H. G. Stever, "A directional Geiger counter," *The Physical Review*, 1941, vol 59,
       9, p. 765 and "The discharge mechanism of fast G-M counters from the deadtime
       experiments," *The Physical Review*, 1942, vol. 61, 1-2, pp. 38–52.

22.   That my experiment showed that a Geiger counter would give a false reading of
       no counts when bombarded by heavy radiation explained perhaps the first major
       scientific discovery of the space age. The first U.S. satellite, *Explorer I*, carried
       James Van Allen's Geiger counter experiment. The data showed the counter in

certain regions around the earth going to zero. That turned out to be the high-radiation area now known as the Van Allen Belt, regions of high-energy particles, mainly protons and electrons held captive by the magnetic influence of the earth.

2   WAR

1. Plaque at the site.
2. *http://www.ibiscom.com/blitz.htm.*
3. This account, and much of the history of the Rad Lab that follows, is taken from the excellent book by Robert Buderi entitled *The Invention that Changed the World,* Touchstone, Simon and Schuster, New York, 1997.
4. Microwaves are part of the electromagnetic spectrum of light, which ranges from very short gamma rays to radio waves a meter or more in length. Visible light is in the 10^{-5} centimeter range, and microwaves are in the centimeter range, making them intermediate between radio and infrared waves. For more, *see http://imagine.gsfc.nasa.gov/docs/science/know_l1/emspectrum.html.*
5. Present, along with the cavity magnetron, were from Britain, Eddie Bowen and John Cockcroft, and from the United States, the host, Alfred Loomis, a retired investment banker who had built a physics laboratory at his home in Tuxedo Park, N.Y.; Carroll Wilson, personal assistant to Vannevar Bush; Karl Compton, president of the Massachusetts Institute of Technology; and Admiral Harold Bowen, director of the Naval Research Laboratory. See Buderi, p. 37.
6. Ibid.
7. Ibid., p. 83.
8. *http://home.zonnet.nl/atlanticwall/radar/#entwick.*
9. Vannevar Bush, president of the Carnegie Institution of Washington and former dean of engineering at MIT, persuaded President Roosevelt in 1940 that a new organization was needed to mobilize the country's scientists for the seemingly inevitable conflict. By Executive Order, Roosevelt created the National Defense Research Committee, NDRC, with Bush as chairman. But the limitations of the NDRC quickly became obvious. It could focus on research but not carry it forward into development and production, having neither the funds nor the authority. A year later, on June 28, 1941, Roosevelt established the Office of Scientific Research and Development (OSRD) with Bush as director, to "serve as a center for mobilization of the scientific personnel and resources of the Nation in order to assure maximum utilization of such personnel and resources in developing and applying the results of scientific research to defense purposes." It became an enormous enterprise, responsible not only for the Rad Lab and the Manhattan Project but also for major efforts on new explosives and propellants, submarine warfare, proximity fuzes, and major medical advances on antimalarials, blood and blood substitutes, and penicillin. The OSRD also lastingly transformed the American research enterprise. Prior to the OSRD, research and development partnerships among government, industry, and academia were rare, and with few exceptions the federal government did not support academic research. OSRD changed all that, creating novel arrangements for work among the

three sectors and enormously increasing the flow of federal funds into academia. Moreover, rather than adopting the military's hierarchical command style (or the German style which apparently trusted only selected industries and military laboratories), the OSRD operated in a highly decentralized and collegial manner, with Bush firm that scientists could do their best work in their own laboratories as civilians. At the same time, where needed, major central laboratories were created. The obvious example is the Rad Lab at MIT, which drew personnel from 69 academic institutions. The enormous success of this style—highly decentralized with the specific technical goals chosen by the scientists—left an imprint that is still with us today in the unique style of American research. See also James Phinney Baxter, *Scientists Against Time,* Little Brown and Co., Boston, 1946; National Science Board, *Science and Engineering Indicators—2000,* National Science Foundation, Washington, D.C, 2000; and G. Pascal Zachary, *Endless Frontier: Vannevar Bush, Engineer of the American Century,* Free Press, New York, 1997, esp. pp. 109ff.

10.   Lee Dubridge was trained as a nuclear physicist. His interests encompassed biophysics, the theory of photoelectric effects, and, of course, radar. "Of course" because in 1940 Dubridge took a leave of absence as chair of the physics department at the University of Rochester to became head of the Rad Lab. He never came back. After serving as Rad Lab director from 1940 to 1945, he went to the California Institute of Technology to serve as its president for 23 years, from 1946 to 1969, doubling during his time the size of the faculty and tripling its physical space. From 1969 to 1970 he was President Nixon's science advisor. Dubridge died in Pasadena on January 24, 1994, at the age of 93. Jesse Greenstein, an astronomer and Cal Tech colleague, captured him well: "He was a modest, eminently likable man, small in stature, but with strong presence. His conversation ranged from reminiscences of great world scientific events and personal friends to the finances of KCET, the Los Angeles PBS station, which he helped found and served as president of its board. He loved opera and made and listened to shelves-full of video recordings of nearly all broadcast performances." See *http://www.nap.edu/readingroom/books/biomems/ldubridge.html.*

11.   That bland phrase ill serves the quality of people who came to the Rad Lab on short notice. Examples include E. O. Lawrence, who invented the cyclotron, a device for accelerating nuclear particles to very high velocities without the use of very high voltages; Kenneth Bainbridge, a noted mass spectroscopist at Harvard; and of course, Lee Dubridge, who, aside from directing Rad Lab throughout World War II, in time became president of Cal Tech. Ultimately, 10 of the people who worked at the Rad Lab won Nobel prizes.

12.   Buderi, p. 105.

13.   Robinson was at the famous cavity magnetron meeting at the old Wardman Park Hotel in Washington that I mentioned earlier. On arriving he was given a quick introduction to American style when, as Robinson described it, the host, Alfred Loomis, "came to the door, opened it himself, wearing trunks and otherwise completely nude. . . . It was the first time I had bourbon, and he probably had the best bourbon anywhere." See *Rad Lab: Oral Histories Documenting World War II Activities at the MIT Radiation Laboratory,* produced by the IEEE Center

for the History of Electrical Engineering, Piscataway, N.J. Principal investigators: John Bryant, William Aspray, Andrew Goldstein, and Frederik Nebeker, p. 288, also see a biography of Alfred Loomis: Jennet Conant, *Tuxedo Park: A Wall Street Tycoon and the Secret Palace of Science that Changed the Course of World War II,* Simon and Schuster, New York, 2002.

14.   E. R. Chamberlin, *Life in Wartime Britain*, Batsford, London, 1972, p. 51.

15.   Brigadier Peter Young, ed., *The World Almanac Book of World War II*, World Almanac Publications, Bison Books Ltd., New York, 1981, p. 198.

16.   Ibid., p. 198.

17.   For me it was a deeply felt privilege and reward to meet and work with the giants of British radar development: A. P. Rowe, W. B. Lewis, P. I. Dee, and J. A. Ratcliffe, at TRE, and J. D. Cockcroft at ADRDE.

18.   Chamberlin, pp. 60–61. Chamberlin comments that only about 9 percent of the population used public shelters and that "the majority of those who used [them] went into an Anderson shelter. And an Anderson shelter on a winter's night was an extremely unpleasant experience. . . . Even official and private ingenuity could make of them nothing more than a steel shell sunk in the damp earth." Later on, there were the so-called Morrison shelters, large steel tables inside houses to protect the inhabitants even if the house itself collapsed. A million were in use when the war ended.

19.   No relation to the American General Electric.

20.   Other members were Lee Dubridge, Rear Admiral Julius Furer, and Brigadier General H. M. McLelland.

21.   Code names for radio grids each with different characteristics, strengths, and weaknesses.

22.   Buderi, p. 212.

23.   *The World Almanac Book of World War II*, p. 527.

24.   Incidentally, when the Rad Lab colloquium crowd, many of them recent hires, heard me referring to the pocket notebook, they asked about it and then asked why don't we have a similar talk on American radar. Lee Dubridge immediately asked Louis Ridenour to do that at the next colloquium.

25.   ALSOS, the Greek word for grove, likely because it was organized by Brigadier General Leslie Groves, head of the Manhattan Project, had several linked tasks: (1) determine the status of Germany's nuclear bomb program (i.e., did it have a bomb?); (2) find the some 1,100 tons of uranium the Germans had taken in 1940 in occupied Belgium; and (3) capture and interrogate German nuclear scientists before the Allies, especially the Russians, did. It succeeded on all three. The ALSOS military leader was Lieutenant Colonel Boris Pash, who interrogated Robert Oppenheimer on his communist affiliations. Sam Goudsmit, a Dutch theoretical physicist who had previously worked at the Rad Lab, was, again, scientific leader of ALSOS. By mid-November 1944, ALSOS found in Strasbourg conclusive evidence that, according to Goudsmit, "Germany had no atom bomb and was not likely to have one in any reasonable form." On April 17, 1945, a British-American strike force found the missing uranium ore in Stassfurt. And Pash in the meantime pursued the German scientists hard, at times getting into firefights with German forces. Pash captured the German scientists in

Hechingen on April 23, all except Otto Hahn and Werner Heisenberg, both of them taken a few days later. (This account is based largely on Richard Rhodes, *The Making of the Atomic Bomb*, Simon and Schuster, New York, 1986, pp. 605–610.)

26. A physicist, he went on to become founding president of Harvey Mudd College and memorialized his experiences in *Harvey Mudd College: The First Twenty Years*, Fithian Press, Santa Barbara, Calif., 1994.

27. Mike Chaffee and Al Bagg.

28. Chamberlin, p. 158.

29. The teams sent by Bowles included Louis Ridenour and David Griggs, as well as shorter stays by others, and became part of Eisenhower's staff at SHAEF.

30. Small bombs with incendiaries carried inside a larger bomb.

31. Mollie Panter-Downes, "Letter from London," *The New Yorker*, May 27, 1944. Reprinted in *The New Yorker Book of War Pieces, 1939–1945*, Schocken Books, New York, 1947, p. 305.

32. Bob had a brilliant career in cosmology and relativity with major contributions in the application of science and mathematics to military issues. He was a professor of mathematical physics at Princeton and served during World War II as scientific liaison officer of the London mission of the OSRD, my home base, and as chief of the Scientific Intelligence Advisory Section of the Allied Forces Supreme Headquarters. During the latter period, I met and traveled with him on technical intelligence matters in France, following hard on the Normandy invasion. After the war he moved to my alma mater, Cal Tech, as a professor of mathematical physics. He died much too soon, tragically after a car accident in 1961.

33. Reginald V. Jones (1911–1997) served as Professor of Natural Philosophy at the University of Aberdeen since 1946 and Professor Emeritus since 1981. He served as Assistant Director of the Royal Air Force Intelligence Section during World War II, where he developed methods to foil the Germans' radar and their radio beam targeting of bomb sites in Britain. In 1993 the Central Intelligence Agency honored him with a perpetual intelligence medal in his name.

34. The Germans' code name for the V-1 was "Kirschkern" (meaning cherry pit), because it was designed to be spit out against England (Theodore von Kármán, with Lee Edson, *The Wind and Beyond*, Little Brown and Co., Boston, 1967, p. 282).

35. "It was a single-stage rocket powered by LOX [liquid oxygen] and alcohol, developing a thrust of 56,000 pounds [cf. 110 pounds for the V-1], a payload of 2,200 pounds, [and] a velocity of 3,500 miles per hour, and inertially guided by gyroscopes and leveling pendulums to its target 200 miles distant" (Walter A. McDougall, *The Heavens and the Earth: A Political History of the Space Age*, Johns Hopkins University Press, Baltimore, 1985, p. 43).

36. Chamberlin, p. 74.

37. The title of the song is Kanaouen ar Vretoned. It was composed in 1836 by Auguste Brizeux (1803–1858), born in Lorient.

38. A. J. Liebling, "Letter from Paris," *The New Yorker*, September 9, 1944. Re-

printed in *The New Yorker Book of War Pieces, 1939–1945*, Schocken Books, New York, 1947, p. 368.

39. See Chapter 3, note 27.

40. The most exhaustive description of the work of the OSRD London Mission is the six-part *Report of OSRD Activities in the European Theater During the Period March 1941 through July 1945*, by Bennett Archambault. The report makes clear the stunning breadth and depth of the work of the office, going beyond radar and guided missiles. The full report is to be found in the MIT Archives (Bennett Archambault Papers, MC 555, Institute Archives and Special Collections, MIT Libraries, Cambridge, Mass.).

41. Theodore von Kármán (1881–1963), born in Budapest, was noted for his work in fluid mechanics, a field he entered in earnest after 1912 when he was Professor of Aerodynamics and Mechanics as well as Director of the Aerodynamics Institute at Aachen, Germany. When World War I broke out in 1914, he became head of research in the Austro-Hungarian Army Aviation Corps. In 1930, he moved permanently to the US to head the Guggenheim Aeronautical Laboratory at the California Institute of Technology. His important theory of boundary layers and his related studies of fluid flow at high subsonic, transonic, and supersonic speeds were of great importance to post-World War II progress in all areas of flight. He left behind a fine memoir, now regrettably out of print: Von Kármán, Theodore, with Lee Edson, *The Wind and Beyond*, Little Brown and Company: Boston, 1967.

42. Theodore von Kármán, with Lee Edson, *The Wind and Beyond*, Little Brown and Co., Boston, 1967, pp. 273–274.

43. Pictures of the site can be seen at *http://www.nasm.edu/galleries/gal114/SpaceRace/sec200/sec210.htm*, which also offers design details on the V-2.

44. *http://www.104infdiv.org/CONCAMP.HTM*.

45. Others at the meeting were Hugh Dryden, Frank Wattendorf, H. S. Tsien, possibly George Schairer, and me.

46. Winston S. Churchill, *The Second World War: Their Finest Hour*, vol. 2, Houghton Mifflin, Boston, p. 381.

3   MIT, MISSILES, AND MARRIAGE

1. James T. Patterson, *Grand Expectations: The United States, 1945–1974*, Oxford University Press, Oxford, 1996, pp. 3–4.

2. Ibid., p. 108.

3. Ibid., p. 111.

4. Walter A. McDougall, *The Heavens and the Earth: A Political History of the Space Age*, Johns Hopkins University Press, Baltimore, 1985, p. 79.

5. John C. Lonnquest and David F. Winkler, *To Defend and Deter: The Legacy of the United States Cold War Missile Program*, study sponsored by the Department of Defense Legacy Resource Management Program Cold War Project, USACERL Special Report 97/01, 1996, p. 19

6. "One witness to the flight was an RAF pilot, who sat that night in the officers'

mess with a puzzled expression on his face. There had been something peculiar in what he had seen, but exactly what had eluded him. Then suddenly he jumped up. 'My God, chaps,' he said, 'I must be going round the bend—it hadn't got a propeller'" (H. Guyford Stever, James J. Haggerty, and the Editors of Life, *Flight*, Life Science Library, New York, 1965, p. 83).

7. Stever, p. 83.

8. McDougall, p. 26.

9. Lonnquest, p. 16

10. *http://www.nap.edu/readingroom/books/biomems/mtuve.html.*

11. James Phinney Baxter, *Scientists Against Time*, Little Brown and Co., Boston, 1946, p. 7.

12. Ibid., p. 12.

13. The Office of the Joint Chiefs of Staff was created several months earlier, in February 1942.

14. Quoted in G. Pascal Zachary, *Endless Frontier: Vannevar Bush, Engineer of the American Century*, Free Press, New York, 1997, pp. 160–161.

15. And the time had also come for the Rad Lab. Here's how Ivan Getting, who received a Medal of Merit for his work at the Rad Lab, described its end: "On the day of President Truman's announcement of the Japanese surrender, Lee Dubridge held the V-J convocation of members of the Radiation Laboratory in the beautiful Eastman Great Court of MIT and proudly announced that the laboratory had completed its work and would close in three months; one-third of its staff would leave the first month, a second third in the second month, and the rest by the end of the third month. A few people would remain to complete the Radiation Laboratory Series of books and others to handle the administrative closeout." At the same time, Vannevar Bush announced that OSRD and NDRC would terminate by the end of the year and there would be no follow-on organization (Ivan Getting, *All in a Lifetime: Science in the Defense of Democracy*, Vantage Press, New York, 1989, p. 204).

16. Zachary, p. 246ff.

17. Ibid., p. 230.

18. Lonnquest, p. 21.

19. Quoted in Zachary, p. 314.

20. Lonnquest, p. 22.

21. Fact Sheet, NASA/JPL, *http://www.jpl.nasa.gov/facts/jpl.pdf.*

22. Sub-, tran-, and supersonic indicate below, at, and above, respectively, the speed of sound, which in turn is shortened to Mach 1. The transonic range actually begins below Mach 1, about Mach .9, as air moving over most of the plane reaches supersonic speeds and shock waves appear.

23. Ramjets are the engines for reaching supersonic speeds but, paradoxically, are simpler than turbojets, lacking both compressor and turbines and dependent on forward motion to force air into carefully designed air intakes for compression. They have no moving parts and work best at Mach 2, or twice the speed of sound. See Stever and Haggerty, p. 85.

24. *http://www.chinfo.navy.mil/navpalib/factfile/missiles/wep-side.html.*

25. NACA was created in 1915 to keep American aviation technologically current. It

eventually transmuted into the National Aeronautics and Space Administration (NASA), but much more about that in Chapter 5.

26.   The first missile under the Bumblebee program left the ground in October 1945, flew 9 miles, and splashed down in Barnegat Bay, off New Jersey and about 30 feet from four presumably very startled fishermen. The program was very successful and quite a number of missiles came out of it Talon, Terrier/Tartar, Typhon, and Triton.

27.   For Ernst Mach (1838–1916), the Austrian physicist who pointed out that an object moving through a space changes the space. In the case of planes, a plane moving into the supersonic range creates shock waves that not only affect the controls on the plane and increase its drag but also radiate out from the plane, announcing their presence with sonic booms.

28.   von Kármán, p. 219.

29.   The chair was Walter McNair of Bell Telephone Laboratories,* whose job then was director of the NIKE missile project. Other members were Francis Clauser from the Douglas Aircraft Corporation, who had been deeply involved early in the development of test airplanes for high-speed flight (he was a product of von Kármán education at Cal Tech); and Robert Gilruth, head of NACA's Pilotless Aircraft Research and Development unit, where the control of rocket-powered guided missiles was studied on test vehicles at a superb, well-instrumented flight test range with excellent telemetering back from the test vehicle to base (Bob later became the head of the Johnson Space Flight Center when NASA was established and a leader of the whole program of manned space flight). Other well-qualified members were Richard Porter, head of guided-missile work at General Electric, which was supported by the Army Ordnance department (he was well acquainted with the rocketry that von Braun and his group of German engineers brought to the United States along with some V-2s for flight testing); Clark Millikan, acting head of the aeronautical department at Cal Tech when von Kármán was away and later to be head of it (a broadly based engineering professor in aeronautics, he was well known for his wind tunnel work at Cal Tech and also for aeronautical design); and Larry Henderson from the RAND Corporation, which had been established with support from the Air Force, growing out of work by Franklin Collbohm and Arthur Raymond of the Douglas Aircraft Company.

30.   Donald Mackenzie, *Inventing Accuracy: A Historical Sociology of Nuclear Missile Guidance*, MIT Press, Cambridge, Mass., 1993, p. 113. By 1951 that requirement had become even tougher—1500 feet for the circular error, or radius within which half the warheads should land. It took 20 years before ICBMs met that target (p. 114).

31.   For further details of our conclusions and recommendations, see RG 218, 1620 series, CCS 334 Guided Missile Committee (Modern Military Records Branch, National Archives, College Park, Md.).

---

*Here and throughout affiliations are at the time of service.

32.   Zachary, p. 336.

33.   Ibid., p. 338.

34.   Vannevar Bush, *Modern Arms and Free Men*, Simon and Schuster, New York, 1949, p. 36. Quoted in Perry, Chapter 2, *Evolution of a Policy*, *http://www.fas.org/spp/eprint/origins/part07.htm*.

35.   But it hedged its bets. "At least partly as a result of *Toward New Horizons*, Army Air Force planners rejected the widely publicized views of the eminent physicist, Dr. Vannevar Bush, who regarded as futuristic the possibility of perfecting intercontinental missiles. Instead, they embraced von Kármán's predictions on the feasibility of ballistic missiles and inserted a missile development program in the five-year R&D projections" (Michael H. Gorn, *Harnessing the Genie, Science and Technology Forecasting for the Air Force, 1944–1986*, Office of Air Force History, Washington, D.C., 1988, pp. 41–42).

36.   MacKenzie, p. 31ff.

37.   The SAB, about which much more later, grew out of the Army Air Force Scientific Advisory Group, formed by Theodore von Kármán. It met for the first time in June 1946, had 30 members, and was chaired by von Kármán (Thomas A. Sturm, *The USAF Scientific Advisory Board: Its First Twenty Years, 1944–1964*, USAF Historical Division Liaison Office, Washington D.C., 1967, p. 15).

38.   Some "temporary"! It lasted 55 years, finally going in 1998. "The building was constructed in . . .1943 as a war building and is of a temporary nature," according to an architect's memo, "the life of said building to be for the duration of the war and six months thereafter." "Its 'temporary nature' permitted its occupants to abuse it in ways that would not be tolerated in a permanent building. If you wanted to run a wire from one lab to another, you didn't ask anybody's permission—you just got out a screwdriver and poked a hole through the wall. Of course this was in the days before the dangers of asbestos were recognized" *http://www-eecs.mit.edu/building/20/*.

39.   The AEC, created by the Atomic Energy Act of 1946, assumed civilian control of "the plants, laboratories, equipment, and personnel assembled during the war to produce the atomic bomb. . . . The transfer list ran to thirty-seven installations in nineteen states and Canada. With the facilities the Army would transfer 254 military officers, 1,688 enlisted men, 3,950 Government workers, and about 37,800 contractor employees" (Richard G. Hewlett, and Oscar E. Anderson, Jr., *A History of the United States Atomic Energy Commission: The New World, 1939/1946, Volume I*, Pennsylvania State University Press, University Park, 1962, pp. 1–2).

40.   Jerome C. Hunsaker (1886–1984) was heavily involved in the development of the science of flight in America for the first three-quarters of the twentieth century. Indeed, he taught the first MIT course in aeronautical engineering and aviation design and was the first head of the MIT Department of Aeronautical Engineering formed in 1939. Among his many contributions, he designed the flying boat NC-4, the first aircraft to fly across the Atlantic Ocean, and supervised the design of the dirigible *Shenandoah*, the first American rigid airship. *http://libraries.mit.edu/archives/mithistory/collections-mc/mc272.html#bio*. Also, see Roger D. Launius, "Jerome C. Hunsaker," in Emily J. McMurray, et al., eds.,

Notable Twentieth-Century Scientists (New York: Gale Research Inc., 1995), pp. 980-81, and William F. Trimble, *Jerome C. Hunsaker and the Rise of American Aeronautics*. Washington, DC: Smithsonian Institution Press, 2002.

41.   These included from MIT Asher Shapiro, an outstanding fluid mechanics student and engineer in mechanical engineering; Clark Goodman, an authority on atomic energy; Fran Friedman, another physicist who was very strong in research in that area and very knowledgeable about nuclear piles, as they were called in those days; Shatswell Ober from the MIT aeronautical engineering department, on my suggestion because he was an airplane performance expert and taught the courses in that area at MIT; and a number of other people who were added to this list to begin to put together some work. Later, Whitman added Wheeler Loomis, who during the war had been associate director of the Rad Lab at MIT and then had returned to his base at the University of Illinois as head of the physics department, and also, Jerrold Zacharias from the physics department, another fellow experienced in radar.

42.   NEPA, a feasibility study, was succeeded in 1951 by a research and development program called Aircraft Nuclear Propulsion, shaped in good measure by the work done by Project Lexington.

43.   Herbert York, *Race to Oblivion: A Participant's View of the Arms Race*, Simon and Schuster, New York, 1970. See also *http://www.learnworld.com/ZNW/LWText. York.RaceToOblivion.html#chapter4*.

44.   Ibid.

45.   RAND (*R*esearch *AND D*evelopment) was created in May 1948, spinning off from the Douglas Aircraft Corporation. General Henry "Hap" Arnold, Army Air Force chief of staff during World War II, had pushed for a place where experts could analyze and propose advanced concepts for the Air Force, and the result was Project Rand established within Douglas in 1945. It spent $640 in its first month, came under Frank Collbohm's direction in 1946, and by 1948 had 200 employees. *http://www.rand.org/50TH/#origins*.

46.   The B-52 Stratofortress was first flown in 1954 and entered service a year later. A total of 744 were built, the last in 1962. It is the primary manned bomber for the United States, flies at 650 miles per hour (Mach 0.86), and at takeoff can carry almost 490,000 pounds. *http://www.af.mil/news/factsheets/B_52_Stratofortress. html.*

47.   James R. Killian, Jr., *Sputnik, Scientists, and Eisenhower*, MIT Press, Cambridge, Mass., 1977, pp. 178–184.

48.   Yes, the "Doolittle" of Doolittle's raid over Tokyo. In March 1942, when it was very dark indeed for the Allies, 16 bombers launched from an aircraft carrier some 800 miles from the Japanese coast, struck Tokyo, Kobe, Nagoya, and Yokohama, with most crash landing afterwards in China and one in the Soviet Union. General Doolittle was awarded the Medal of Honor. Jim Doolittle (1896–1993) was commissioned as a first lieutenant in the Signal Corps Aviation Section in 1920 and two years later was the first to fly cross country, from Palm Beach, Fla. to San Diego. He made one stop, did it in 21 hours and 19 minutes, and earned the Distinguished Flying Cross. A year later, in 1923, he went to MIT, emerging two years later with a doctorate in aeronautics, one of the first in

the country to do it. His considerable talents were recognized in the many assignments he was given and did admirably, including commanding the Eighth Air Force in Europe and the Pacific.

49. For more on what it was like, not only for my colleagues but for many others as well, see "Red Scares Abroad and at Home" in James T. Patterson, *Grand Expectations: The United States, 1945–1974*, Oxford University Press, Oxford, 1996, pp. 165–205.

50. Reinstated in 1963.

51. It was properly a "Hearing in the Matter of J. Robert Oppenheimer" before the Personnel Security Board of the Atomic Energy Commission, held from April 12 to May 6, 1954. Whatever it was called, it was a trial.

52. From June 1948 to May 1949, U.S., British, and French planes flew 278,000(!) flights to deliver 2 million tons of supplies to the residents of West Berlin.

53. Both quotes are from the complete official text of Churchill's speech released to the press by the MIT News Office ("For Release After 10:00 P.M. [EST] News Service, Massachusetts Institute of Technology, Thursday, March 31, 1949."

54. Patterson, p. 135.

4   DEFENSE

1. The SAB was then under the Army Air Force, carried over to the new U.S. Air Force in 1947, and in 1948 placed directly under the Air Force chief of staff, and subsequently under both the Air Force Chief of Staff and the Secretary of the Air Force.

2. Quoted by Thomas A. Sturm, in *The USAF Scientific Advisory Board: Its First Twenty Years, 1944–1964*, USAF Historical Division Liaison Office, Washington D.C., 1967, p. 14.

3. September 24, 1945, memorandum from Brigadier General Alden R. Crawford to Major General E. M. Powers, "Necessity for an Army Air Forces Air Defense Program."

4. There are essentially two types of long-range missiles: an unmanned airplane that is jet propelled and aerodynamic, a cruise missile; and a ballistic missile, carrying its own oxygen, which is boosted above the atmosphere and into space by a rocket engine; once it reaches its apogee, the engine cuts off and it follows a ballistic path (i.e., controlled only by gravity and drag).

5. Jacob Neufeld, *Ballistic Missiles in the United States Air Force, 1945–1960*, Office of Air Force History, Washington, D.C., 1990, p. 27.

6. Ivan Getting, *All in a Lifetime: Science in the Defense of Democracy*, Vantage Press, New York, 1989, p. 350.

7. George E. Valley, Jr., "How the SAGE Development Began," *Annals of the History of Computing*, vol. 7, no. 3, July 1985, p. 204.

8. Ibid., p. 197.

9. Other ADSEC members included Charles Stark Draper, also from MIT, Allen Donovan of the Cornell Aeronautical Institute, John Marchetti of the Air Force Cambridge Research Laboratories, George C. Comstock of Airborne Instruments

Laboratory, Inc., Henry Houghton of the MIT Department of Meteorology, and William R. Hawthorne of the MIT Department of Mechanical Engineering.

10. Like most visiting scientists, he would ride the Atlantic in the plane's bomb bay.

11. Valley, p. 204.

12. Yes, the transistor was invented in 1947, but it didn't come into use for computing devices until the late 1950s and wasn't packaged into integrated circuits until the late 1960s.

13. *http://acomp.stanford.edu/siliconhistory/Olds/ROIonBasicResearch.html.*

14. Afterburners are second combustion chambers immediately in front of the engine's exhaust nozzle. Very fuel inefficient, they are generally used only for supersonic military aircraft.

15. Donald Mackenzie, *Inventing Accuracy: A Historical Sociology of Nuclear Missile Guidance*, MIT Press. Cambridge, Mass., 1993, pp. 115, 120.

16. Getting, p. 351.

17. Sturm, p. 40.

18. Valley, Jr., p. 204.

19. Ibid., p. 214.

20. Ibid., p. 213.

21. SAGE and Project Whirlwind had an intimate and symbiotic relationship that powered early advances in computing, especially digital computing, helped launch the computer in the United States, and, not least, set a management style that was to be copied by electronics and computer firms in the Route 128 corridor around Boston and Silicon Valley. Whirlwind grew out of initial work for the navy in 1944 on a computer to enable simulated flight testing, work that led to the conception and development of a general-purpose digital computer with applications far beyond flight testing. As Project Whirlwind's technical goals grew, so did its budget until it was consuming 10 percent of the annual budget of the Office of Naval Research (ONR). Budget cuts were inevitable, until the Air Force, seeing the potential for Whirlwind through George Valley, took over the major part of the costs, enabling the project to move forward toward developing a ferrite core with a nine-second access time, remarkable then. "All told, ONR spent roughly $3.6 million on Whirlwind, the Air Force, $13.8 million. In return, Whirlwind and SAGE generated a score of innovations. On the hardware side, Whirlwind and SAGE pioneered magnetic-core memory, digital phone-line transmission and modems, the light pen (one of the first graphical user interfaces), and duplexed computers. In software, they pioneered use of real-time software; concepts that later evolved into assemblers, compilers, and interpreters; software diagnosis programs; time-shared operating systems; structured program modules; table-driven software; and data description techniques. Five years after its introduction in Whirlwind, ferrite-core memory replaced every other type of computer memory, and remained the dominant form of computer memory until 1973. Royalties to MIT from nongovernment sales amounted to $25 million, as MIT licensed the technology broadly. SAGE accelerated the transfer of these technologies throughout the nascent computer industry. SAGE was a driving force behind the formation of the American computer and electronics industry"

(National Research Council, *Funding a Revolution: Government Support for Computing Research*, National Academy Press, Washington, D.C., 1999, pp. 93–94).

22. James T. Patterson, *Grand Expectations: The United States, 1945–1974*, Oxford University Press, Oxford, 1996, p. 210ff.

23. *http://www.boeing.com/companyoffices/history/bna/f86.htm*.

24. Sturm, p. 35.

25. Louis Ridenour, professor of physics at the University of Illinois, was the first chief scientist of the Air Force, the overall editor of the Radiation Laboratory series, and at the time of his sudden death when he was only 48, vice president of Lockheed Aircraft and general manager of its avionics and electronics division.

26. Other members, in addition to myself, were Al Donovan, Francis H. Clauser from Douglas Aircraft, and Allen V. Astin from the National Bureau of Standards.

27. Quoted in Sturm, p. 41.

28. Ibid., p. 43.

29. Vannevar Bush, *Science, the Endless Frontier*, National Science Foundation, Washington, D.C., July 1945. For the text online see *http://www.nsf.gov/od/lpa/nsf50/vbush1945.htm*.

30. Bush, pp. 8–9.

31. Ibid., p. 5.

32. William Blanpied, of the National Science Foundation, pointed out to me that "Bush's Letter of Transmittal of *Science, the Endless Frontier* to President Truman did mention the social sciences, but in a way that can only be regarded as dismissive. Bush's letter emphasized that when, in November 1944, President Roosevelt asked for his recommendations about a postwar program for science, he assumed that the president was referring to the natural and engineering sciences, not the social sciences, partially on the grounds that social science research was already adequately funded by private sources relative to the natural sciences and engineering."

33. G. Pascal Zachary, *Endless Frontier: Vannevar Bush, Engineer of the American Century*, Free Press, New York, 1997, p. 369.

34. See Chapter 1 for details and references.

35. John von Neumann (1903–1957), a mathematical prodigy as a child in Budapest, became a brilliant mathematician who was the youngest (at age 23) to lecture at the University of Berlin; one of the first professors, with Albert Einstein, of the Institute for Advanced Studies at Princeton; and a key figure in the development of digital computers, so much so that his architecture is called von Neumann processors. He developed a theory of automata, coinvented game theory, contributed importantly to quantum mechanics, and, most significant to my life, was very influential in military science and technology in fields such as ICBMs, nuclear weaponry, and military strategy. He enjoyed life enormously and looked at it with some bemusement, once commenting that, "If people do not believe that mathematics is simple, it is only because they do not realize how complicated life is." See *http://ei.cs.vt.edu/~history/VonNeumann.html*.

36. Two valuable sources for much fuller discussions of these issues are by Donald Mackenzie, *Inventing Accuracy: A Historical Sociology of Nuclear Missile Guid-*

ance, MIT Press, Cambridge, Mass., 1993 (esp. pp. 105–113) and Jacob Neufeld, *Ballistic Missiles in the United States Air Force, 1945–1960*, Office of Air Force History, Washington, D.C., 1990. In addition, a concise summary of the history and lessons to be drawn is given by Stephen B. Johnson in "The Organizational Roots of American Economic Competitiveness in High Technology," a paper presented at the Conference on R&D Investment and Economic Growth in the 20th Century, Berkeley, Calif., March 26–28, 1999. See *http://ishi.lib.berkeley.edu/cshe/r%26d/papers/johnson.html.*

37. Quoted in Mackenzie, p. 106.

38. The name reportedly came via Simon Ramo, a member of the committee who wanted a simple working name and suggested the Tea Garden Committee, after Trevor Gardner. That was rejected as too obvious for a classified committee in favor of the Tea Pot Committee. See Reference 10 in *http://www.spacecom.af.mil/HQAFSPC/history/gardner.htm.*

39. This recommendation echoed similar recommendations of a RAND report issued two days earlier, not surprising since there was steady communication between the two groups.

40. Neufeld, p. 255.

41. General Bernard A. Schriever (1910–) was the first commander of the Western Development Division, subsequently the Ballistic Missile Division, which pioneered the development of the US ICBMs. He was born in Bremen, Germany, raised in Texas, and educated at Texas A&M and Stanford University, taking, respectively, undergraduate and master's degrees in aeronautical engineering. He entered the Army Air Corps in 1932, was commander by the end of World War II of Advanced Headquarters, Far East Air Service Command, and in 1946 went to the Pentagon, where he worked closely with Theodore von Kármán, then working on the blueprint for the future Air Force, *Toward New Horizons.* He became through that association and with others, including Louis Ridenour, a strong advocate for development and deployment of missile for American security. Trevor Gardner was instrumental in putting him in charge of the ballistic missile program, a confidence that was fully met in the development on schedule of a succession of missile families, from Atlas to Titan to Minuteman. He retired in 1966. Schriever brought to his awesome responsibilities a very special style which the *New York Times* in December 1957 (Quoted in Neufeld, p. 108) described as one of "relaxed precision" that "gives the suggestion that he has seen a lot of Jimmy Stewart films." *http://www.spacecom.af.mil/hqafspc/history/schriever.htm.*

42. Stephen B. Johnson, "The Organizational Roots of American Economic Competitiveness in High Technology," a paper presented at the Conference on R&D Investment and Economic Growth in the 20th Century, Berkeley, Calif., March 26–28, 1999. See *http://ishi.lib.berkeley.edu/cshe/r%26d/papers/johnson.html.*

43. Donald L. Putt (1905–1988) was a career U.S. Air Force officer who specialized in the management of aerospace research and development. Trained as an engineer, he entered the Army Air Corps in 1928 and worked in a series in increasingly responsible posts at Air Materiel Command and GHQ Air Force. In 1948-1952 he was director of research and development for the Air Force, and

between 1952 and 1954 he was first vice commander and then commander of the Air Research and Development Command. Thereafter, until his retirement in 1958, he served as deputy chief of the development staff at Headquarters USAF.

44. Von Kármán had earlier established in NATO the Advisory Group for Aeronautical Research and Development (AGARD) and had chaired this together with the SAB. AGARD proved an exceptional success in reviving aeronautics in Europe. Its meetings tapped the best in NATO and the other European countries. Von Kármán felt a missionary duty to help all those countries now that he was the world's leading aeronautical engineer and scientist. AGARD and SAB eventually proved to be too much as he aged. The AGARD revered him as much as we did.

45. And in fact that's what happened. The prototype for the KC-135, the Boeing 367-80, led to the Boeing 707, the first U.S. commercial jet transport. The 707 production model made its maiden flight in December 1957, and Pan American World Airways put it into transoceanic service less than a year later, on October 26, 1958. It was of course not the first commercial jet transport. That was the ill-fated de Havilland Comet I.

46. One reason the aerodynamics was well tended to may have been the problems with Convair's F-102 interceptor, where the estimates of the aerodynamic drag on a delta-winged aircraft, as the B-58 was to be, had proved too optimistic. See *http://www.csd.uwo.ca/~pettypi/elevon/baugher_us/b058-01.html.*

47. *http://www.fas.org/nuke/guide/usa/bomber/b-58.htm.*

48. James R. Killian, Jr., *Sputnik, Scientists, and Eisenhower*, MIT Press, Cambridge, Mass., 1977, p. 71.

49. Confirmation to Gardner's post was held up in the Senate for several months "because of his steady support for Dr. J. Robert Oppenheimer, who had been accused of disloyalty to the United States." In 1953, at the height of anticommunist feelings in the United States, Oppenheimer, who had served as scientific director of the Manhattan Project, which built the first atomic bomb, was accused of having communist sympathies and deprived of his security clearance. It effectively ended his government service. See *http://www.pbs.org/wgbh/aso/databank/entries/baoppe.html* and *http://www.spacecom.af.mil/HQAFSPC/history/gardner.htm.*

50. He got there by both understanding the bureaucracy he had to deal with and finding ways to co-opt and defeat it. That story is well told in Jacob Neufeld's *Ballistic Missiles in the United States Air Force, 1945–1960*, Office of Air Force History, Washington, D.C., 1990. Regrettably, the book at last look is out of print.

51. Killian, p. 68.

52. Quoted in Neufeld, p. 96.

53. Formally titled "Meeting the Threat of Surprise Attack."

54. Killian, p. 68.

55. Ibid., pp. 72–73.

56. Ibid., p. 74.

57. Ibid., p. 77.

58. The U.S. membership of our SAB committee was broadly based, fairly representing many interested and capable institutions. We had two members from Bell Telephone Laboratories, Hendryck Bode and S. E. Miller, important because Bell was developing the Nike Zeus missile, which was specifically designed for point defense against ballistic missiles using a ground-based detection, fire control, and missile launch system; from the RAND Corporation we had Ed Barlow and Bill Graham, both of whom had spent a great deal of time on the subject of antiballistic missiles; from industry we had two representatives, Rube Mettler from TRW and former Air Force Chief Scientist Chalmers Sherwin; and from Raytheon we had Ivan Getting. Ivan was a major stalwart of SAB as well as vice president of Raytheon, where he had started a number of missile programs, some of which eventually led to the Hawk missile and then to the Patriot missile of Desert War fame. From the Lincoln Laboratories beginning to get involved in this antiballistic missile work, we had Al Hill and Dan Dustin; from the major nuclear laboratories of the Department of Energy, Wolfgang "Pief" Panofsky from the Stanford Linear Accelerator and Herb York from the Livermore Laboratories, both men quite expert in nuclear explosives and armament generally; from the National Advisory Committee on Aeronautics, Bob Gilruth, who had been doing a great deal of launching of rocketry, guidance, and so on; from General Electric and from the Killian Committee, Andy Longacre, who was very knowledgeable about long-range radars; and from Cal Tech, Homer Joe Stewart, a distinguished aerospace scientist.

59. Lindbergh was a faithful participant in our meetings and even came to a cocktail party Bunny and I gave. He was much anticipated by our young children. As I was talking with him, one child tugged at my leg to ask where the famous flyer was. I pointed to Lindbergh, and she began to tip her eyes from his shoes slowly up to his face, for her a giant. I experienced a flashback to the Lindbergh baby kidnapping, when he was being hounded by reporters and their flash cameras. While sitting opposite and talking to Lindbergh at breakfast prior to a committee meeting in California, a flash camera went off at a secretary's birthday party. He went rigid, and his eyes went wild for a moment. He regained control and apologized and began to explain. I cut him off saying, "I understand," and he thanked me.

60. The U-2 was a very high-flying photoreconnaissance plane built by Lockheed with a wingspread so wide and light that the tips draped on the ground. The U-2 flights continued until May 1960, when the Soviets finally shot one down.

61. Walter A. McDougall, *The Heavens and the Earth: A Political History of the Space Age*, Johns Hopkins University Press, Baltimore, 1985, p. 117.

62. That thinking, then being explored by Army Ordnance and Bell Telephone Laboratories, eventually led to the development of the Nike Zeus A missile system, the first weapon designed to attack incoming missiles. It had a number of problems and was never deployed, and the next Nike model succeeded it.

63. In fact, I stayed almost six months longer. An unwritten job of a chief scientist is to help get his successor. Mine was Courtland D. Perkins, a former SAB member. He was a "real airplane man," heading the aeronautical program at Princeton. But he couldn't come till the summer, so I extended my appointment an extra half-year.

64. A fine history of the Chief Scientist's Office, including my tenure, is: Day, Dwayne A., *Lightning Rod: A History of the Air Force Chief Scientist's Office.* Washington, D.C.: Chief Scientist's Office, United States Air Force, 2000.

5 INTO SPACE

1. See Chapter 4.

2. Quoted in James T. Patterson, *Grand Expectations: The United States, 1945–1974*, Oxford University Press, Oxford, 1996, p. 286.

3. Quoted in Walter A. McDougall, *The Heavens and the Earth: A Political History of the Space Age,* Johns Hopkins University Press, Baltimore, 1985, p. 114.

4. Indeed, federal defense spending declined in fiscal years 1953–1956. Patterson, p. 289.

5. The Western Development Division was created in 1954 to build the *Atlas* ICBM, following on the recommendation of the Tea Pot Committee (see Chapter 3). Sited in Inglewood, California, it was led by General Bernard A. Schriever and placed organizationally for a time under the Air Research and Development Command.

6. *Atlas* was fueled by kerosene oxidized by liquid oxygen. Unlike later missiles in which several stages were fired serially, all the *Atlas* engines fired at liftoff. The "special structures" included pressurized fuel tanks made of thin sheets of stainless steel. *http://www.pawnee.com/fewmuseum/atlas.htm.*

7. First called a medium-range missile, but the same thing.

8. See Chapter 2, note 9 for details on the OSRD.

9. The name was changed in 1992 to the Department of Civil and Environmental Engineering.

10. This became in 1974 part of the Department of Materials Science and Engineering.

11. For example, the Office of Naval Research and the Air Force Office of Scientific Research.

12. The name was changed to the Department of Aeronautics and Astronautics in January 1959, about which more later.

13. To hear what we all heard that October, go to *http://www.hq.nasa.gov/office/pao/History/sputnik/index.html* and click on the wave file.

14. Both quotes are from James R. Killian, Jr., *Sputnik, Scientists, and Eisenhower*, MIT Press, Cambridge, Mass., 1977, p. 9.

15. Ibid., p. 8.

16. Ibid., p. 2.

17. Others at the meeting included Sidney Chapman, S. Fred Singer, and Harry Vestine.

18. This account of the origins of the IGY and the notion of a space satellite to do science is based substantially on McDougall, p. 118.

19. Note that two months later Eisenhower made the ICBM a "research program of the highest priority."

20. Air Force Scientific Advisory Board, *Report of the SAB Ad Hoc Committee on Advanced Weapons Technology and Environment,* October 9, 1957.

21. Cargill R. Hall, and Jacob Neufeld, eds., *The U.S. Air Force in Space—1945 to the Twenty-first Century,* Proceedings of the Air Force Historical Foundation Symposium, September 21–22, 1995, U.S. Air Force History and Museum Programs, Washington, D.C., pp. 15–16.

22. Edward Teller (1908–), born in Hungary, came to the United States in 1935. From 1935 to 1941 he was professor of physics at George Washington Univ. and during World War II worked on atomic bomb research at a number of facilities, including the laboratory at Los Alamos that built the first atomic bomb. Later (1946–1952) he was professor of physics at the University of Chicago. He was also associated (1949–1951) with the thermonuclear research program of the Los Alamos National Laboratory. Teller was instrumental in the first successful U.S. hydrogen bomb explosion, November 1, 1952. From 1952 to 1960, Teller was professor of physics at the University of California and director of the Livermore division of its radiation laboratory. For his contributions to the development, use, and control of nuclear energy, Teller received the 1962 Enrico Fermi Award. He, too has written a memoir, and a very fine one at that: Edward Teller (with Judith Shoolery), *Memoirs: A Twentieth-Century Journey in Science and Politics.* Cambridge, Mass.: Perseus Publishing, 2001.

23. Quoted in Thomas A. Sturm, *The USAF Scientific Advisory Board: Its First Twenty Years, 1944–1964,* U.S. Air Force Historical Division Liaison Office, Washington, D.C., 1967, p. 81.

24. Members included all members (except Dr. Ramo, who was not at the meeting) of the Cislunar Committee plus Edward Teller and David Griggs, the new chief scientist of the Air Force.

25. Killian, pp. 27–28.

26. In retrospect, public beliefs about the relative strengths of the United States and the Soviet Union in space were almost 180 degrees wrong. The United States was very strong, with robust but classified programs in missile technology and reconnaissance satellites, whereas the Soviet Union, despite *Sputnik,* was weak on the military side of missiles and space.

27. Dr. Killian was the first truly presidential science advisor, in the sense that his only boss was the president. The closest approximation prior to his appointment was the chair of the Science Advisory Committee of the White Office of Defense Mobilization, organized subsequent to the onset of the Korean War. The first chair was Oliver Buckley, retired president of Bell Telephone Laboratories, appointed on April 19, 1951, by President Truman. Science Advisory Committee members included James Killian, the only nonscientist in the group. Truman apparently made little use of the new organization during the remaining months of his presidency. *http://www.aaas.org/spp/cstc/golden.*

28. McDougall, p. 143. The first operational reconnaissance satellite project was CORONA. The first CORONA satellite was launched August 18, 1960, followed by hundreds more. It operated for almost 12 years during the Cold War. The August flight yielded more photographic coverage of the Soviet Union than all of the U-2 flights to that date, those having begun in 1956. It delivered 3,000

feet of film covering 1.65 million square miles of Soviet territory. *http://www.nro.odci.gov/corona/cor-ab.html.* See also Dwayne A. Day, John M. Logsdon, and Brian Lattell eds., *Eye in the Sky: The Story of the Corona Spy Satellites,* Smithsonian History of Aviation Series, Washington, D.C., 1998.

29.   *http://www.hq.nasa.gov/office/pao/History/sputnik/16.html.*

30.   The White House also had an unstated reason for its strong role in space organization. The intelligence programs of the country were strongly centered in White House groups—the National Security Council and the Foreign Intelligence Committee, both of which were deeply involved in the use of satellites for space reconnaissance ("spy satellites") and military communications, all super secret programs.

31.   *http://www.hq.nasa.gov/office/pao/History/Timeline/1958.html.*

32.   A good background to this, especially the legislative history, is provided by McDougall, esp. Chapter 7. For many of the relevant documents, time lines, and other information on the birth of NASA, see the excellent NASA history page at *http://history.nasa.gov/history.html.*

33.   Hugh L. Dryden (1898–1965) had a distinguished career in research and administration. He published more than 100 technical articles on his work in high-speed aerodynamics, fluid mechanics, and acoustics. He served as Director of NACA from 1947 to 1958, and as Deputy Administrator of NASA from 1958 until his death in 1965. In 1950 he received the Daniel Guggenheim Medal for "outstanding leadership in aeronautical research and fundamental contributions to aeronautical science." In 1955 he received the Wright brothers memorial trophy for "significant public service of enduring value to aviation in the United States." Hugh was honored by the National Civil Service League with the Career Service Award for 1958. As his biographer, Michael Gorn, put it: "Indifferent to self-advancement, he nonetheless rose to the pinnacle of the aeronautics profession and subsequently assumed a pivotal role in the initial period of space exploration." *http://www.dfrc.nasa.gov/History/Publications/Monograph_5/Gorn_Intro.html.*

34.   We had a very good membership, including Werner Von Braun, pioneer rocketeer, first for the Germans in World War II and then for the U.S. Army; Bill Pickering, leader of the Jet Propulsion Laboratory and a pioneer in the electronics field for guidance and control and instruments; Jim Van Allen, a physicist from Iowa State University and a leading user of scientific rockets for payloads to explore the region of the space directly around the earth; Sam Hoffman, leader of the development of big rockets from North American Aviation; and, ex officio, Hugh Dryden, Smitty DeFrance (director of the NACA Ames Laboratory), Abe Silverstein (director of NACA Lewis Propulsion Laboratory in Cleveland), and Bob Gilruth (director of the NACA Wallop's Island Test Firing Range). In looking over the membership I felt that we needed some added starters in the field of electronics and guidance, radar particularly, so Hendrik Bode of Bell Laboratories and head of the Nike guided-missile program and a colleague of mine in earlier anti-ICBM days, and Charles Stark Draper, director of MIT's Instrumentation Laboratory and the leading inertial guidance developer, were added.

35. Jacob Neufeld, *Ballistic Missiles in the United States Air Force, 1945–1960,* Office of Air Force History, Washington, D.C., 1990, p. 151.

36. For much more on Johnson's work and style in the Senate, see the third volume of Robet Caro's biography of him, *Master of the Senate: The Years of Lyndon Johnson,* Knopf, New York, N.Y. 2002.

37. McDougall, p. 165.

38. T. Keith Glennan (1905–1995) served as the first NASA Administrator from August 19, 1958 to January 20, 1961, when James Webb succeeded him. Trained as an electrical engineer he had a strong career in industry, including the then new sound motion picture industry. During World War II he served as Director of the U.S. Navy's Underwater Sound Laboratories in New London, Connecticut. After the war, he went from an industrial position to the presidency of Case, which he transformed from mainly a local institution to one competitive with the country's top engineering schools. He transformed NASA during his 2 _ year tenure as Administrator." *http://www.hq.nasa.gov/office/pao/History/Biographies/glennan.html.*

39. *http://www.hq.nasa.gov/office/pao/History/report58.html.*

40. Ibid.

41. Indeed, the day NASA was signed into law the director of the Bureau of the Budget signed off on a budget giving more money to the Advanced Research Projects Agency, coordinating military space programs, than to NASA. However, that didn't last long, and soon NASA grew spectacularly. McDougall, p. 191.

42. The committee had 20 members, including many from the "Ridenour Committee" that had examined the Air Force's research and development needs and structure immediately after the end of World War II. Among the members was Bennett Archambault, board chairman of the Stewart-Warner Corporation in Chicago and my boss during World War II, when I was part of the London mission of the OSRD.

43. Quoted in Sturm, p. 84.

44. Neufeld, pp. 170–171.

45. Edward Teller, who, with his wife, was at the ranch many times, for business, recovery from medical problems, or simply relaxation was left "with the clear impression that everything was oversized, even the dates on what seemed to be thousands of palm trees. The main hall and dining room of the house were almost but not quite big enough to serve as a football field" (Edward Teller, with Judith Shoolery, *Memoirs: A 20th Century Journey in Science and Politics,* Perseus Publishing, Cambridge, Mass., 2002, p. 458).

46. These included J. C. R. Licklider from MIT; Leo Goldberg from the University of Michigan; and Charles Townes from Columbia University. Townes would later share the Nobel Prize in physics.

47. Successor to the Air Research and Development Command.

48. Sturm, p. 98.

49. Ibid., p. 117.

50. Ibid., p. 110.

51. Ibid., p. 111.

52. The United Aircraft Corporation become the United Technologies Corporation in 1975.

53. Not that the corporation was a slouch. Its Pratt & Whitney Division produced jet engines in 1948, its Otis Division the first operator-less elevators in 1950, its Sikorski Division the first turbine-powered helicopter in 1957, and not too soon after I arrived the Pratt & Whitney jet engine for the Boeing 707, the first American commercial jet.

54. C. S. Lewis, *The Lion, the Witch, and the Wardrobe*, Geoffrey Bles, London, 1950.

55. *http://cslewis.drzeus.net/books/fiction.html.*

6   GOING PUBLIC

1. John "Jake" Christian Warner (1897–1969), armed with a doctorate in chemistry earned in 1923 at Indiana University, came to Carnegie Tech as an instructor in 1926. With others he pushed hard for a stronger research program, rose to successively more demanding and broader positions with Carnegie Tech, and became its president in 1950 when he was 53. He retired in 1965, having transformed Carnegie Tech. His achievements included the Graduate School of Business Administration, the firm and very strong grounding in the emergent computer sciences, an independent board of trustees, major new buildings, and finally, and most critical, recruitment of exceptional faculty, including Herb Simon. He died in 1969 at age 92, leaving much of his estate to the University. [I'm indebted for this account to Edwin Fenton, *Carnegie Mellon 1900–2000, A Centennial History*. Pittsburgh: Carnegie Mellon University Press, 2000, pp. 149ff.]

2. Andrew Carnegie wanted a technical and trade school in Pittsburgh and created the school in 1900 with a gift of $1 million in gold bonds and a promise of a site by the city of Pittsburgh. The site was in Oakland, in the east end of Pittsburgh, away from the soot, smoke, and dirt of its industry. Carnegie wanted to change Pittsburgh's sorry national reputation, and he expressed in his diary his hope that "not only our own country, but the civilized world will take note of the fact that our Dear Old Smoky Pittsburgh, no longer content to be celebrated only as one of the chief manufacturing centres, has entered upon the path to higher things" (*http://www.post-gazette.com/newslinks/Oakland.asp*).

3. "R. K." was the son of Richard Beatty Mellon, brother of Andrew Mellon. Andrew Mellon had two children, Paul and Alisa.

4. I should emphasize that Edwin Fenton's excellent *Centennial History* reminded me of things I had forgotten, distorted with time, or simply didn't know. Edwin ("Ted") Fenton is now emeritus professor of history at Carnegie Mellon, having joined the faculty in 1954. He received the university's highest honor for his teaching and educational leadership.

5. Edwin Fenton (p. 45) quotes the 1908 course catalog: "The courses of instruction offered in this school are planned to develop womanly attributes and give a foundation on which to build a career in distinctly feminine fields. Its emphasis is primarily laid upon the home, which is esteemed the important and logical sphere for educated women."

6. Herbert A. Simon (1916–2001) came to Carnegie Tech in 1949 to join its new

Graduate School of Industrial Administration after establishing a reputation for his work on organizational decision-making. He developed in the late 1940s his ideas of "bounded rationality," that rather seeing entrepreneurs simplistically as purely rationalists intent on maximizing profits, they instead operate as cooperating decision makers whose capacities for purely rational actions are bounded by limits of knowledge and by personal and social ties. At Carnegie, he led in the use of computer simulations of human cognition. He focused on discovering the symbolic processes that people use to think, and with his colleagues enlarged the computer as simply a machine for arithmetic work to a processor of symbols and hence human action and thoughts. He received Nobel Prize in Economics in 1978 for "for his pioneering research into the decision-making process within economic organizations." For more, see: *http://www.nobel.se/economics/laureates/1978/simon-autobio.html.*

7. Simon and his colleagues pushed "computer science" beyond its restrictive definition of the "theory and design of computers" and redefined it to include, in Newell's words, "the study of all the phenomena arising from them." The department of computer sciences was established in July 1965, or shortly after I became president, and was one of the first such departments in the country. In 1988 the department became the school of computer science. See *http://www.cs.cmu.edu/aboutscs/mission.html.*

8. The field took a sharp turn with the development in 1959 of the integrated circuit, which replaced separate and heat-producing parts—transistors, resistors, etc.—with one much cooler "chip" that was much easier to package in a machine and cost a lot less than individual components. The Fairchild Semiconductor Corporation produced the first commercially available chips in 1961, and in 1962 the Air Force used chips in its computers and *Minuteman* missiles.

9. Arthur Kennedy, who won a Tony for his starring role in the 1949 production of *Death of a Salesman*; Andy Warhol; William Ball, founder of the American Conservatory Theater; and Stephen Schwartz, lyricist/composer of "Pippin" and "Godspell," which received Grammy awards, and "Pocahontas," which got him two Academy Awards for best song and best original score.

10. Fenton, p. 102ff.

11. John, I knew, would be hard to keep and indeed left soon after I arrived to become president of Haverford College, serving as president until 1968 when, true to his beliefs, he resigned after the Haverford Board voted to accept women only as transfer students. See *http://www.haverford.edu/publications/fall99/century3.htm.*

12. Fenton, p. 156.

13. Roger L. Geiger, *Research and Relevant Knowledge: American Research Universities Since World War II*, Oxford University Press, New York, 1993, p. 147.

14. Ibid., pp. 147–149.

15. Ibid., p. 150.

16. Casals first visited the United States in 1901. It was to have been an extensive series of engagements, with performances in 80 different locations planned, but midway through the tour Casals seriously injured his left hand while hiking in California. He had been climbing Mount Tamalpais, near San Francisco, when a

large rock somehow became dislodged and fell on his hand, crushing some fingers. Casals said that the first thought that came to his mind was, "Thank God, I'll never have to play the cello again!" *http://www.cello.org/casals/page10.htm*). Fortunately, after about four months of treatment in San Francisco, his left hand and fingers regained their strength and agility, and he was able to continue his career. Those who heard him perform then said that his long vacation from performing somehow had added an emotional depth to his interpretation that had not been there before.

17.   Quoted in Fenton, p. 145.

18.   Ibid. Fenton also adds: "How appropriate! The founders of the three family fortunes supporting Carnegie Mellon and the Heinz School—Andrew Carnegie, Thomas Mellon, and H. J. Heinz—had been friends when they were still vigorous young men" (p. 145).

19.   This summary is based on Fenton, esp. pp. 138–139 and 144–145.

20.   Aiken Fisher (1908–1996) became Chairman of the Board of the Fisher Scientific Company in 1965, the year I came to Carnegie. The Company was established by Aiken Fisher's father in 1902, and become one of the major suppliers of scientific products, tools, and service in the world. Aiken Fisher was the first board chairman of Carnegie Mellon and was awarded an honorary degree by the University. He died at 88, of cancer.

21.   Both quotes are from Fenton, p. 162.

22.   A quarter of a century later, Paul Mellon, in *Reflections in a Silver Spoon: A Memoir* (written with John Baskett and published in 1992 by Morrow) described this merger: "The last President of Carnegie Tech, H. Guyford Stever went on to become Carnegie Mellon's first President. Stever, resigned in 1972 to become Director of the National Science Foundation. . . ."

23.   See *http://www.wild-trout.co.uk/leven.htm* for a fine watercolor of this splendid trout and its history, which notes in part: "Following the retreat of the ice sheets from northern Britain some 10,000 years ago, a massive block of ice was left stranded just north of the Firth of Forth in what is now southeast Scotland. As the climate slowly warmed, the ice melted, and its water cut a river into the Forth at Largo Bay. The lake that now fills the depression made by this gigantic block of ice has become the most famous trout water in the world: Loch Leven."

24.   Geiger, p. 198.

25.   See *http://www.census.gov/prod/99pubs/99statab/sec04.pdf*.

26.   Geiger, p. 213.

27.   Ibid., p. 201.

28.   Ibid., p. 202.

29.   Ibid., p. 245.

30.   Ludwig F. Schaefer, *Evolution of a National Research University, 1965–1990: The Stever Administration and the Cyert Years at Carnegie Mellon,* Carnegie Mellon University Press, Pittsburgh, Pa., 1992, p. 111. Indeed, Dr. Schaefer offers a detailed accounting of Carnegie Mellon's financial problems, notably in "Management: 1965–1972."

31.   Ibid., p. 106.

32.   Fenton, p. 158.

33. Ibid., p. 179.

34. Herbert Simon, *Models of My Life*, MIT Press, Cambridge, Mass., 1996, p. 279.

35. Quoted in Fenton, p. 169.

36. James T. Patterson, *Grand Expectations: The United States, 1945–1974*, Oxford University Press, Oxford, 1996, p. 595.

37. Ibid., p. 597.

38. Fenton, p. 177.

39. The inevitable shorthand for this group was PEACHY. Members included Carnegie Mellon, the University of Pittsburgh, Chatham College, Mount Mercy College, Duquesne University, and later Point Park College.

40. Patterson, p. 598ff.

41. As Edwin Fenton reminds me (p. 182), the 1970 issue of *The Thistle* was especially offensive, and when they sent me page proofs in advance I said "it was in bad taste, unrepresentative of campus life and scurrilous." But I didn't stop it, although I did move to shield the student sponsor from liability.

42. Simon, p. 287.

43. Ibid., p. 282.

44. Ibid., p. 287.

45. It included Frederick Seitz, who had gone to Rockefeller University to be president, having been president of the National Academy of Sciences; Phillip Handler, now president of the National Academy of Sciences and also chairman of the National Science Board; William O. Baker on the Carnegie Mellon board and president of Bell Labs; Simon Ramo, who was one of the founders of TRW; General Bernard Schriever, who had recently retired from being the first commander of the Air Force Systems Command and before that was the first commander of the Western Division, which built the nation's first intercontinental ballistic missiles; and Edward Teller, the so-called father of the H-bomb.

46. Fenton, p. 172.

47. Ibid., p. 169.

48. Simon, p. 257.

49. These included the Roy Hunt family; Vera Heinz; the Fisher brothers and Fisher Scientific; the Fred Foys; Fletch and Peg Byrom and Koppers Corporation; the Carnegie Mellon units; the board; the faculty; the alumni; the administration and staff; the PCHE presidents; and a host of other acquaintances and friends who had made us welcome and our stay rewarding in Pittsburgh.

7   TO WASHINGTON

1. Story as told by Toby A. Appel, *Shaping Biology: The National Science Foundation and American Biological Research, 1945–1975*, Johns Hopkins University Press, Baltimore, 2000, Foreword. I'm not going to tell you the "sausage-making" details of how the NSF was created. However, there are many places to turn for that story, such as "The Birth of NSF," *Mosaic*, Nov./Dec. 1975, pp. 20–27. The *Mosaic* piece is from a longer work also making for profitable reading: Milton Lomask, *A Minor Miracle: An Informal History of the National Science Founda-*

tion, U.S. Government Printing Office, Washington, D.C., 1976. Other fine sources are George T. Mazuzan, *The National Science Foundation: A Brief History*, esp. pp. 6–13, available online at *http://www.nsf.gov/pubs/stis1994/nsf8816/nsf8816.txt*; William A. Blanpied, *Impacts of the Early Cold War on the Formulation of U.S. Science Policy: Selected Memoranda of William T. Golden, October 1950–April 1951*, American Association for the Advancement of Science, Washington, D.C., 1995 (*http://www.aaas.org/spp/cstc/golden/*); J. Merton England, *A Patron for Pure Science*, National Science Foundation, Washington, D.C., 1982; Daniel J. Kevles, "Scientists, the military, and the control of postwar defense research: The case of the Research Board for National Security, 1944–46," *Technology and Culture*, Jan. 1975, vol. 16, pp. 20–47; Daniel J. Kevles, "The National Science Foundation and the debate over postwar research policy, 1942–1945," *Isis*, Mar. 1977, vol. 68, pp. 5–26.

2.   Mazuzan, p. 4.

3.   *http://www.nsf.gov/od/lpa/nsf50/history.htm*.

4.   *http://ntalpha.bfa.nsf.gov/NSFHist.htm*.

5.   The social sciences in time joined the foundation's portfolio and were given "statutory stimulus" in July 1968 in a bill amending its charter that became Public Law 90-407. Michael D. Reagan, *Science and the Federal Patron*, Oxford University Press, New York, 1969, p. 191.

6.   He called it the National Research Foundation.

7.   Reagan, p. 190.

8.   Ibid., p. 190.

9.   Interview of Don Price by Milton Lomask, January 13, 1973, Historians' Files, 1945–1985, Record Group 307, Row 30 (unprocessed), National Records and Archives Administration, College Park, Md.

10.   Ibid.

11.   Jeffrey K. Stine, *A History of Science Policy in the United States, 1940–1985*, Science Policy Study, Background Report No. 1, Task Force on Science Policy, Committee on Science and Technology, House of Representatives, 99th Congress, Second Session, September 1986, p. 57.

12.   ONR had the political space to offer generous terms to the universities. "Senior naval officers did not approve of ONR's role, but they were too preoccupied with first the demobilization following World War II and then the start of the Cold War and the Korean mobilization to take the steps necessary to alter it. But outside the Navy, ONR was viewed as conducting a military mission, one apparently buttressed by a vital national security rationale. . . . Scientists were encouraged to propose their own projects and were not required to submit progress reports—mere publication in the open literature, refereed literature was considered sufficient. . . . ONR [also followed] the special wartime policy of paying full overhead costs, generously defined, thus reimbursing universities for sponsored research at rates that government agencies in the 1930s shunned and private foundations to this day refuse to do. Literally hundreds of millions of dollars have been provided to research universities for allocation by their administrators via this mechanism" (Harvey Sapolsky, "Financing science after the Cold War," in *The Fragile Contract: University Science and the Federal Government*, David H.

Guston and Kenneth Keniston, eds., MIT Press, Cambridge, Mass., 1994, pp. 166–167).

13. I was the fourth director, preceded by Alan T. Waterman, a physicist who served for 12 years; Leland J. Haworth, also a physicist, 1963–1969; and, my immediate predecessor, William McElroy, a biologist, 1969–1971. Apparently, according to Jim Killian, Alan Waterman was the third choice after Karl Compton and James Conant. He was a great choice, in part because, as Don Price observed, "the Office of Naval Research had made a great hit with the universities by doing what they didn't think any military organization would do, to have a program of comparatively unrestricted money for basic research. Now, Waterman was not in terms of personal glamour and personality a Vannevar Bush by any means. He was a rather diffident and neutral-appearing type of man. But I suspect that in the post-war atmosphere with all the weakness in the position of the Science Foundation that he was just as effective as a more dynamic-appearing man would be" (interview of James Killian by Milton Lomask, January 16, 1973, Historians' Files, 1945–1985, Record Group 307, Row 30 (unprocessed), National Records and Archives Administration, College Park, Md., and Lomask interview with Don Price).

14. Of course, NSF wasn't unique in its use of peer review. Other agencies—the National Institutes of Health, the Office of Naval Research, and the research programs of the Atomic Energy Commission—all used the principle to varying extent.

15. These included Senators Jacob K. Javits (R-N.Y.), Hugh Scott (R-Pa.), Edward Kennedy (D-Mass.), Peter H. Dominick (R-Colo.), Richard S. Schweiker (R-Pa.), Harrison Williams (D-N.J.), Russell Long (D-La.), and Allan J. Ellender (D-La.). Senator Williams chaired the Committee on Labor and Public Welfare that conducted my nomination hearing on November 30, 1971.

16. The commitment in the early 1960s by the federal government and the aerospace industry to build a commercial aircraft capable of operating beyond the speed of sound ran into very severe criticisms in the late 1960s. These included the sonic booms created by the aircraft and its potential for destroying the ozone layer. Congress, after several hearings, terminated the program in 1971. "Although the SST conflict really dealt with the nation's policy for technology rather than science, it signified new involvement for scientists in the political process, through Presidential science advice and Congressional testimony, as well as through involvement in citizen groups." (Stine, p. 59.)

17. He was best known for his research on luciferase, an enzyme responsible for the "twinkle" in fireflies.

18. Mazuzan, p. 19.

19. It was a rare moment for Dubridge, having had a rough time of it with Nixon, even though they had known each other for over two decades, when Nixon was governor of California and Dubridge was president of the California Institute of Technology. Among the more visible problems was of course Nixon's veto of Dubridge's nomination of Franklin Long as director of the NSF; but even during the transition, Dubridge was "pointedly not included among members of Nixon's

inner circle called to the president-elect's suite in New York City's Pierre Hotel to draft position papers on issues for the transition" (Herken, p. 167).

20.   *http://ntalpha.bfa.nsf.gov/NSFHist.htm.*

21.   Stine, p. 57.

22.   The late Philip Handler, a biochemist, served as president of the National Academy of Sciences from 1969 to 1981, and was a member of the National Science Board from 1962 to 1970, the last four years as chair.

23.   Don K. Price, "Money and influence: The links of science to public policy," *Daedalus*, Summer 1974 (Science and Its Publics: The Changing Relationship), vol. 3, no. 3, *Proceedings of American Academy of Arts and Sciences*, p. 97.

24.   Quoted in Milton Lomask, *A Minor Miracle: An Informal History of the National Science Foundation*, U.S. Government Printing Office, Washington, D.C., 1976, p. 233.

25.   Some saw this then and today still do as smacking of ignorance of the value of fundamental research. I think a better perspective was offered by Don Price, who suggested that the emphasis on applied science "does not cast doubt on the fundamental utility of science as a means of acquiring knowledge, or even on the expectation that basic knowledge will be translated into useful applications that will further human progress. On the contrary, it shows an impatient confidence in these traditional ideas, and a determination to take political shortcuts to make them effective. The fundamental philosophical basis for this reaction is not any new alternative cognitive system; it is old-fashioned Jeffersonian confidence that science will be furthered better by very pragmatic and applied approaches than by scholastic theorizing" (Price, p. 99).

26.   Price, p. 100.

27.   Albert H. Teich, "Federal Support of Applied Research: A Review of the United States Experience," available online at *http://www.ulib.org/webRoot/Books/National_Academy_Press_Books/federal_role/fedl051.htm.*

28.   For example, in the 1960s the foundation made a number of sizeable institutional development grants to enable them to raise the quality of their research programs.

29.   And the money was also to be used to offset the impact of the Mansfield Amendment. This amendment, actually introduced by Senator J. William Fulbright (D-Ark.) but associated with Senator Mike Mansfield (D-Mont.), made it unlawful for the Department of Defense to pay for basic research projects unless they were clearly related to a "military function or operation." And more tellingly for NSF, it ordered the foundation to support a larger share of such projects. Even before the Amendment became law on October 7, 1970, it was applied not only by the Department of Defense but also by other agencies supporting research, such as the Atomic Energy Commission. By the end of 1970, NSF had to cope with $40 million laid on its doorstep by the amendment, and Bill McElroy projected that NSF would have to pay an additional $74 million for projects dropped by other agencies because of the amendment (Lomask, p. 240).

30.   A much fuller account, from which this is abstracted, is provided by Lomask (Chapter 14).

31.   Alfred J. Eggers, Jr., was appointed permanent director of the Research Applica-

tions Directorate on March 2, 1971. Eggers was an aerospace engineer who contributed important experimental and theoretical research in supersonic and hypersonic aerodynamics and in aerospace vehicle technology.

32. Lomask, pp. 246–247.

33. Ibid., p. 248.

34. Ibid., p. 249.

35. That convenient and politically weighty location came to an end when in 1993 the foundation was moved to an office building in Ballston, in Arlington, Virginia.

36. Ehrlichman directed the White House "plumbers" unit. He also approved the break-in at the office of the psychiatrist of Daniel Ellsberg, the defense analyst who leaked the Pentagon Papers to the press. He was convicted of conspiracy to obstruct justice and perjury in the Watergate case and of conspiracy in the Ellsberg case. Haldeman was Nixon's chief of staff and spent 18 months in prison for his role in Watergate.

37. We weren't close colleagues professionally, since he was an economist in the School of Humanities and Social Studies, and I was in the engineering school. However, we did occasionally work together—for example, when he was a member of the MIT Staff-Administration Committee that I chaired.

38. The federal budget process operates under a fiscal year cycle that is 12 months in length. The federal fiscal year begins on the October 1 preceding the calendar year for which the fiscal year is named (e.g., fiscal year 1973 began on October 1, 1972, and ended on September 30, 1973). The president, of course, submitted the FY 1973 budget in January 1972, and the preparation of that budget and its climb upwards from offices to division to directorate to director to OMB, with countless planned and unplanned side excursions along the way, began early in the summer of 1971.

39. It can also function as a radar telescope.

40. Frank Press and Raymond Siever, *Understanding Earth*, W. H. Freeman, New York, 1994, p. 450.

41. National Academy of Sciences, *Astronomy and Astrophysics for the 1970s, Vol. 1: Report of the Astronomy Survey Committee*, National Academy of Sciences, Washington, D.C., 1972, pp. 1, 3.

42. Ibid., pp. 4, 54–55.

43. For an account of this extraordinary experiment, see Norman Metzger, *Men and Molecules*, Crown Publishers, New York, 1972, pp. 84–100. The experiment rested on the very rare conversion by a neutrino of a chlorine isotope to argon, Cl $37 \rightarrow$ Ar 37. Since then, an array of neutrino detectors have been built, embodying different detection schemes. For example, the SAGE and GALLEX detectors in Russia and Italy, respectively, both use vast amounts of Gallium, looking for the conversion of a Gallium atom to Germanium, $^{71}Ga \rightarrow {}^{71}Ge$; while the Super-Kamiokande detector in Japan is a pure water detector, depending on a neutrino hitting an electron in a water molecule, knocking it out of orbit, and making it detectable with photomultiplier tubes.

44. National Academy of Sciences, *Physics in Perspective, Volume 1*, National Academy of Sciences, Washington, D.C., 1972, p. 185.

45. *http://www-odp.tamu.edu/glomar.html.*

46. Ibid.

47. It was built by Cornell University under contract with the Air Force Cambridge Research Laboratories funded by the Advanced Research Projects Agency. The Department of Defense was at first mildly interested in this project but became very interested when it was pointed out that it could detect the ionization trails put down by rockets and *Sputniks.*

48. Bunny and I on a visit to Arecibo realized that the dish was about 20 acres, about the area of our Randolph property. The site wasn't chosen because astronomers prefer pleasant places but rather because, if planets were to be studied, the instrument had to be sited near the equator. The Arecibo site also had large sinkholes of limestone that offered natural shapes needed for the large reflector. Nevertheless, some 270,000 yards of solids and rock had to be excavated or blasted out. See Daniel R. Altschuler, and Chris Salter, "Arecibo: 36 years ago," available online at *http://www.naic.edu/about/history/35years.htm.*

49. The foundation had the lead responsibility for the United States for this program, "initially organized to study the biological structure and function of ecosystems and determine man's relation to them. One of the goals is to predict the consequences of possible natural or man-induced changes brought on specific ecosystems. . . . The research work was organized into large projects within five distinct kinds of life zones called biomes—grassland, desert, coniferous forest, deciduous forest, and tundra" (*National Science Foundation Annual Report—1972*, p. 19).

50. National Science Foundation, "Tundra: The cold ecosystem," *Mosaic*, Winter 1974, p. 3.

51. Toby A. Appel, *Shaping Biology: The National Science Foundation and American Biological Research, 1945–1975*, Johns Hopkins University Press, Baltimore, 2000, pp. 259–261.

52. The United States and other nations had research on the continent since 1956, although Britain claims to have been doing research and exploring in the Antarctic for about 200 years (!) The program has for three decades operated under the Antarctic Treaty, "whose primary purpose is to ensure 'in the interests of all mankind that Antarctica shall continue forever to be used exclusively for peaceful purposes and shall not become the scene or object of international discord.' The Treaty provides for freedom of scientific research in Antarctica and promotes international cooperation toward that end. The original Parties to the Treaty were the 12 nations that were active in conducting research in the Antarctic during the International Geophysical Year (IGY) of 1957–1958. Throughout the years, additional countries have become Consultative Parties, until there are now 26". The treaty joined the advent of space satellites as another legacy of the International Geophysical Year. During the IGY, July 1957–December 1958, "12 countries—Argentina, Australia, Belgium, Chile, France, Japan, New Zealand, Norway, South Africa, the UK, the USA and the USSR—established more than 40 bases on the Antarctic continent and a further 20 on the subantarctic islands" (*http://ast.leeds.ac.uk/haverah/spaseman/index.shtml*).

53. Logistical support for the foundation's Antarctic program came from, and still does, the U.S. Navy and the U.S. Coast Guard.

54. *National Science Foundation Annual Report—1972*, p. 41.

55. Alan T. Waterman did go to Antarctica after he left office.

56. Amundsen left a note for Scott in a tent he set up at the pole.

57. Amundsen's advice to Byrd on the attempt was apt: "Take a good plane, take plenty of dogs and only the best men" *(http://www.south-pole.com/p0000107.htm)*.

58. The exposition was held in 450 acres at League Island near the U.S. Navy Yard in South Philadelphia. Thirty foreign nations participated and 7 million people visited. Participation lagged expectations, however, and financial problems dogged the project from beginning to end. The association passed into receivership in 1927, and it was years before the claims of the organization's many creditors were resolved in U.S. district court.

59. Siple went on to a career in Antarctic exploration and wrote several books, including *A Boy Scout with Byrd*. His association with Byrd was continuous and strong until the admiral's death in 1957. See *http://www.south-pole.com/p0000111.htm*.

60. A major figure in instrumentation and control for airplanes and missiles, Seamans served as Secretary of the Air Force, the position he had when we jointly "circumnavigated" the world at the South Pole; as deputy administrator of NASA; and as the first administrator of the Energy Research and Development Administration, the predecessor to the Department of Energy. He was president of the National Academy of Engineering and is now professor emeritus in the Department of Aeronautics and Astronautics at MIT.

61. Robert Scott called his 1902 hut "The Discovery Hut" and wrote that "it was obvious that some sort of shelter must be made on shore before exploring parties could be sent away with safety, as we felt that at any time a heavy gale might drive the ship off her station for several days, if not altogether. With the hut erected and provisioned, there need be no anxiety for a detached party in such circumstances. . . . We found, however, that its construction was no light task as all the main and verandah supports were designed to be sunk three or four feet in the ground. We soon found a convenient site close to the ship on a small bare plateau of volcanic rubble, but an inch or two below the surface the soil was frozen hard, and many an hour was spent with pick, shovel, and crowbar before the solid supports were erected and our able carpenter could get to work on the frame." Several members of Earnest Shackleton's 1915 effort to cross the continent from sea to sea, assigned to put down depots for the main expedition, died trying to reach the Discovery Hut. See *http://www.theice.org/historicguide.html*.

62. Since my visits to the Cape Royds and Evans huts, they have been fully recognized as historical sites under the Antarctic Treaty. Visitors cannot take away artifacts.

63. Indeed, the station built in 1956 for the International Geophysical Year had become unusable for science and habitation by the end of the 1960s when the decision was made to build a new station.

64. Bob Seamans when he got home offered to send about half of the ice to the

foundation for scientific work-up, but I declined because a single sample from one depth wasn't that helpful; a continuous core from the top down was needed. Rather, I urged Bob to "continue your own line of investigation—possibly by conducting a series of small experiments (about the size of a drinking glass) to determine how the ice interacts with a good brand of Scotch" (Stever to Seamans, January 15, 1973, Historians' Files, 1945–1985, Record Group 307, Row 30 (unprocessed), National Records and Archives Administration, College Park, Md.).

8   TUMULT

1.   Congressman Melvin Laird reportedly agreed to be Nixon's secretary of defense only if the science advisor was kept out of national security matters. See Bruce L. R. Smith, *The Advisers: Scientists in the Policy Process*, The Brookings Institution, Washington, D.C., 1992, p. 169.

2.   Quoted in Gregg Herken, *Cardinal Choices: Presidential Science Advising from the Atomic Bomb to SDI*, Stanford University Press, Stanford, Calif., 2000, p. 169. See pp. 166–183 for a detailed accounting of the fall and eventual rebirth of a science advisory apparatus in the White House.

3.   Memorandum from Henry Simmons to Jim Cannon, March 18, 1975, Office of Science and Technology Policy, 1975, Box 32, James M. Cannon Collection, Gerald R. Ford Library, p. 2.

4.   See Box 8-1 for a list of my predecessors as White House science advisors.

5.   For more details, see Herken, p. 180.

6.   Until science was cast out of the White House, David Z. Beckler served as PSAC's executive secretary and principal assistant to six science advisors.

7.   The Office of Emergency Preparedness was the successor of several agencies dating back to 1947 and had principal responsibility for dealing with civil emergencies and disasters, investigating imports that might impair national security, and oil policies. These functions were dispersed to various line agencies. The National Aeronautics and Space Council was set up in 1958, as part of the response to *Sputnik*. It was abolished, and none of its functions were transferred.

8.   Philip M. Smith had been head of polar programs at NSF. When responsibility for the U.S. activities in Antarctica shifted from the Navy to the NSF, Phil had worked for about a year at OMB to get it done. Hugh Loweth wanted Phil to stay on at OMB and work on the preparation of the next year's science and technology budget, which he did. All that was excellent experience for the assistant to the director job. For the rest of my time in the two-hat period and the later operation at the White House, Phil was my right-hand man. His inside experience at OMB was invaluable in establishing close working relationships with that powerful organization. Phil and I worked together on many other ventures in the past three decades as he continued his outstanding career.

9.   Many other talented people joined me, and I simply can't credit them all. For example, I needed help in the speechwriting area, a growth industry for me. There I was blessed by my friend, Glenn Seaborg, who mentioned that he had a

great speechwriter when he was head of the Atomic Energy Commission, Stanley Schneider. Stan and I hit it off right off the bat, and he joined me and proved to be a great boon. There is nobody in Washington who can succeed in his job without having a good speechwriter, and I had one of the best.

We were fortunate to have as deputy director Ray Bisplinghoff. Ray was invaluable with his broad experience as a teacher of engineering in the aeronautics and astronautics area, as a consultant to many major aerospace companies of the country, and as a strong participant in the space activities of the National Aeronautics and Space Administration. Alfred Eggers and his colleagues in the RANN (see Chapter 7) program often helped when the science advisor had jobs to do in many applied science and technology areas of interest to government.

10. The hookup of the U.S. and Soviet spacecraft occurred on July 17, 1975. For a thorough description of the entire program, formally known as the *Apollo-Soyuz Test Project*, see *http://www.hq.nasa.gov/office/pao/History/SP-4209/cover.htm*.

11. Responsible for NSF congressional liaison.

12. See Chapter 7.

13. See Chapter 2. Dave Langmuir headed the two-person team (I was the second person) in the OSRD Science Liaison Office in London during Word War II.

14. The Russians originally intended them for my predecessor, Ed David, an avid rock collector.

15. *Time*, June 5, 1972. See *http://www.cnn.com/SPECIALS/cold.war/episodes/16/1st.draft/* for longer text.

16. Vadim Aleksandrovich Trapeznikov was an Academician with the Division of Physical-Technical Sciences of the Soviet Academy of Sciences. He died in 1994. See *http://www.icp.ac.ru/RAS_1724-1999/CD_PAH/ENG/27/2783.HTM*.

17. The number one was ill.

18. Nikolai Viktorovich Podgorny was deposed in 1977 in favor of Leonid Brezhnev, who wanted to combine the posts of presidium chairman (i.e., Soviet president) and party secretary-general. He died in 1983.

19. See *http://www.camk.edu.pl/info/index.html*.

20. Nicolae Ceaucescu ran Romania from over 30 years, until he was executed in a coup in 1989.

21. For details, see *http://www.bsf.org.il/*.

22. These international comparisons have grown in time in both sophistication and scope. Nowadays, in addition to the percentage of the gross national product going into science and scientific research and development, comparisons include, inter alia, the different fields of emphasis in the efforts of science and technology in the different countries, the efficiency and effectiveness of science education fields, and the outputs of various countries in terms of patents and intellectual properties.

23. The social and behavioral sciences were added in 1980. For more details, see *http://www.nsf.gov/nsb/awards/nms/start.htm*.

24. The award citation was on target: "For his leadership in the science and engineering basic to aeronautics, for his effective teaching and related contributions in many fields of mechanics, for his distinguished counsel to the Armed Services,

and for his promoting international cooperation in science and engineering." See Chapters 2 and 3 for more on this splendid man and my time with him.

25.   Agnew was investigated by the U.S. attorney in Baltimore for allegedly receiving payoffs from engineers seeking contracts when Agnew was Baltimore county executive and governor of Maryland. Agnew asserted his innocence, but he resigned on Oct. 10, 1973, and pleaded nolo contendere, or no contest, to a single charge that he had failed to report $29,500 of income received in 1967. He was fined $10,000 and placed on three years' probation. See *http://gi.grolier.com/presidents/ea/vp/vpagnew.html.*

26.   For a more detailed Watergate time line, see *http://washingtonpost.com/wp-srv/national/longterm/watergate/chronology.htm.*

27.   The Bohemian Grove is a 2,700-acre redwood forest, located in Monte Rio, California, with accommodations for 2,000 people to camp. The Bohemian Club, a private, all-male club, owns it.

28.   Paul Donovan became its director and Paul Craig the deputy director. We also got considerable help from Richard E. Balzhiser, who before it was abolished had been associate director in the Office of Science and Technology responsible for energy, environment, and natural resources. I tried to lure Dick to the foundation, but he preferred to go into the private sector, where he had a splendid career.

29.   John W. Tukey, who died in 2000, was one of the seminal contributors to statistics in the twentieth century, credited for the invention of many graphical and numerical methods now commonly used in statistics, including the fast Fourier transform algorithm. He received the National Medal of Science in 1973.

30.   Inevitably, the government apparatus for energy and the environment expanded and became more complex. For example, the White House created a new Office of Energy Policy in June 1973 to be directed by John Love, former governor of Colorado. The following December it became the Federal Energy Office headed by John Sawhill and in May the Federal Energy Administration. The Atomic Energy Commission was split in 1974 into the Energy Research and Development Administration and the Nuclear Regulatory Commission. And in 1977 the Energy Research and Development Administration transmuted into the Department of Energy. The White House Council on Environmental Quality came into being in 1970 led by Russell Train. The Environmental Protection Agency opened for business in December 1970, with William D. Ruckelshaus the first administrator, succeeded in 1973 by Russell Train. And in October 1972 the Congress had established its own Office of Technology Assessment (OTA) "as an aid in the identification and consideration of existing and probable impacts of technological application." And it had wisely mandated close coordination by the OTA with NSF enjoining both to maintain "promotion of coordination in areas of technology assessment, and the avoidance of unnecessary duplication or overlapping of research activities in the development of technology assessment techniques and programs." Alas, the Congress—unwisely—abolished the OTA in 1995.

31. Of course, I got the question many times that every science advisor gets: "Do you often see the president?" Tiring of the question and with tongue firmly in cheek, I would sometimes reply: "Yes, I see him almost everyday out of my office window when he walks from the Oval Office to his limousine."

32. See *http://www.nsf.gov/nsb/awards/waterman/start.htm*. In addition to a medal, the awardee receives a grant of $500,000 over a three-year period for scientific research or advanced study in the mathematical, physical, medical, biological, engineering, social, or other sciences. The first award, in 1976, was given to Charles L. Fefferman, Professor of Mathematics at Princeton University "for his research in Fourier analysis, partial differential equations and several complex variables which have brought fresh insight and renewed vigor to classical areas of mathematics and contributed signally to the advancement of modern mathematical analysis."

33. There are two others: (1) always write your minutes in such a way that your master can come to only one decision, and (2) don't drink before the evening. I often violated the first but never the second.

34. Nixon was empowered to appoint Ford through the 25th Amendment, which provides that when the vice presidency becomes vacant the president shall appoint a successor, subject to majority approval by both houses of Congress. Gerald R. Ford, sworn in as vice president on December 6, 1973, had served in the House since 1949, winning 12 consecutive elections and became minority leader of the House in 1965, holding that position for eight years until he became vice president. For more, see *http://www.ford.utexas.edu/grf/fordbiop.htm*.

35. Of course, the political reality was that the office, whatever its congressional pedigree, could be ignored, appointments delayed, etc.

36. Stever to Donald Rumsfeld et al., October 24, 1974, Science and Technology Policy, Legislation, Glenn R. Schleede Papers, Box 35, Gerald R. Ford Library. Donald Rumsfeld was assistant to the president.

37. James Reston, "Calling all scientists," *New York Times*, October 10, 1974, p. 39. Reston flattered me by calling me in the same column an "able and talented man," but I couldn't really quarrel with his judgment that I "wasn't at the center of policy-making at a time when science is central to the problems of the nation and the world." While I was science advisor, I still sat in the National Science Foundation, whose clout across the government was limited and hence limited in coordinating much more powerful agencies.

38. John F. Burby, "Science Report/Congress ready to move on new federal R&D structure," *National Journal*, Dec. 14, 1974, pp. 1871–1876.

39. James R. Killian, Jr., *Sputnik, Scientists, and Eisenhower*, MIT Press, Cambridge, Mass., 1977, p. 257. See also the National Academy of Sciences, *Report of the Ad Hoc Committee on Science and Technology, Science and Technology in Presidential Policymaking: A Proposal*, National Academy of Sciences, Washington, D.C., 1974.

40. Rockefeller was asked to take on restoration of science to the White House as one of his principal responsibilities, no doubt in part because of his work with the Commission on Critical Choices for America, which he organized and funded for the purpose of developing national policy alternatives. After extended con-

gressional inquiries into his financial resources (considerable), he was confirmed by a vote of 287 to 128 in the House and 90 to 7 in the Senate. He was sworn in as the forty–first vice president on December 19, 1974.

41. Memorandum of the vice president to the president, March 3, 1975, Michael Raoul-Duval Collection, OSTP folder, Box 23, Gerald R. Ford Presidential Library.

42. Robert C. Cowen, "Research notebook: Is science out of fashion?" *Christian Science Monitor*, June 4, 1975.

43. The vice president, who had had a long public service career dating back to the FDR administration, had a keen interest in science and technology. He and his chief of staff proved to be critical allies in getting control over those who were reluctant to see the science office reestablished.

44. Letter from the vice president to Senator Kennedy, December 3, 1975, and Senator Kennedy to the vice president, December 8, 1975, Glenn R. Schleede Collection, Science and Technology Policy, 1975, Senate and House Positions, Box 38, Gerald R. Ford Presidential Library.

45. For more on Dick Atkinson, see *http://www.ucop.edu/ucophome/pres/atprofil.html*.

46. George T. Mazuzan, "The National Science Foundation: A brief history," available online at *http://www.nsf.gov/pubs/stis1994/nsf8816/nsf8816.txt*; p. 4.

47. Don K. Price, "Money and influence: The links of science to public policy," *Daedalus*, Summer 1974 (Science and Its Publics: The Changing Relationship). Vol. 3, no. 3. *Proceedings of the American Academy of Arts and Sciences*, p. 101.

48. John Walsh, "NSF: Congress takes hard look at behavioral science course," *Science*, May 2, 1975, p. 426.

49. "NSF gets a record $768 million," *Science*, Sept. 20, 1974, p. 1030.

50. I thank William Blanpied, of NSF, for noting that Doris McCarn was secretary to Alan Waterman, when he was chief scientist at the Office of Naval Research before he came to the foundation as its first director. Doris was sworn in first, Waterman second. So she was NSF's first employee and Waterman its second.

51. That included substantial support by the foundation for graduate students through fellowships and similar means. For example, NSF invested $256 million in fellowships and traineeships between 1952 and 1967. Memo from Stever to Schleede, January 29, 1976, Glenn R. Schleede Collection, National Science Foundation, 1976, Appropriations for FY 1977, Box 22, Gerald R. Ford Presidential Library.

52. Much of this summary is adapted from my January 29 memo to Glenn Schleede referenced above.

53. Letter from Jack Kratchman to the Honorable Joe L. Evins, June 12, 1975, Historians' Files, 1945–1985, Record Group 307, Row 30 (unprocessed), National Records and Archives Administration, College Park, Md.

54. MACOS was developed by the Education Development Center in Cambridge, Massachusetts, a well-regarded organization that had developed other curricula with foundation support, notably in the physical sciences. Extremely distinguished consultants guided the project, including Jerome Bruner and Irven Devore of Harvard and Asen Balikci of the University of Montreal.

55. Kratchman to Evins.

56. This committee was chaired by Robert E. Hughes, NSF assistant director for national and international programs, and included two members of the National Science Board, Grover E. Murray, president of Texas Tech University and Texas Tech University School of Medicine, and L. Donald Shields, president of California State University at Fullerton.

57. Edward Patrick Boland (D-Mass.) served in the Congress for 18 successive congresses, from January 3, 1953, to January 3, 1989. He was enormously influential as an appropriator in the rising fortunes of the National Science Foundation and other research agencies, such as the National Institutes of Health. I always enjoyed my dealings with him. He was tough and questioning but always fair, and when he felt you were right on a hard issue, he was a powerful ally. Not least, he was resolute with another strong supporter of good science—William Natcher (D-Ky.) —in fighting "pork-barrel" spending.

58. Quoted in Walsh, p. 426. The reference to "competition from other publishing houses" was another arrow in the attack on the NSF; specifically, charges of impropriety in the manner in which the foundation had found a publisher for the MACOS course, the disposition of royalties, and the like. We responded to these additional charges very effectively I thought, and rather than go into detail on these, I refer anyone interested to the *Science* article by John Walsh, especially pp. 427–428.

59. The story of course had many twists and turns, cut short by this summary. The reader who wants all the gory details can find them in the excellent series of articles by John Walsh in *Science*. In addition to the cited May 1975 article, others by Walsh in *Science* worth reading are "NSF and its critics in Congress: New pressure on peer review," June 6, 1975, pp. 999–1001; "Peer review: NSF faces changes, the question is how extensive," Oct. 17, 1975, pp. 253–256; and "NSF: How much responsibility for course content, implementation?" Nov. 14, 1975, pp. 644–646.

60. President Ford was well on his way to stopping and reversing the decline in research funding. The Carter administration went on to seek real growth in research support by the National Science Foundation and other agencies, such as the Departments of Agriculture, Defense, and Energy. Overall, support grew for federal defense and nondefense research during the Carter administration, embodying as well growth for the support of basic research. With the arrival of the Reagan administration, nondefense research and development support declined markedly, whereas defense research and development increased sharply. Total support for basic research began rising again in 1984 and continued to do so through the remainder of the twentieth century. See *http://www.aaas.org/spp/dspp/rd/trendtot.pdf.*

61. Ford, the "accidental president" now seeking the nomination for presidency on his own, had a strong challenge from the right by the two-term governor of California, Ronald Reagan. Ford countered by picking a conservative, Senator Robert Dole from Kansas, as his running mate, dropping Nelson Rockefeller. Even so, Ford won the nomination by only 60 votes. And he lost the general election to Jimmy Carter, but in bittersweet vindication made it a close race after being badly behind in the opinion polls at the start of the campaign. For more

on the 1976 presidential campaign, see *http://www.americanpresident.org/KoTrain/Courses/GF/GF_Campaigns_and_Elections.htm.*

62. Senators Jesse Helms (R-N.C.), James McClure (R-Ind.), Clifford Hansen (R-Wyo.), and Carl Curtis (R-Neb.).

63. For this correspondence and related materials, see the Glenn R. Schleede Collection, the Folder on Science and Technology Policy, 1976: Nomination of Dr. Stever (1), Box 40, Gerald R. Ford Presidential Library.

64. Other members were Otis Bowen, Glen Campbell, Ed David, Elizabeth LeDuc, Paul O'Neill, Fritz Russ, Charlie Slichter, Charlie Townes, Caspar Weinberger, and Brad Wiley. I was an ex-officio member.

65. Don, a distinguished neurobiologist, stayed past the end of the Ford presidency to head the Food and Drug Administration under President Carter for two and one-half years. He returned to Stanford in 1979 to serve as provost and then for 12 years as its president. In June 2000 he became editor-in-chief of *Science*. Bill Nierenberg, who died of cancer in September 2000, was a physicist who served as director of the Scripps Institution of Oceanography for 21 years, from 1965 to 1986. He was instrumental in founding JASONS, a group of physicists who conduct studies on military issues. As a memorial tribute put it, Bill had a very dynamic style and "held his views strongly, and showed no fear of making a mistake. He would teach Chinese to the China-born and was heard to lecture on French chateaux to a president of a society devoted to restoring them. Although an impatient listener, Bill still 'heard' you. To use an analogy from Antisubmarine Warfare, in exchanging information he used the principles of active rather than passive sonar, bouncing his ideas off you and observing your reactions." (*http://www-senate.ucsd.edu/assembly/nierenberg.htm*).

66. By 1.7 million votes out of 81 million cast.

9   END AND START

1. Quoted in National Research Council, *US Industry in 2000: Studies in Competitive Performance*, National Academy Press, Washington, D.C., 2000, p. 1.

2. National Research Council, *Securing America's Industrial Strength*, National Academy Press, Washington, D.C., 1999, p. 4.

3. Richard R. Nelson and Paul M. Romer, "Science, economic growth, and public policy," *Technology, R&D, and the Economy*, Bruce L. R. Smith and Claude E. Barfield, eds., The Brookings Institution and The American Enterprise Institute, Washington, D.C., 1996, pp. 54–55.

4. That $80 billion equated to about $62 billion in constant 2001 dollars. About $30 million of that was for defense research and development, the dominant portion of that going to prototyping and weapons testing. Of the nondefense funds, about $10 billion went to basic research. The distortions in representing federal research and development as a single number—grouping weapons testing with basic research—were addressed in the mid-1990s by a committee of the National Academy of Sciences, which argued that a new measure was needed that set out what the federal government was actually spending on the provision

of new knowledge, what it called the "Federal Science and Technology Budget." The idea was somewhat radical and thus took time to gain acceptance. But acceptance came, notably in the FY 2003 budget, which explicitly used the measure. The report itself can be found at *http://books.nap.edu/html/federal_funds*. See the next chapter for further details on the work of this committee, of which I was a member.

5.   Quotes taken from Davis Dyer, *TRW: Pioneering Technology and Innovation Since 1900,* Harvard Business School Press, Boston, 1998, pp. 233–234. Indeed, this volume provides a thorough account of the history of the companies, both in its originally separate parts—Ramo-Wooldridge (R-W), which had built its reputation on system engineering for the nascent Air Force ICBM programs, and Thompson Products, a mass producer of parts and components for the automotive, aircraft, and aircraft engine industries. Merging them was a challenge: they were geographically apart, R-W in Southern California and Thompson in Cleveland; and more critically the cultures of the two organizations were polar, R-W people more at home in the highly specialized and intense world of aerospace and advanced electronics and Thompson people more "old-line" running a business that by sticking to a basic formula had been successful for decades. That the merger came off is quite a story, and Dyer tells it well, especially pp. 225–249.

6.   Rube Mettler was a fellow Cal Tech graduate, earning all his degrees there, including a doctorate in electrical and aerospace engineering. He began his career with Hughes Aircraft in 1949, moved to TRW in 1955, and retired as chairman and CEO of TRW in 1989. He was elected to the National Academy of Engineering in 1965 as an "outstanding creative missile and systems engineer."

7.   At the same time, Stanley Pace became president and chief operating officer. Certainly not least, Simon Ramo reached his sixty fifth birthday and chose to retire from his vice chairmanship of the board, but stayed as a member. He kept his California office right next to that of the chairman and frequently consulted with and led special studies for the chairman, remaining a force to reckon with.

8.   Dyer, p. 341.

9.   Richard Delauer (1918–1990) earned a doctorate in aeronautics and mathematics from Cal Tech, pioneered work on nuclear rocket propulsion, and in 1958 joined the new TRW Space Technology Laboratories as assistant laboratory director, eventually rising to responsibility for all of TRW's defense, space, electronics, and information systems. Dick Delauer was elected to the National Academy of Engineering in 1969 for the "design of spacecraft and ballistic missile weapons systems; application of systems engineering methodology in defense, space, and civil systems." He became under secretary of defense for research and engineering in 1981. He left government service in 1984, founded a consulting group, and from 1989 until his untimely death was chief executive officer of Fairchild Space and Defense.

Simon Ramo, cofounder of TRW, has been honored in many ways, including the highest civilian honor, the Presidential Medal of Freedom. He was elected to the National Academy of Engineering in 1964. He was born in Salt Lake City on May 13, 1913, to Lithuanian immigrants who owned a clothing store. Ramo earned a B.S. in electrical engineering at the University of Utah and, at 23, a

Ph.D. in electrical engineering and physics at Cal Tech. From 1936 to 1946 Ramo worked for General Electric (GE) in Schenectady, New York, developing microwave transmission and detection equipment and GE's electron microscope. In 1946 he accepted a position with Hughes Aircraft in Culver City, California, where he developed fire control, radar, navigation, computer, and other aircraft electronics systems. Ramo and fellow engineer Dean E. Wooldridge left Hughes Aircraft in 1953 to form the Ramo-Wooldridge Corporation, obtaining financial support from Thompson Products, Inc. For the next four years, Ramo-Wooldridge had primary responsibility for guiding the development of the *Atlas*, *Titan*, and *Minuteman* ICBMs. In 1958 Ramo-Wooldridge merged with Thompson Products to form Thompson Ramo-Wooldridge, Inc., a name later shortened to TRW, Inc.

George E. Solomon was elected to the National Academy of Engineering in 1967 for the "design and development of space and weapon systems." After earning a Ph.D. at Cal Tech, Solomon joined the Ramo-Wooldridge Corporation, rising to executive vice president and general manager of its electronics and defense sector.

John S. Foster, Jr., after earning his physics doctorate from the University of California at Berkeley, began his career with the Radio Research Laboratory at Harvard University in 1942. He spent 1943 and 1944 as an advisor to the 15th Air Force on radar and radar countermeasures in the Mediterranean theater of operations. He was director of the Lawrence Livermore National Laboratory and for eight years director of research and engineering before joining TRW. His many awards include the Enrico Fermi Award of the Department of Energy and the James Forrestal Memorial Award. He was elected to the National Academy of Engineering in 1969 for "technological leadership in defense research and engineering."

10. Pierce Angel, an old-timer in the Cleveland operations, became the focus of these information exchange groups, though his health-related retirement cut short his participation.

11. Formally, the Redondo Beach Space Center, for which ground was broken in December 1960. It included several major buildings, including one that Rube Mettler called "a tremendous gamble," $7 billion for a facility to manufacture spacecraft. It included a 50-foot-high manufacturing bay, a 10-ton overhead crane, and a 30-foot space simulator that could handle spacecraft much larger than those built at the time. The gamble paid off, as both NASA and the Air Force selected TRW in the 1960s to manufacture major and very large spacecraft programs—the Orbiting Geophysical Observatory in the case of NASA and for the Air Force Project 823, or "Vela Hotel," to detect nuclear explosions. That was only the beginning. Out of this building and other facilities at Space Park emerged *Pioneer 10*, the first spacecraft to leave the solar system; the tracking and data relay satellites (TDRS) for communications between the earth and low-earth orbit satellites, including the Space Shuttle; the Compton Gamma Ray Observatory; the Defense Support Program satellites to monitor missile launches; and the engine for the lunar descent module that enabled six astronauts to land

on the moon and, certainly not least, was key in the safe return of *Apollo 13*. Dyer, pp. 2, 235–236.

12.   Bement worked with William Perry, director of defense research and engineering in the Carter administration, on the materials research programs that led to the stealth technologies that demonstrated their effectiveness in the Gulf War. Arden then became David A. Ross Distinguished Professor of Nuclear Engineering and head of the School of Nuclear Engineering at Purdue University. In 2002 he returned to public service as director of the National Institute of Standards and Technology.

13.   Richard J. Fruehan, Dany A. Cheij, and David M. Vislosky, "Steel," in National Research Council, *U.S. Industry in 2000*, pp. 75–76.

14.   Indeed, the Bethlehem Steel story has a sad end, for the company filed for bankruptcy on October 15, 2001. Its financial problems worsened a lot after September 11, and according to its chairman, "what was a declining market has become a free fall" (*Washington Post*, October 16, 2001, p. E1). The impact of the steel tarriffs imposed by the Bush administration in 2002 remains to be seen.

15.   That cultural difference is still evident in TRW today, nearly a quarter of a century later.

16.   Schering-Plough was created in 1971 by a merger of the Schering Corporation and Plough, Inc. Each had modest beginnings. "The Schering concern started in 1852 as a local pharmacy called the Green Apothecary, but soon began to produce the new wonder drugs chloroform and cocaine. Schering later made a fortune by pioneering the use of synthetic drugs to avoid importing raw materials" (Alexandra Richie, *Faust's Metropolis: A History of Berlin*, Carol and Graf Publishers, New York, 1998, p. 144). Plough, Inc., was created in 1908 by 16-year-old Abe Plough. With $125 borrowed from his father, he bought a horse and wagon and began selling remedies to town folk and farmers around Memphis, Tennessee.

17.   Comprehensive analyses of the worldwide automobile industry were developed at MIT by Jim Womack, Dan Jones, and Daniel Roos, my former colleagues, who started the MIT International Motor Vehicle Program in the early 1980s. See Alan Altshuler, Martin Anderson, Daniel T. Jones, and Daniel Roos, *The Future of the Automobile*, MIT Press, Cambridge, Mass., 1984, and James P. Womack, Daniel T. Jones, and Daniel Roos, *The Machine that Changed the World*, Rawson Associates, New York, 1990. They are classic studies of engineering and management.

18.   We met with President Suharto and Minister for Research and Technology B. J. Habibie among others. Habibie was an aeronautical engineer by training and since 1986 a foreign associate of the National Academy of Engineering. In the 1980s Indonesia was stable and a good place for business, although the ruling Suharto family had already begun to get into trouble because of nepotism. Habibie became president in 1998 but was forced out after 17 months of an extremely turbulent time for the nation.

19.   Even now, at the start of a new millennium, Japan has relatively few venture capital investors other than banks and established companies, few university-industry joint research programs, and still fewer industrial and incubator parks than we have in the United States.

10   ENDING THE CENTURY

1. Gregg Herken, *Cardinal Choices: Presidential Science Advising from the Atomic Bomb to SDI,* Stanford University Press, Stanford, Calif., 2000, p. 208. See also pp. 208–211 for a detailed account from which I've drawn for the origins of the March 23 address and its repercussions.

2. For the complete text of the March 23 speech, see *http://www.fas.org/spp/starwars/ offdocs/rrspch.htm.*

3. See Chapter 4 for more on the ABM study I chaired in 1954.

4. G. A. Keyworth, II, "Federal science and technology policy—The Reagan years," paper presented at 25th Anniversary Symposium of the Office of Science and Technology Policy, Massachusetts Institute of Technology, May 1, 2001, pp. 5–6.

5. Herken, p. 213. The official was Victor Reis, then Office of Science and Technology Policy (OSTP)'s assistant director for national security and space." Laetrile was at the time one of the better-known supposed "cures" for cancer, made from substances in the pits of apricots and other fruits. No reliable evidence for its efficacy has yet turned up.

6. Ibid., p. 210.

7. Keyworth, p. 6.

8. *http://www.fas.org/spp/starwars/offdocs/m8310017.htm.*

9. I was also mindful that OTA committees have a sharply different role than those of the National Research Council. NRC reports are the responsibility of the committee, and NRC procedures stem from that. For OTA reports the committees are advisory only, and the reports are the responsibility of project staff and OTA. Thus, an OTA report can be issued even, say, if most if not all of the advisory committee members disagree with it; that's impossible with NRC reports.

10. In addition to me as chairman, there was Solomon Buchsbaum, executive vice president of AT&T Bell Labs; Ashton Carter, Kennedy School of Government, Harvard; Robert Clem, director of systems sciences at Sandia National Labs; Sidney D. Drell, deputy director of the Stanford Linear Accelerator Center; Daniel J. Fink, president of D. J. Fink Associates, Inc.; Richard Garwin, IBM fellow, Thomas J. Watson Research Center; Noah Guyler, Admiral, U.S. Navy, American Committee on East-West Accord; Colin Gray, president of the National Institute for Public Policy; George Jeffs, president of North American Space Operations, Rockwell International; General David Jones, U.S. Air Force (ret.), former chairman, Joint Chiefs of Staff; Robert S. McNamara, former president of the World Bank; Michael M. May, associate director at large, Lawrence Livermore National Laboratories; H. Allen Pike, program manager, Space Station at Lockheed, Missiles in Space; Fred Seitz, president emeritus, Rockefeller University; Robert Seldon, associate director for theoretical computational physics, Los Alamos National Laboratory; Marshall D. Shulman, director of the Herrman Institute for Advanced Study of the Soviet Union, Columbia University; Ambassador Gerard C. Smith, president of Consultants International Group, Inc.; Sayre Stevens, vice president of Systems Planning Corporation;

Major General John Toomay, U.S. Air Force (ret.), consultant; and Seymour Ciberg, vice president, research and engineering operations, Martin Marietta Aerospace.

11. Office of Technology Assessment, U.S. Congress, *Ballistic Missile Defense Technologies*, Princeton University Press, Princeton, N.J., 1986.

12. Ibid., p. 34.

13. Ibid., p. iii.

14. That of course turned out to be wrong, and the conflict over the technical feasibility and strategic wisdom of building an ABM system is now a major issue for the second Bush administration.

15. Office of Technology Assessment, U.S. Congress, *SDI: Technology, Survivability, and Software*, vol. 2, U.S. Government Printing Office, Washington, D.C., 1988, p. 5

16. Universities Research Association is now one of several consortia that sit in effect between major facilities and the government agencies that support them, and whose job it is to assure and manage the national use of these facilities. Thus, the association now has 89 member universities and through its governing board of university representatives serves as the primary contractor to the Department of Energy for operation of the Fermi National Accelerator Laboratory. See *http://www.ura-hq.org/about/index.html.*

17. It became Fermilab after the great Italian physicist, Enrico Fermi, on May 11, 1974.

18. Of course, there were (and are) other centers of high-energy physics in the United States and other countries, at which momentous discoveries were made. For example, the first evidence for point-like quarks came from the Stanford Linear Accelerator Center, confirming speculation by Richard Feynman and providing strong support for the Standard Model.

19. *http://www.news.cornell.edu/releases/Jan00/RRWilson_obit.hrs.html.*

20. *http://www.fnal.gov/pub/about/whatis/history.html.*

21. With Melvin Schwarz and Jack Steinberger.

22. The lengthy political battles about the Superconducting Supercollider facility construction, about its partial completion followed by an orderly closing down, and, not least, about helping very able people find new positions, were difficult chapters in the SSC story. The load fell on the shoulders of John Marburger, chairman of URA, John Toll, president of URA, and John Peoples, director of Fermilab. But the loss of the SSC by no means spelled the end of U.S. high-energy particle physics in the United States. Fermilab and several of the other principal laboratories received support for new facilities and strong research programs. The field, however, being a prime example of expensive "big science," always presents difficult budget decisions.

23. The Gang of Four was a group of four hard-core communists, including Mao's wife, who in the mid-1960s pushed for total destruction of traditional Chinese culture to be replaced by textbook communist ideology and culture. They became the leading forces in Mao's cultural revolution. For more, see *http://www.europeaninternet.com/china/underst/gangfour.php3.*

24.   Phil Smith, who worked with me when I was science advisor to President Ford
      and then stayed to work with Frank Press as associate director of the Office of
      Science and Technology Policy aptly described U.S.-China dialogue at the time
      of my visit: "As the administration was getting underway, China was also just
      beginning its renewed modernization drive, an undertaking of enormous magni-
      tude then and now and the genesis of China's interest in acquiring technology
      from the West. Trade, science, technological, and academic contacts had been
      taking place since the Nixon initiatives, but they were private, non-governmental
      exchanges and contacts. The president and National Security Adviser Zbigniew
      Brzenzinski asked Frank to develop a possible program of government-to-
      government cooperation in S&T. After a few months of preparation, Frank led a
      large delegation, including the heads of NSF, NASA, NOAA, NIH, USGS, the
      research directors of DOE and the Department of Commerce, and other govern-
      ment R&D leaders on an extended visit to China. At the time it was, and may
      still be, the largest official R&D delegation of government S&T officials to go
      abroad. They went in their own Air Force plane, which added prestige to the visit
      from the perspective of their Chinese hosts. The discussions in China led to the
      formal agreement on cooperation in science and technology signed by the Presi-
      dent and Vice Premier Deng Xiaoping in early 1979, initiation of government-
      to-government cooperation in R&D, and the start of the influx of Chinese
      students studying in American universities. There's an interesting anecdote re-
      lated to Chinese student study in U.S universities. In discussing student study
      with Deng Xiaoping, Frank cautioned that it was possible that some students
      would choose to remain in the U.S. to pursue their scientific careers. After a
      pause, the Vice Premier responded: 'Dr. Press, we have plenty more!'" (Philip M.
      Smith, "Science and Technology in the Carter Presidency," paper presented at the
      25th Anniversary Symposium of the Office of Science and Technology Policy,
      Massachusetts Institute of Technology, May 1, 2001).

25.   The National Academy of Engineering is part of the National Academies, the
      other parts being the National Academy of Sciences, the Institute of Medicine,
      and the National Research Council. The National Academy of Sciences came
      first, in 1863. The National Research Council, an organizational vehicle for con-
      ducting studies, mainly for the government, was created in 1916. A major reason
      for bringing it into being, and the subsequent and eternal confusion on its role
      vis-à-vis the National Academy of Sciences, was to provide a means for creating
      committees that included non-Academy members. The National Academy of
      Engineering arrived in 1964 and the Institute of Medicine in 1970. Oversimpli-
      fying just a tad, the latter two were created because of the restrictive criteria for
      election to the NAS—"distinguished contributions to science." That often made
      it tough for people who made major contributions to engineering science and
      technology and to the advancement of health care to be recognized by election to
      the Academy. All three organizations—NAS, NAE, IOM—annually elect new
      members to their respective academies and getting elected was and remains very
      tough, making it one of the highest honors for a professional in the fields of
      science, engineering, and medicine. For more, see *http://nationalacademies.org*.

26.   Chapter 3 has more on Tsien's brilliant career and political troubles.

27.   National Research Council, *Engineering Education in the People's Republic of China: Report of a Visit*, National Academy Press, Washington, D.C., 1983.

28.   On the other hand, we enjoyed that culture when we were shown the antiquities of every region visited and the top political leader gave us an unsurpassable Chinese banquet, culminating in Beijing when we were guests of the minister of education in the Great Hall on the occasion of the twenty-ninth anniversary of the communist revolution.

29.   This was a project of the NRC's Board on Science and Technology for International Development sponsored by the Agency for International Development.

30.   National Research Council, *U.S. Science and Technology for Development: A Contribution to the 1979 UN Conference*, U.S. Government Printing Office, Washington, D.C., 1978, p. iii.

31.   Ibid., p. 7.

32.   *http://www.iiasa.ac.at/docs/history.html?s*

33.   *http://www.iiasa.ac.at/.*

34.   Francis R. Scobee, commander; Michael J. Smith, pilot; Judith A. Resnik, mission specialist 1; Ellison S. Onizuka, mission specialist 2; Ronald E. McNair, mission specialist 3; Gregory B. Jarvis, payload specialist 1; and Sharon Christa McAuliffe, payload specialist 2.

35.   Formally, the Presidential Commission on the Space Shuttle Challenger Accident. Its chair, William P. Rogers, secretary of state under President Nixon, and I were both Colgate alumni, Bill Rogers having graduated five years ahead of me. We received honorary Colgate degrees at the same occasion, his twenty-fifth anniversary and my twentieth. We had several interactions, when Bill was at the State Department and I was at the National Science Foundation, for example, on NSF's role in the law of sea negotiations and in establishing the United States-Israel Bi-national Science Foundation mentioned in the previous chapter.

36.   *http://science.ksc.nasa.gov/shuttle/missions/51-l/docs/rogers-commission/Appendix-F.txt.*

37.   The members of the Panel on Technical Evaluation of NASA's Redesign of the Space Shuttle Solid Rocket Booster, in addition to me, were Laurence J. Adams, Martin Marietta Corporation; David Altman, United Technologies Corporation; Robert C. Anderson, TRW, Inc.; Jack L. Blumenthal, TRW, Inc.; Robert C. Forney, E. I. DuPont Nemours & Co.; Alan N. Gent, University of Akron; Dean K. Hanink, General Motors; James W. Mar, Massachusetts Institute of Technology; Edward W. Price, Georgia Institute of Technology; and Robert D. Watt, Stanford Linear Accelerator Center.

38.   This was Myron F. Uman, perhaps the most critical selection for what turned out to be an incredibly arduous task. Then a member of the professional staff of the National Research Council for over a decade, he came armed with a doctorate in electrical engineering and plasma physics from Princeton University, academic time as a former faculty member of the University of California at Davis, and experience in running many programs and studies. He was indispensable.

39.   National Research Council, *Collected Reports of the Panel on Technical Evaluation*

of NASA's Redesign of the Space Shuttle Solid Rocket Booster, National Academy Press, Washington, D.C., 1988.

40. John Thomas was selected by NASA to be responsible for the redesign, and Morton Thiokol selected Alan McDonald to lead the effort. They soon recognized the expertise of our panel members and that we could be a positive and supporting team.

41. The panel's concern was that flaws might arise in manufacturing or assembling the rocket that could not be detected by quality control processes. So flaws should be intentionally introduced into test articles to be sure the design is resilient.

42. National Research Council, p. 38.

43. STS is Space Transportation System, the formal name for the shuttle.

44. The surgery included two laminectomies, excision of posterior arches of vertebrae, and several foraminotomies, removing some of the foramina or passages between vertebrae to ease nerve compression.

45. This was one of several so-called White Papers (a term misappropriated from the British who would call them "Green Papers," intended for discussion) done by the Academies for the new president. The other topics were AIDS, global environmental change, science and technology advising in the White House, and federal science and technology budget priorities.

46. These are more fully summarized in H. Guyford Stever, and David L. Bodde, "Space policy: Deciding where to go," *Issues in Science and Technology*, Spring 1989, pp. 66–71.

47. Ibid., p. 70.

48. The president was a psychiatrist, David Hamburg, who before going to the Carnegie Corporation served as president of the Institute of Medicine. Earlier he had been a professor at Stanford University. Hamburg had very broad interests in science, technology, and health policy, and this motivated his interest in establishing the Carnegie Commission. He became president emeritus of the Carnegie Corporation in 1997 after a 14-year tenure.

49. John R. Steelman, *Science and Public Policy: A Report to the President*, U.S. Government Printing Office, Washington, D.C., 1947.

50. Of course the immediate predecessor to the Steelman Report, Vannevar Bush's *Science, the Endless Frontier*, was also influential but more as a hard-hitting call to arms—national strength in research in peacetime—than as a detailed battle plan, which the Steelman Report provided.

51. The national research and development effort has been well above Steelman's one percent mark since the early 1950s, while the federal portion climbed to almost 2 percent in 1961 but has been closer to 0.5 percent since then. See *http://www.aaas.org/spp/dspp/rd/trendusg.pdf*.

52. William A. Blanpied, *Science and Public Policy: The Steelman Report and the Politics of Post-World War II Science Policy*, available online at *http://www.aaas.org/spp/yearbook/chap29.htm*.

53. Jeffrey K. Stine, *A History of Science Policy in the United States, 1940–1985*, Science Policy Study, Background Report No. 1, Task Force on Science Policy, Committee on Science and Technology, House of Representatives, 99th Congress, Second Session, September 1986, pp. 30–32.

54. A full listing and text of the Carnegie Commission's products can be found at *http://carnegie.org/sub/research/index.html#science.*

55. Joshua Lederberg received a Nobel Prize in 1958 for his work in bacterial genetics. William T. Golden, among his many contributions, served as a special consultant to President Truman on mobilizing the nation's scientific resources.

56. Vannevar Bush, *Science, the Endless Frontier*, National Science Foundation, July 1945, Washington, D.C. See also *http://www.nsf.gov/od/lpa/nsf50/vbush1945.htm.*

57. Both reports are available through the Carnegie Commission's website at *http://carnegie.org/sub/pubs/ccstfrep.htm* and also through the State Science and Technology Institute's site at *http://www.ssti.org/Publications/carnegie.htm.*

58. "We the people of the United States, in order to form a more perfect union, establish justice, insure domestic tranquility, provide for the common defense, promote the general welfare, and secure the blessings of liberty to ourselves and our posterity, do ordain and establish this Constitution for the United States of America."

59. This is obviously cursory. For more details, see National Research Council, *Linking Science and Technology to Society's Economic Goals*, National Academy Press, Washington, D.C., 1996.

60. National Research Council, *Harnessing Science and Technology for America's Economic Future*, National Academy Press, Washington, D.C., 1996.

61. It was a very distinguished and able committee: Frank Press, Carnegie Institution of Washington, chair; Lew Allen, Jr., Charles Stark Draper Laboratory, Inc.; David H. Auston, Rice University; Forest Baskett, Silicon Graphics Computer Systems; Barry R. Bloom, Albert Einstein College of Medicine; Daniel J. Evans, Daniel J. Evans & Associates; Baruch Fischhoff, Carnegie Mellon University; Marye Anne Fox, University of Texas at Austin; Shirley A. Jackson, U.S. Nuclear Regulatory Commission; Robert I. Levy, Wyeth-Ayerst Research; Richard J. Mahoney, Monsanto Co. (ret.); Steven L. Mcknight, Tularik, Inc.; Marcia K. McNutt, Massachusetts Institute of Technology; Paul M. Romer, University of California at Berkeley; Luis Sequeira, University of Wisconsin; Harold T. Shapiro, Princeton University; H. Guyford Stever, trustee and science advisor; and John P. White, Department of Defense.

62. National Academy of Sciences et al., *Allocating Federal Funds for Science and Technology*, National Academy Press, Washington, D.C.,1995.

63. *http://www.nap.edu/readingroom/books/fedfunds/.*

64. *http://www.greatachievements.org/greatachievements/index.html.*

11 AFTERWORD

1. Walter A. McDougall, *The Heavens and the Earth: A Political History of the Space Age*, Johns Hopkins University Press, Baltimore, 1985, p. 75.

2. Roger L. Geiger, *To Advance Knowledge: The Growth of American Research Universities, 1900–1940*, Oxford University Press, New York, 1986, p. 183.

3. Ibid., p. 86.

4. Daniel J. Kevles, *The Physicists: The History of a Scientific Community in Modern America,* Alfred A. Knopf, New York, 1978, p. 290.

5. Ibid., p. 289.

6. Ibid., p. 307.

7. Quoted in James Phinney Baxter, *Scientists Against Time,* Little Brown and Co., Boston, 1946, p. 9.

8. Which, of course, became the Carnegie Mellon University during my tenure.

9. An umbrella term for the National Academy of Sciences, National Academy of Engineering, and Institute of Medicine.

Appendix A

Chronology

1916   Born October 24 in Corning, N.Y.

1938   A.B., summa cum laude, physics, Colgate University

1941   Ph.D., magna cum laude, physics, California
              Institute of Technology

1941–1942   Staff member, Radiation Laboratory, Massachusetts
              Institute of Technology

1942–1945   Scientific liaison officer, London Mission, Office of
              Scientific Research and Development

1945–1965   MIT Faculty:
                  Assistant professor, 1946–1951, aeronautical
                      engineering;
                  Associate professor, 1951–1956;
                  Professor, 1956–1965, aeronautics and
                      astronautics;
                  Associate dean of engineering, 1956–1959;
                  Department chair, mechanical engineering, and
                      naval architecture and marine engineering,
                      1961–1965

1946	Married Louise Risley Floyd. Four children: Horton Guyford Stever, Jr., August 25, 1947; Sarah Stever, February 25, 1949; Margarette Stever Weed, June 18, 1952; Roy Risley Stever, August 17, 1954
1946–1972	Member or chair of numerous committees, principally to advise the military services, especially the U.S. Air Force, on melding science and technology with military goals and needs. Advised on the goals and missions for the newly formed National Aeronautics and Space Administration, for the president on the patent system, and for the Congress on science and technology policy.
1955–1956	Chief Scientist, U.S. Air Force
1956–1972	Corporate board member and/or consultant to United Aircraft Corporation, Koppers Corporation, and Fisher Scientific Company
1956–1961	Vice chairman, Air Force Scientific Advisory Board
1962–1968	Chairman, Air Force Scientific Advisory Board
1965	Elected to the National Academy of Engineering
1965–1972	President, Carnegie Institute of Technology (in 1967, Carnegie Mellon University)
1970–1976	Member, National Science Board
1972–1976	Director, National Science Foundation
1973	Elected to the National Academy of Sciences

1976–1977   Director, Office of Science and Technology Policy

1977–1989   Corporate board member and/or consultant to TRW,
               Schering-Plough, Caterpillar, and Goodyear

1977–present   Advisor through committee memberships and
               chairmanships on matters including the Strategic
               Defense Initiative, responding to the Challenger
               accident, science and technology policy, and other
               issues for the National Academies, Office of
               Technology Assessment, and Carnegie Commission
               on Science, Technology, and Government; trustee of
               the Universities Research Association, Inc., Woods
               Hole Oceanographic Institution, and Science Service

Acronyms

ABM	antiballistic missile
ADRDE	Army Defense Research and Development Establishment
ADSEC	Air Defense Systems Engineering Committee
AGARD	Advisory Group for Aeronautical Research and Development
AICBM	anti-intercontinental ballistic missile
APL	Applied Physics Laboratory
C³	communications, command, and control
CIOS	Combined Intelligence Operations Section
CMAP	Carnegie Mellon Action Program
CVD	Committee on Valve Development
GAAP	guided antiaircraft projectile
ICBM	intercontinental ballistic missile
ICSU	International Council of Scientific Unions
IGY	International Geophysical Year
IIASA	International Institute for Applied Systems Analysis
IRBM	intermediate-range ballistic missile
JCS	Joint Chiefs of Staff
LORAN	long-range navigation system
MACOS	Man: A Course of Study

MAD	mutually assured destruction
MEW	Microwave Early Warning system
MIT	Massachusetts Institute of Technology
NACA	National Advisory Committee for Aeronautics
NASA	National Aeronautics and Space Administration
NEPA	Nuclear Energy for the Propulsion of Aircraft
NRDC	National Research Defense Council
ODM	Office of Defense Mobilization
OPEC	Organization of Petroleum Exporting Countries
OSRD	Office of Scientific Research and Development
OST	Office of Science and Technology
OSTP	Office of Science and Technology Policy
OTA	Office of Technology Assessment
PSAC	President's Science Advisory Committee
RAF	Royal Air Force
RANN	Research Applied to National Needs
SAB	Scientific Advisory Board
SAC	Strategic Air Command
SAGE	semi-automated ground environment
SHAEF	Supreme Headquarters, Allied Expeditionary Force
STL	Space Technology Laboratories
TRE	Telecommunications Research Establishment
URA	Universities Research Association, Inc.

Appendix C

Suggested Readings

Appel, Toby A. *Shaping Biology: The National Science Foundation and American Biological Research, 1945–1975*. Baltimore: Johns Hopkins University Press, 2000.

Barfield, Claude E. *Science Policy from Ford to Reagan: Change and Continuity*. Washington, D.C.: American Enterprise Institute for Public Policy Research, 1982.

Baxter, James Phinney. *Scientists Against Time*. Boston: Little Brown and Co., 1946.

Beckler, David Z. "The precarious life of science in the White House," *Daedalus*, summer 1974 (Science and Its Publics: The Changing Relationship), vol. 3, no. 3 of the Proceedings of the American Academy of Arts and Sciences, pp. 115–134.

Blanpied, William A., ed. *Impacts of the Early Cold War on the Formulation of U.S. Science Policy*. Selected Memoranda of William T. Golden: October 1950–April 1951. Edited with an Appreciation by William A. Blanpied. Conversation with Kenneth Pitzer, Director, Research Division, Atomic Energy Commission. Nov. 1, 1950. Available online at *http://www.aaas.org/spp/cstc/golden/title.htm*.

Buderi, Robert. *The Invention that Changed the World*. New York: Touchstone, Simon and Schuster, 1997.

Bush, Vannevar. *Modern Arms and Free Men*. New York: Simon and Schuster, 1949.

Bush, Vannevar. *Science, the Endless Frontier*. Washington, D.C.: National Science Foundation, July 1945. See also *http://www.nsf.gov/od/lpa/nsf50/vbush1945.htm*.

Chamberlin, E. R. *Life in Wartime Britain*. London: Batsford, 1972.

Conant, Jennet, *Tuxedo Park: A Wall Street Tycoon and the Secret Palace of Science that Changed the Course of World War II,* Simon and Schuster, New York, 2002.

Day, Dwayne A., *Lightning Rod: A History of the Air Force Chief Scentist's Office.* Washington, D.C: Chief Scientist's Office, United States Air Force, 2000.

Dupree, A. Hunter. *Science in the Federal Government: A History of Policies and Activities.* Baltimore: Johns Hopkins University Press, 1986.

Dyer, Davis. *TRW: Pioneering Technology and Innovation Since 1900.* Boston: Harvard Business School Press, 1998.

Fenton, Edward. *Carnegie Mellon 1900–2000, A Centennial History.* Pittsburgh: Carnegie Mellon University Press, 2000.

Geiger, Roger L. *Research and Relevant Knowledge: American Research Universities Since World War II.* New York: Oxford University Press, 1993.

Geiger, Roger L. *To Advance Knowledge: The Growth of American Research Universities, 1900–1940.* New York: Oxford University Press, 1986.

Getting, Ivan. *All in a Lifetime: Science in the Defense of Democracy.* New York: Vantage Press, 1989.

Golden, William T., ed. *Science and Technology Advice to the President, Congress, and Judiciary.* Washington, D.C.: AAAS Press, 1993.

Gorn, Michael H. *Oral Interview of H. Guyford Stever.* Washington, D.C.: Office of Air Force History, 1989.

Gorn, Michael H. *Harnessing the Genie, Science and Technology Forecasting for the Air Force, 1944–1986.* Washington, D.C.: Office of Air Force History, 1988.

Guston, David H., and Kenneth Keniston, eds. *The Fragile Contract: University Science and the Federal Government.* Cambridge, Mass.: MIT Press, 1994.

Hall, Cargill R., and Jacob Neufeld, eds. *The U.S. Air Force in Space—1945 to the Twenty-first Century.* Proceedings, Air Force Historical Foundation Symposium, September 21–22, 1995. Washington, D.C.: USAF History and Museum Programs, 1998.

Herken, Gregg. *Cardinal Choices: Presidential Science Advising from the Atomic Bomb to SDI.* Stanford, Calif.: Stanford University Press, 2000.

Hershberg, James G., *James B. Conant: Harvard to Hiroshima and the Making of the Nuclear Age.* Stanford, CA: Stanford University Press (Reissue edition), 1955.

Hewlett, Richard G., and Oscar E. Anderson, Jr., *A History of the United States Atomic Energy Commission: The New World, 1939/1946*, vol. 1. University Park: Pennsylvania State University Press, 1962.

Kevles, Daniel J. *The Physicists: The History of a Scientific Community in Modern America.* New York: Alfred A. Knopf, 1978.

Killian, James R., Jr. *Sputnik, Scientists, and Eisenhower.* Cambridge, Mass.: MIT Press, 1977.

Jones, R. V. *Most Secret War: British Scientific Intelligence, 1939–1945.* London: Hamish Hamilton Ltd., 1985.

Johnson, Stephen B. "The organizational roots of American economic competitiveness in high technology." Paper presented at the Conference on R&D Investment and Economic Growth in the 20th Century. Available online at *http://ishi.lib.berkeley.edu/cshe/r%26d/papers/johnson.html.*

Lomask, Milton. *A Minor Miracle: An Informal History of the National Science Foundation.* Washington, D.C.: U.S. Government Printing Office, 1976.

Lonnquest, John C., and David F. Winkler. *To Defend and Deter: The Legacy of the United States Cold War Missile Program.* USACERL Special Report 97/01, Nov. 1996.

Mackenzie, Donald. *Inventing Accuracy: A Historical Sociology of Nuclear Missile Guidance.* Cambridge, Mass.: The MIT Press, 1993.

Mazuzan, George T. *The National Science Foundation: A Brief History*, esp. pp. 6–13. Available online at **http://www.nsf.gov/pubs/stis1994/nsf8816/nsf8816.txt.**

McDougall, Walter A. *The Heavens and the Earth: A Political History of the Space Age.* Baltimore and London: The Johns Hopkins University Press, 1985.

National Academy of Sciences. *Science and Technology in Presidential Policymaking: A Proposal.* Washington, D.C.: National Academy of Sciences, 1974.

National Academy of Sciences. *Astronomy and Astrophysics for the 1970s, Vol. 1: Report of the Astronomy Survey Committee.* Washington, D.C.: National Academy of Sciences, 1972.

Neufeld, Jacob. *Ballistic Missiles in the United States Air Force, 1945-1960.* Washington, D.C.: Office of Air Force History, 1990.

Patterson, James T. *Grand Expectations: The United States, 1945–1974.* Oxford: Oxford University Press, 1996.

Perry, Robert L. *Origins of the USAF Space Program, 1945–1956, Volume V, History of Deputy Commander (AFSC) for Aerospace Systems, 1961.* Available online at *http://www.fas.org/spp/eprint/origins/part07.htm.*

Press, Frank, and Raymond Siever. *Understanding Earth.* New York: W. H. Freeman, 1994.

Price, Don K. "Money and influence: The links of science to public policy," *Daedalus,* summer 1974 (Science and Its Publics: the Changing Relationship), vol. 3, no. 3 of the Proceedings of the American Academy of Arts and Sciences, pp. 97–113.

Rad Lab: Oral Histories Documenting World War II Activities at the MIT Radiation Laboratory. Produced by the IEEE Center for the History of Electrical Engineering, Piscataway, N.J. Principal investigators John Bryant, William Aspray, Andrew Goldstein, and Frederik Nebeker.

Reagan, Michael D. *Science and the Federal Patron.* New York: Oxford University Press, 1969.

Rhodes, Richard. *The Making of the Atomic Bomb.* New York: Simon and Schuster, 1986.

Schaefer, Ludwig F. *Evolution of a National Research University, 1965–1990: The Stever Administration and the Cyert Years at Carnegie Mellon.* Pittsburgh, Pa.: Carnegie Mellon University Press, 1992.

Simon, Herbert. *Models of My Life.* Cambridge, Mass.: MIT Press, 1996.

Smith, Bruce L. R., and Claude E. Barfield. *Technology, R&D, and the Economy.* Washington, D.C.: Brookings Institution and American Enterprise Institute, 1996.

Smith, Bruce L. R. *The Advisers: Scientists in the Policy Process.* Washington, D.C.: Brookings Institution, 1992.

Smith, Philip M. *The National Science Board and the Formulation of National Science Policy.* Washington, D.C.: National Science Foundation, 1981.

Stine, Jeffrey K. *A History of Science Policy in the United States, 1940–1985.* Science Policy Study, Background Report No. 1, Task Force on Science Policy, Committee on Science and Technology, House of Representatives, 99th Congress, Second Session, September 1986.

Sturm, Thomas A. *The USAF Scientific Advisory Board: Its First Twenty Years, 1944–1964.* Washington D.C.: USAF Historical Division Liaison Office, 1967.

Teich, Albert H. "Federal support of applied research: A review of the United States experience." Available online at *http://www.ulib.org/webRoot/Books/National_Academy_Press_Books/federal_role/fedl051.htm.*

Teller, Edward, with Judith Shoolery. *Memoirs: A 20th Century Journey in Science and Politics.* Cambridge, Mass.: Perseus Publishing, 2002.

Trimble, William F., *Jerome C. Hunsaker and the Rise of American Aeronautics.* Washington, DC: Smithsonian Institution Press, 2002.

Valley, George E., Jr., "How the SAGE development began," *Annals of the History of Computing,* July 1985, vol. 7, no. 3.

von Kármán, Theodore, with Lee Edson. *The Wind and Beyond.* Boston: Little Brown and Co., 1967.

Young, Peter, ed. *The World Almanac Book of World War II.* London: World Almanac Publications, Bison Books Ltd., 1981.

York, Herbert. *Race to Oblivion: A Participant's View of the Arms Race.* New York: Simon and Schuster, 1970. See also *http://www.learnworld.com/ZNW/LWText.York.RaceToOblivion.html#chapter4.*

Zachary, G. Pascal. *Endless Frontier: Vannevar Bush, Engineer of the American Century.* New York: Free Press, 1997.

Index

361

C

Q

R

S